11/14

Eisenhower Public Library
4613 N. Oketo Avenue
Harwood Heights, IL. 60706
708-867-7828

Confronting Contagion

Our Evolving Understanding of Disease

MELVIN SANTER

OXFORD
UNIVERSITY PRESS

Oxford University Press is a department of the University of Oxford.
It furthers the University's objective of excellence in research, scholarship,
and education by publishing worldwide.

Oxford New York
Auckland Cape Town Dar es Salaam Hong Kong Karachi
Kuala Lumpur Madrid Melbourne Mexico City Nairobi
New Delhi Shanghai Taipei Toronto

With offices in
Argentina Austria Brazil Chile Czech Republic France Greece
Guatemala Hungary Italy Japan Poland Portugal Singapore
South Korea Switzerland Thailand Turkey Ukraine Vietnam

Oxford is a registered trade mark of Oxford University Press
in the UK and certain other countries.

Published in the United States of America by
Oxford University Press
198 Madison Avenue, New York, NY 10016

© Oxford University Press 2015

Library of Congress Cataloging-in-Publication Data
Santer, Melvin, author.
  Confronting contagion : our evolving understanding of disease / Melvin Santer.
      p.   cm.
  Includes bibliographical references and index.
  ISBN 978–0–19–935635–5 (alk. paper)
  I. Title.
  [DNLM: 1. Communicable Diseases—history. 2. Communicable Diseases—etiology.
3. Disease Outbreaks—history. WC 11.1]
  RA643
  616.9—dc23
                                   2014010298

9 8 7 6 5 4 3 2 1

Printed in the United States of America on acid-free paper

*This book is dedicated to Ursula Victor Santer
(1932–2003) and Emily Miriam Santer,
Ruth Irene Santer, and Lewis Aaron Santer.*

The god Apollo,

. . . . unleashes his arrows on animals and soldiers for nine days
black bolts of plague fly on the Argives

HOMER, The Iliad

The human body contains blood, phlegm, yellow bile and black bile. These
are the things that make up its constitution and cause its pains and health.

THE HIPPOCRATIC TREATISE, The Nature of Man

Now what is the cause of plagues, and whence on a sudden the force of
disease can arise and gather deadly destruction for the race of men and
herds of cattle, I will unfold. First I have shown before that there are seeds
(semina, atoms) of many things which are helpful to our life, and on the
other hand it must needs be that many fly about which cause disease and
death.

LUCRETIUS, De rerum natura

An inland lake . . . there were . . . very many little animalcules.

Letter from Leeuwenhoek to Henry Oldenburg,
First Secretary of the Royal Society

. . . . . for the smallest sort of animalcules, which come daily to my view,
I conceive to be more than 25 times smaller than one of those blood-globules
which make the blood red.

Leeuwenhoek to Oldenburg

From hence it appears that the Art of Physic (Medicine) was anciently
established (1) by a faithful Collection of Facts Observed, whose Effects
were (2) afterwards explained, and their Causes assigned by the Assistance
of Reason; the first carries Conviction along with it, and is indisputable;
nothing more certain than Demonstration from Experience, but the latter is
more dubious and uncertain; since every Sect May explain the Causes of
particular Effects upon different Hypotheses.

HERMAN BOERHAAVE, Academical Lectures on the Theory of Physics,
1751, 2nd edn, vol. 1; Eighteenth Century Collections Online,
http://galenet.galegroup.com/servlet/ECCO

That man has long and persistently been concerned to achieve some
understanding of the enormously diverse, often perplexing, and sometimes
threatening occurrences in the world around him is shown by the manifold
myths and metaphors he has devised in an effort to account for the very
existence of the world and of himself, for life and death, for the motion of
the heavenly bodies, for the regular sequence of day and night, for the

changing seasons, for thunder and lightning, sunshine and rain. Some of these explanatory ideas are based on anthropomorphic conceptions of the forces of nature, others invoke hidden powers or agents, still others refer to God's inscrutable plans or to fate

CARL HEMPEL, *Philosophy of Natural Science*, 1966, Prentice-Hall

CONTENTS

ACKNOWLEDGMENTS

This book had its beginnings when my wife, Ursula Victor Santer, and I taught a course for non-scientists at Haverford College designated "History of Microbiology." Such "Histories" generally concentrated on the role of microorganisms in disease, and therefore their narratives began in the nineteenth century. Disease theory, however, begins millennia before the nineteenth century; the written materials provide a rich source of speculation about the cause of infectious/contagious diseases. That theorizing is an integral part of the speculations about the content and operation of the physical and biological world. I encourage the reader to read "Notes to the Reader" and the prologue to gain an introduction to the form and content of the book.

I want to thank a number of people who read portions of the book and encouraged me to continue. They are Jim Joyce, Sidney Axinn, Ruth Rothman, Lewis Santer, Karl Johnson, Madelyn Gutwirth, Marcel Gutwirth, Jim Dahlberg, and Tracy Kosman. I want to thank Lewis Santer, who has contributed to the book during the final stages of its preparation. I love you.

My thanks go to Dora Wong, science librarian at Haverford College. I am indebted to Sasha Santer Hill for work on Permissions to quote copyrighted material. I want to thank Professor K. Codell Carter for providing me with the original J. Henle reference used in Chapter 12.

NOTES TO THE READER

Walter Pagel opens his book on Paracelsus, *An Introduction to Philo-sophical Medicine in the Era of the Renaissance,* with the follow-ing paragraph:

> Much of modern medicine developed in the xvi[th] and xvii[th] centuries against a background of trends of thought that were not purely or mainly scientific. The main purpose of the present writer's historical enquiries since 1926 has been to place scientific and medical discoveries in the to us less comprehen-sible philosophical and religious setting in which they first appeared
>
> PAGEL, p. 1, 1958, Karger

Pagel is stating the essential fact that to understand the origins of dis-ease theories during any historical period, one must understand the con-temporary philosophical and religious ideas about the composition and the operation of the world. It is my intention to adopt and extend this approach to understand the cause of infectious/contagious disease starting in antiq-uity and going forward to the twentieth century.

I want to request of you, dear reader, to temporarily suspend judgment about the causes of contagious disease that you know to be the initiated by the invasion of a host by various microscopic agents. I request the suspen-sion of this knowledge so that you can place yourself in the particular period, prior to the last quarter of the nineteenth century, that will be de-scribed and can find yourself, like the writers of the time, struggling to understand a phenomenon that is visible but incomprehensible because its underlying causes are invisible and impossible to know. Despite this dif-ficulty, it was essential for humans to try to explain the causes of these diseases in order to treat them. This crucial philosophical assumption is in the Hippocratic treatise *The Science of Medicine*, which was composed in the fifth century BCE, 2,500 years ago:

> "To know the cause of a disease and to understand the use of the various methods by which disease may be prevented amounts to the same thing in effect as being able to cure".
>
> *Hippocratic Writings*, p. 145, 1978, ed. by G. E. R. Lloyd, Penguin

In this book we shall initially use the term "infectious" to characterize disease apparently acquired from some source external to a host. During this period it was not recognized that such a disease could be passed from one susceptible host to another. When this fact was acknowledged, such diseases were classified as "contagious."

Contagious disease is a general biological phenomenon. It occurs in plants, humans, and other animals and among microorganisms. Thus, each category of diseases is a special case of the general phenomenon. However, the development of our understanding of the causes of disease of the various biological groups did not develop as an integrated whole. For example, the experimental work on the causes of plant diseases and human diseases evolved separately, so the information gained in the study of plant disease had little impact initially on the work on human diseases.

Obviously and understandably, human diseases have assumed importance in the lives of people, and the preponderance of the histories of contagious disease has consequently focused exclusively on human diseases. Our goal is to include all contagious disease works in this book.

| *The Why of This Book: Explaining Infectious/Contagious Diseases*

T o introduce what this book is about, it will be useful to compare the description of an infectious disease given by Thucydides in his famous history of the Peloponnesian war initiated in 431 BCE, and a comment about such epidemic diseases by William Harvey in 1653 CE, some 2,000 years later. In Thucydides' narrative it was the second year of the war, the first days of summer, and the Spartan forces had reached Attica, the area surrounding Athens, when a deadly "plague" broke out. Thucydides wrote as a historian and as someone who contracted and survived the disease. He presented the symptoms and progress of the disease as it devastated the population in brilliant technical detail, equal to the descriptions of diseases in the Hippocratic writings:

"though no pestilence of such extent nor any scourge so destructive of human lives is on record any where. For neither were physicians able to cope with the disease, since they at first had to treat it without knowing its nature, the mortality among them being greatest because they were most exposed to it, nor did any other human art prevail. And the supplications made at sanctuaries, or appeals to oracles and the like were futile, and at last men desisted from them, overcome by the calamity".

And he went on to write,

"Now any one, whether physician or layman, may, each according to his personal opinion, speak about its probable origin and state the causes which, in his view, were sufficient to have produced so great a departure from normal conditions".

THUCYDIDES, *History of the Peloponnesian War*, 1969, p. 343, Harvard

From the description of the symptoms it is not possible to give a definite diagnosis of the disease. It was most likely caused by a virus. Once the patient had recovered from the disease, there appeared to be immunity from another infection. During this time a disease that infected a number of people was described as "infectious"—that is, the disease was caused by some external carrier such as air or food. There was no explicit acknowledgment that diseases could pass from person to person and therefore be labeled contagious. However, Thucydides offered the vague suggestion that this could occur when he commented that physicians who tended the sick had the greatest mortality.

Thucydides stated that it was a new illness so neither physicians, presumably Hippocratic doctors, nor temple healing was of any use. There appeared to be no antecedent conditions that could be designated as a cause. It was apparently imported from North Africa, brought in during the summer months by the invading Lacedaemonians (the Spartan army), first arriving in the port of Piraeus and then spreading to Athens. Thucydides, a contemporary of Hippocrates and the developing Hippocratic Corpus, was reluctant to offer a cause of such a calamitous disease or was, perhaps, skeptical of the many causes proposed for infectious diseases in the various treatises of the Corpus and by practitioners of temple medicine.

On November 30, 1653, some 2,000 years later, the great William Harvey, who firmly established the circulation of blood in humans, and certainly knew many things, confessed to one John Nardi of Florence that there were two issues he found difficult to explain. He wrote,

"How the idea, or form, or vital principle should be transfused from genitor [male parent] to the genetrix [female recipient] and from her transmitted to the conception of the ovum, and thence to the foetus, and in this produce not only an image of the genitor . . . but also various peculiarities or accidents, such as disposition . . . hereditary diseases . . . etc. All of these accidents must inhere in the geniture and semen . . . from which an animal is not only produced, but by which it is afterwards governed, and to the end of its life preserved".

Harvey simply acknowledged that it was hard for him to account for heredity. He went on to state that he also did not understand the cause of contagious diseases, particularly those leading to epidemics:

"So do I hold it scarcely less difficult to conceive how pestilence or leprosy should be communicated to a distance by contagion, by a zymotic element contained in woolen or linen things, household furniture, even the walls of a house . . . as we find stated in the 14th chapter of Leviticus. How, I ask, can contagion, long lurking in such things . . . after a long lapse of time produce its like nature in another body? Nor in one or two only, but in many, without respect of strength, sex, age, temperament, or mode of life, and with such violence that the evil can by no art be stayed or mitigated".

WILLIAM HARVEY, *The Works of William Harvey*, 1965, pp. 610–611, Johnson

Neither Thucydides, a contemporary of Hippocrates, nor Harvey, living 2,100 years later, with access to Hippocratic–Galenic medicine and all contemporary theories, appeared to believe in the various causes that had been proposed over the millennia, including bad air, the configuration of the planets, and the anger of God. Thus, one confronts the following problem. After thousands of years and much speculation about the cause of disease, an individual like Harvey, and certainly Thucydides thousands of years before him, remained an agnostic, if not an atheist, about the various explanations for disease, including the idea that god(s) cause them.

We now know that these diseases are the result of the interaction of microscopic biological entities with the human body. These agents, for example viruses and bacteria, were impossible to observe, and so were the processes leading to the development of disease. Consequently, the variety of causes that were proposed were speculations, hypotheses created by thinkers who had to formulate ideas about a complex, invisible process on the basis of visible outcomes. The invisible nature of the events leading to disease was not revealed during the 2,000 years from the time of Thucydides to Harvey in the seventeenth century and beyond. Nevertheless, during that period, many theories were proposed. They changed over time, although there was no evidence in the modern sense of the term to suggest changes were required. Nothing was ever proved or disproved. Gods were the cause of disease in Homeric times. Gods ceased to be the cause of disease for the Hippocratic writers, whereas God returned as the cause of disease in Christian Europe—although not the gods of the Greeks.

This raises an interesting question. Since there was no way to judge whether a theory was correct, why did theories change? We will discuss the source of these various theories and why they changed and why it took 2,000 years—and 200 more years after Harvey—to arrive at a proper account of the causes of contagious disease among humans, animals, plants, and microorganisms.

This book does not provide a linear narrative history of our understanding of the cause of infectious/contagious disease. Instead the storyline focuses on historical episodes, and most importantly the foundations of belief, to understand the origins of the arguments that justify the various causes. We begin our coverage with a history of theories from antiquity to Galen and then concentrate on theories from the fourteenth century CE to the seventeenth century, the revolutionary era in science, and finally move forward to the twentieth century. We shall follow the transition of explanations from traditional beliefs in gods to explanations based on metaphysical discourses about the composition of the world and the way it works, that include naturalistic explanations, to the use of experiments as the way to understand the cause of natural phenomena including disease. Our coverage of theories from the seventeenth century will be more detailed and will attempt to fairly present claims to certain and not-so-certain knowledge. In presenting these various episodes in the intellectual history of the cause of infectious/contagious disease we will uncover numerous insights into the methodology of this branch of biological science.

Where Do Hypotheses Come From?

Illness and pain are among the earliest experiences of humans. These unfortunate episodes provide examples of the immediate acquisition of knowledge through our senses—that is, knowledge acquired "passively," as John Locke (1632–1704) put it, or "without any operation of our will" (D'Alembert, 1717–1783). It is quite clear that ancient civilizations had intimate experience of disease and epidemics. One example of such an event, terrifying and incomprehensible, was provided by Thucydides in the fifth century BCE.

How was one to explain this sequence of events? The modern scientific investigation of a phenomenon involves two steps: the construction of a hypothesis and the testing of the hypothesis. The first step is crucial to the development of understanding: without an identified hypothesis, the search has no focus. To appreciate this logic let us examine how a hypothesis may be formulated.

A hypothesis emerges as an inductive process, a model advocated by Francis Bacon in the seventeenth century. It works like this. All the data related to a phenomenon are collected. Once this is accomplished the cause will emerge via a procedure labeled "inductive inference." Under this epistemological model, the study of the cause of an infectious disease (the plague of Athens, for example) would proceed in this way. After the disease arises in a population, all data associated with the disease are gathered. People suddenly exhibit symptoms: headaches, bleeding, coughing, fever. They break out in pustules. The body is being destroyed. Some die, some live. One can add, as antecedent data, the season of the year, the weather, the direction of the wind, and the configuration of the planets. From this evidence the cause of the disease somehow emerges from the information collected. It all seems so straightforward. With such a procedure it would appear that discovery of the cause would be inevitable.

But obviously the outcome from this method of proceeding is not inevitable. In fact, it is a program almost impossible to carry out. One fatal weakness is the inability to distinguish important antecedent data from data that are irrelevant; in short, there is no way of knowing which data are necessarily connected to the cause and which are not. There is no way to distinguish among the variety of "evidence" unless you already have a hypothesis, which means that you already suspect the cause and look for evidence that supports that cause.

A plausible account of hypothesis formation is described by Carl Hempel:

> "There are, then, no generally applicable "rules of induction" by which hypotheses or theories can be mechanically derived or inferred from empirical data. The transition from data to theory requires creative imagination. Scientific hypotheses and theories are not derived from observed facts, but invented in order to account for them".

> CARL HEMPEL, *Philosophy of Natural Science*, p. 15, 1966, Prentice-Hall

Hempel makes the important point that there is nothing in the method of simple induction that can provide explanations of unobservable causes. The process of discovery is not a logical process; it is a psychological event in the mind of the investigator. These insights do occur, which brings us to the following important questions: What factors determine the creation of hypotheses? What might be the objective conditions in the intellectual environment that influence, if not determine, how the creative imagination of individuals will work?

Remember, until the early decades of the nineteenth century there were no experimental ways to objectively differentiate among hypotheses about the cause of contagious disease, but nevertheless there were changes over time in disease theory. This makes it important to understand the processes by which different hypotheses are proposed in the absence of experimental evidence. What we will encounter is a consistent theme: theories derive their legitimacy, their acceptance, because they conform to contemporary religious or philosophical beliefs. If these contemporary beliefs change, it logically follows that disease theories would change as well. In short, change in disease theory did not result from a conflict over experimental evidence but evolved as the philosophical grounding was replaced.

Confronting Contagion

CHAPTER 1 | Homer–Hesiod–Torah–Greek Playwrights

Diseases Are Almost Always Caused by God(s)

EARLY THEORIES ABOUT THE cause of infectious diseases in the "Western" world can be found in the literary works of Homer and Hesiod, the Greek playwrights, and the Tanakh (Torah, Prophets, and Writings) of the Jewish tradition.

The writings of the early Greeks and Hebrews are not the only source of disease theory; there are materials from the Egyptian and Babylonian civilizations. Their speculations about the cause of phenomena, such as disease, weather, earthquakes, and the movement of the sun, moon, and stars, attributed these events to gods who appear with many names and are responsible for the origin of the world and the processes within it. But since the Homeric–Hesiod epics achieved such status they were credited or blamed by Herodotus (484–after 430 BCE) for the creation of the role of gods in the world. In "The Histories," Book 2, he wrote,

> "But it was only—if I may so put it—the day before yesterday that the Greeks came to know the origin and form of the various gods and whether or not all of them had always existed; for Homer and Hesiod are the poets who composed our theogonies and described the gods for us, giving them all their appropriate titles, offices, and powers, and they lived, as I believe, not more than four hundred years ago".
>
> HERODOTUS, *The Histories*, 1954, p. 117, Penguin

To reveal the role that god(s) played in causing disease we shall examine examples of such speculations in the Homeric–Hesiod poems, in Greek plays, and in material from Israelite writings.

Homer

The control of events by gods, their behavior, and their interaction with mortals are expressed in two stories in the "Iliad," in Books 1 and 4 (*ca.* the eighth century BCE) and in Sophocles' play "Oedipus the King," written in the period 430 to 420 BCE. In Book 1 the god Apollo inflicts a plague on King Agamemnon's army. The preceding events are described in the following way. Agamemnon has deeply offended a man of prayer, Khryseis, by taking his daughter. Khryseis asks for her return:

. . . "but let me have my daughter back for ransom".

Agamemnon's troops advise the king to

"Behave well to the priest. And take the ransom!

But Agamemnon would not . . . and brutally he ordered the man away: Let me not find you here by the long ships . . . if I do the staff and ribbons of the god will fail you.

The old man retreated, and prayed to the god Apollo:

let my wish come true: your arrow on the Danaans for my tears".
HOMER, *Iliad*, 1975, pp. 12–13, trans. by Robert Fitzgerald, Anchor-Doubleday

Apollo responds and unleashes his arrows on animals and soldiers for nine days.

On the 10th day, the god Hera takes pity on the dying soldiers and moves Achilles to assemble the troops to determine why there is such anger on the part of Apollo. Achilles arranges an assembly and, addressing Agamemnon, stresses the imminent catastrophe and suggests they consult a seer (diviner or priest) to tell them why Apollo is so angry. A man, Kalchas, comes forward, characterized by Homer as a person with a divine gift.

Kalchas tells Achilles what he already knows: that Agamemnon is responsible for the god's anger because of his behavior toward Khryseis. Kalchas relies on Achilles to protect him for having spoken the truth to the assembly in the presence of Agamemnon. Agamemnon replies irritably to Kalchas, accusing him of never having given him a good prophecy.

Nevertheless, he returns the daughter on one condition, that he keeps Achilles' woman, Briseis. Khryseis now prays to Apollo to end the plague. The plague is terminated.

Thus, a god brings the plague and a god can end the plague. The god does this at the request of a priest. In each case there is good reason for the priest to make the request and for the prayer to be answered. In this case it is punishment for a particular deed by the king that is unacceptable to the priest and to the deity.

In "Oedipus the King" by Sophocles, the priest addresses Oedipus:

"A blight is on the fruitful plants of the earth, a blight is on the cattle in the fields, a blight is on our women that no children are born to them, a God that carries fire, a deadly pestilence, is on our town, strikes us . . . and the house of Cadmus is emptied of its people while black Death grows rich in groaning and lamentation".

SOPHOCLES, *Oedipus the King*, 1954, p. 12, University of Chicago Press

In the play the king sends Creon to Apollo in the Pythian temple to learn how to save the city. Creon relates that he has heard from the god that the plague is the result of "pollution" in the land. It is punishment for a murder committed by the king. The punishment is on the whole city, including plants and cattle.

In Book 4 of the "Iliad" the war between the Akhaians and Troy is proceeding, while the gods are drinking wine and observing the siege of the city of Troy. The tension is great among the gods since it involves the issue of the termination or continuation of the conflict. Meanwhile there is a personal battle between Menelaus and Hector. Each has his adherents among the gods: Hera and Athena are for Menelaus, and Aphrodite has protected Hector. There is the possibility that the conflict can be resolved and Troy would remain intact and Helen returned to Menelaus, but that was not to be: Hera and Athena were "devising evil for the Trojans." How was this being done? Hera has confronted Zeus contending that ending the war would render in vain all their efforts to destroy Troy. Although Zeus questions her why such animosity against Priam and his sons, he is persuaded and offers a bargain to allow the battle to continue, although Troy is one of his favorite cities, if in the future the goddesses will not contest his will to destroy some city. Hera agrees and appeals to Zeus to order Athena to "visit horrible war again on Akhaians and Trojans." Athena is sent to earth to induce the Trojans to fire the first arrow at Menelaus; however, the arrow is deflected by Zeus' daughter and only wounds Menelaus. Zeus sends his "sacred herald" to summon Makhaon, who is the son of Asclepius, to treat Menalaus. Makhaon removes the arrow and "he sucked it clean of blood" and treated it with "balm a medicine that Kheiron [Chiron] gave his father."

These storylines vividly illustrate the way Homer portrays the power and behavior of gods. They bring disease to whole populations. They are endowed with human characteristics: argumentative, passionate, vindictive, and unpredictable. They are willing to sacrifice a city. To maintain civility among the gods, the citizens are doomed to war and destruction. In all this maneuvering there is the interaction between gods, and between gods and humans. In the latter case Zeus is the one who summons a physician to repair Menalaus' wounds. In addition in this episode we get a quick history of the genealogy of the physician Makhaon from Asclepius and Chiron.

Hesiod

Hesiod, in "Works and Days," introduces a system where justice is to prevail and there are punishments for violations. The poem opens with a hymn to Zeus, who dwells in the "uppermost palace" and controls the lives of men and holds "us to justice." Hesiod writes that an unjust society is subject to famine, disaster, and disease; in contrast, where there is justice in a society, there is health and prosperity. The origin of these evils, which include disease, is revealed in the story of Prometheus. Prometheus has stolen fire from Zeus and secretly gives it to humans. Zeus is furious over the deception and promises "great pain" now and in the future to men. He orders Hephaestus, a god, to create a woman who obtains skills from Athena, beauty from Aphrodite, and "shamelessness and . . . deceit" from Hermes. Her name is Pandora and she will be a "plague for men."

Before the creation of Pandora the

"tribes of men had . . . lived on earth free . . . from evils . . . and from painful diseases the bringer of death to men".

A place comparable to the garden of Eden.

All of these evils, including the primary ones, disease and famine, were contained in a jar. When Pandora raised the lid, all these calamities were released and

"the earth is abounding in evils and so is the sea.
And disease come upon men by day and by night, everywhere moving at will".

HESIOD, *The Poems of Hesiod*, 1983, pp. 99–100, trans. by R.M. Frazer, Oklahoma

In another version by West:

"Countless troubles roam among men: full of ills is the earth, and full the sea. Sicknesses visit men by day, and others by night, uninvited, bringing ill to mortals".

HESIOD, *Theogony and Works and Days*, 1988, p. 40, trans. by M. West, Oxford

Hesiod's final comment is

"Thus in no way can anyone escape the purpose of Zeus".

Comparison of Greek and Israelite Writings

Under a model where gods control the physical and biological world in both cultures, the human being is subject to these divine laws and an infringement results in some punishments. There are instances of such behaviors that elicit identical responses in both literatures. In Exodus 15:26,

"If you will heed the Lord your god diligently, doing what is upright in His sight and keeping all his laws, all the diseases that I brought upon the Egyptians then I will not bring upon you, for I the Lord am your healer".

An almost identical threat and promise is made in "Works and Days." Hesiod states that if justice is not present in a city, evil will befall men. Zeus will punish,

"Often even a whole city pays for the wrong of one person. Zeus will send famine together with plague . . . nor will their women bear children. If justice is present the city will flourish".

HESIOD, *The Poems of Hesiod*, 1983, p. 109, trans. by R.M. Frazer, Oklahoma University Press

The Exodus 15:26 story is explicitly in the form of a covenant, an if/then contract. The Hesiod paragraph can be interpreted in this way.

Childlessness is also a recurring theme in the Torah. In Leviticus 20:20–21 it is a punishment from God, and in Exodus 23:25–26, God is the cause of inability to conceive and can prevent miscarriage and barrenness. The punishment that Zeus sends in "Oedipus Rex" contains the entire package of punishments: famine, plague, and childlessness. In various chapters of the Torah, in addition to childlessness, the conjunction of famine and disease also appears. In Leviticus 26:3–4,

> "If you follow my law and faithfully observe my commandments I will grant you rains in their season so that the earth shall yield its produce and the trees their fruit".

"In Deuteronomy 28:22, disease and famine appear following the injunction "if you will not obey the Lord":

> "the Lord will smite you with consumption and with fever, . . . and with drought, . . . and with blasting, and with mildew and every plague"

In Jeremiah 14:12 there is an almost identical threat from God:

> "I will consume them by the sword, and by the famine, and by the pestilence".

Again, in 1 Kings 8:37,

> "If there be in the land famine, if there be pestilence, if there be blasting or mildew, locust or caterpillar".

Not only are there similar punishments, they are expressed in similar terms.

In the "Iliad" the priest Khryseis implores Apollo to punish Agamemnon and his army:

> "O hear me master of the silver bow . . .
> let my wish come true:
> your arrows on the Danaans for my tears"!

In Deuteronomy 32:42, God states

> "I will render vengeance to mine adversaries, I will make mine arrows drunk with blood".

All quotes from *The Holy Scriptures*, 1955, 1989, Jewish Publication Society

Here we have almost identical imagery, although retribution in the "Iliad" is due to the evil of one man while in the latter story it is punishment of the persecutors of the Israelites. However, in the Israelite sources, Chronicles 2, there is an example of punishment of a whole people due to the transgression of a king: "There will occur an incurable disease on thy people, children, wives." This theme of retribution is present in "Works and Days." Zeus sends a terrible suffering from heaven. "Often even a whole city pays for the wrong of one person who is a doer of evil."

A Series of Questions

In the Greek and Hebrew literature, is it the case that all diseases are caused by god(s)? Can we detect in these ancient writings instances where diseases can occur without the intervention of a deity? Does the role of a physician or "healer" diminish the power of a god? That is, can the intention of a deity be contained by human intervention? If diseases (some?) are not the action of god(s), then the questions that arise are: why are they present, what is their cause, and how can we characterize such causes?

Replies in the "Iliad"

Let us return to Book 1 of the "Iliad," where Agamemnon has taken the daughter of the priest and a disease has been inflicted on his army. It all appears straightforward: a human commits an apparently unacceptable act and is punished by a god. However, a number of issues are raised in the poem that disturb this one-to-one relationship.

First of all, it is obvious that Apollo did not respond immediately to the injustice committed by the king. Homer did not create a story in which a misdeed automatically triggers a punishment, nor did he suggest such a relationship exists. After Apollo hears the prayer of the priest he does send a disease. The onset of the disease appears to be a cooperative effort of a human and a god.

In the initial encounter between Agamemnon and the priest the king tells the priest to leave and threatens him with unnamed consequences if he does not leave, underscoring the threat with the statement that the god will not protect him. Agamemnon believes that the vengeance of a human king can override the protection of a god. The priest certainly believes in the power of the king for, as Homer expresses it, the old man feared and obeyed him. This was probably the experience of people. A human king rules arbitrarily. Prayers were not answered. Furthermore, the disease continues for nine days but not everyone dies, neither the main actors, nor, most obviously, the king, whose actions brought on the plague. One can speculate that this reflects Homer's experience with epidemics; not everyone contracts a disease nor does everyone die. There is no explanation offered. Perhaps there is the unstated belief that it is determined by gods.

Let us look at another book of the "Iliad," Book 4:183. There is a dispute between Zeus, Hera, and Athena, the latter two calling for the destruction of Troy. Athena descends to the troops and encourages one of them to shoot an arrow at Menelaus. Athena, however, deflects the arrow so that Menelaus is wounded. Then, Makhaon, son of Asclepius, is called, removes the arrow, cleans the blood from the wound and covers the wound with "balm." Here is a complicated episode that includes the apparent desire of a god to kill a human and the intervention of the same god to prevent immediate death; this intervention causes a wound, which is treated and cured by a human physician. If it was the intent of the goddesses to inflict death or more likely injury on Menelaus, the physician has the power to counteract the action of the god. Indeed, it appears that only a human can prevent further injury.

In Book 19 there is a remarkable narrative with interesting biological content whose "science" seems wholly out of place at this time. In this account Patroklos is killed. The grief-stricken Achilles is comforted by his mother, Thetis, who advises him to wear Patroklos' armor and return to battle. Achilles is concerned that if he leaves the body,

> "black carrion flies may settle on Patroklos' wounds, . . . and I fear they may breed maggots to defile the corpse, now life is torn from it.
> His flesh may rot".

But Thetis assures Achilles that she will shield the body from the flies, "though he should lie unburied a long year":

> "I'll find a way to shield him from the black fly hordes that eat the bodies of men killed in battle. Though he should lie unburied a long year, his flesh will be intact and firm".

HOMER, *Iliad*, 1975, p. 458, trans. by R. Fitzgerald, Anchor-Doubleday Books

She accomplishes this in two ways, by introducing "red nectar and ambrosia in his nostrils to keep his body whole," and preventing flies from landing on his body. This scene is so arresting I will present another translation, this one by Richmond Lattimore, whose content agrees with the translation by Robert Fitzgerald. Achilles says,

> "Yet I am sadly afraid that flies might get into the wounds beaten by bronze in his body and breed worms in them, and these make foul the body, seeing that the life is killed in him, and that all his flesh may be rotted".

Achilles' mother responds,

> "My child, no longer let these things be a care in your mind.
> I shall endeavor to drive from him the swarming and fierce things, those flies which feed upon the bodies of men who have perished; and although he lie here till a year has gone to fulfillment, still his body shall be as it was, or firmer than ever".

> *The Iliad of Homer*, 1951, pp. 392–393, trans. by Richmond Lattimore, University of Chicago Press

Homer does not ascribe to the gods the destruction of Patroklos' body; flies cause the destruction. Remarkably, Homer appears to understand that the flies are not the products of rotting flesh but are the cause of the decay. This is an exceptional conclusion because this phenomenon is a contentious issue some 2,500 years later. Decay can be countered by human intervention, by a physician as well as a half-god, using different compounds.

In the "Odyssey," Book 11:167, Odysseus speaks with his dead mother in the underworld. He asks what caused her death: was it Artemis' arrows (a god's arrows) or dying from consumption? He is contrasting the two possibilities.

> "Odysseus: Mother, I came here, driven to the land of death in want of prophecy from Teiresias' shade . . . But come now, tell me this, and tell me clearly, what was the bane that pinned you down in Death? Some ravaging long illness, or mild arrows a-flying down one day from Artemis?

> Mother: So I, too pined away, so doom befell me, not that the keen-eyed huntress with her shafts had marked me down and shot to kill me; not that

illness overtook me—no true illness wasting the body to undo the spirit; Only my loneliness for you Odysseus . . . took my own life away".

HOMER, The Odyssey, 1963, p. 190, trans. by Robert Fitzgerald,
Anchor-Doubleday

Death appears to be the consequence of three different causes: arrow from gods, wasting illness, and loneliness, longing for a lost loved one. The latter two are not sent by gods.

Replies in Hesiod and Greek Plays

In "Works and Days" Hesiod writes that the earth is full of evils, "everywhere moving at will" . . . . . . "sickness visit men. . . . . .uninvited, bringing ills to mortals". . . It is possible that Hesiod is suggesting that the gods have lost control of the spread of disease just as they have lost control of fire, which Prometheus stole and gave to humans. The gods are unable to reverse this act, nor apparently can they reverse the spread of disease, traveling according to its own free will. It is evident that gods create diseases, but disease is now a something, an entity, that has an existence independent of the gods, and as such travels the world, unpredictable in its behavior, striking some and not others. As such it is not a punishment from the gods for some unacceptable act, but rather it strikes in arbitrary ways, thus behaving in a way that conforms to the experience of humans with disease.

Contractual Relations Between Humans and God(s) That Appear to Limit the Power of the Deity

There are examples of interactive, cooperative behavior between gods and humans to achieve a particular goal that appear to limit the power of the god. In Sophocles' play "Philoctetes," produced late in the fifth century BCE, Philoctetes is meant to fight in the Trojan War but is bitten by a snake, which leaves him with a diseased leg and in great pain. As a consequence Odysseus leaves him on an island. The snakebite has not been some chance encounter but is punishment for the violation of a sanctuary, and thus he is prevented from going to Troy to participate in the war. During his stay on the island he treats the wound with herbs, but there is no cure. To win the war the Greeks must have Philoctetes and his bow, which came from the god Heracles. Philoctetes refuses to go to war but does so when the god Heracles sends Asclepius from heaven to cure him. The deed, the cure, is accomplished as a result of a contractual agreement between a human and a god: if the god sends a cure, then the human agrees to participate in the conflict. Such an instrumental interplay between an apparently all-powerful deity and a human is also presented in Genesis 28:20–21 following Jacob's dream. When he awakes he makes a vow in the form of an if/then oath:

> "If God remains with me, If He protects me on this journey that I am making, and gives me bread to eat and clothing to wear, and I return safe to my father's house, the Lord shall be my God"

There is another such contractual suggestion in the "Iliad," Book 1:204. Athena visits Achilles to convince him to cool his rage against Agamemnon. Achilles consents in the following words:

> "Honor the gods' will, they may honor ours".

Gods, Physicians, Healers

It is likely that soldiers wounded in battle would die due to the destruction of the body from diseases (sent by the gods?) following the wounding process. The power of a divinity appears to be diminished by the necessary use of physicians to counter these effects. It is clear that the healer saves many men from the consequences of battlefield wounds. We have encountered the physicians Makhaon and Podalirius in the "Iliad"; they are healers, and their value is explicitly expressed in Book 11:514, p. 261:

> "a surgeon is worth an army full of other men at cutting shafts out, dressing arrow wounds".

The role of healer may be assumed by others who know how to administer drugs. In the "Odyssey," the daughter of Zeus, Helen, alleviates suffering and sadness with a drug she adds to wine. Although she is a god, she has learned these skills from an Egyptian woman. That is not surprising since Egypt is described as a land where every man is a healer, wise above humankind.

In Book 4 of the "Iliad," Menelaus is wounded by a master bowman. Agamemnon calls upon Makhaon, "son of Asklepios," who extracts the arrow and purifies the wound. There is the understanding that if this procedure is not carried out, the normal course would lead to disease, with consequences as serious as death. Homer recognizes the consequences of wounds, and why not? Such wounds occurred thousands of times, and thousands of times soldiers died, with fever, decay of the body, and death. At the time of the wounds no prayers are proposed to ask a god for a cure; a human healer will provide the cure.

An example of a human healer, a legendary figure, is Epimenides, who according to Diogenes Laertius and Aristotle lived about 600 BCE and had extraordinary powers "prophetic and cathartic." The most famous story about him is told by Diogenes and has to do with a pestilence that attacked Athens. The Athenians sent a ship to Crete to request help from Epimenides, who came to Athens. He was able to reveal the cause of the plague and was able to rid the city of the disease. He is portrayed as a seer, a healer with knowledge of herbs and their value as medicines, a person equivalent to Asclepius. Diogenes reports, without further detail, that he is the first person who purified homes and fields. A purifier is one who carries out some action that removes a pollution. Disease is an instance of a pollution, and consequently Epimenides is one who can cure disease.

Quarantine and its Meaning

Another human-instituted procedure that affects the consequence of disease is the system of quarantine. Leviticus 13:1–14, 57 present a comprehensive explanation of the method. This text combines the elements of a theoretical discourse on infectious-contagious diseases and a how-to manual to control their spread. Although it is generally accepted in this tradition that diseases come from a deity, God does not appear as an actor in these stories: it is always a priest who diagnoses the disease and offers solutions to a condition that is judged a ritual impurity.

There are three diseases described, one on the skin of humans, another a disease of cloth, fabric, or leather, and a disease of the plaster or mud that covers the surfaces of stone houses. In each case the same Hebrew word is used for the disease, *tsara'at*. If the human disease was considered acute, as determined by the priest, it falls under the category of an impurity and the individual was removed from the community and remained under observation for seven days. If the disease had not receded, another seven days of quarantine would follow, at which time the diagnosis by the priest might be that the disease was no longer acute and the person was pronounced free of the disease and returned to the community.

In a disease on fabric, an entity was present on the material that is described as being "streaky green or red." Again, the priest was consulted; he would separate the material from the community for seven days. After that period there were a number of options. If the colored substance had spread, the fabric was to be burned. If it had not spread, the material would be washed two times or the affected area could also be cut out. It is highly likely that the colored substances—for there appear to be more than one—are fungi. Fungi are also indicated when considering diseases of houses. Leviticus 14:34 reports lesions that discolor plaster or mud. The visible material is similar to the substance described on fabrics. Once again, the priest was to be consulted, and he decided whether the house should be evacuated for seven days. If after that time the disease had spread, those areas of the house were to be removed and placed outside the community. The inside of the house was washed. If after all these procedures the disease was still present the house was destroyed and the remains were removed from the community.

One way to understand the reason for quarantine is to assume that it was understood that something passed from an infected person to an uninfected individual when they were in close contact but could not bridge the distance provided by quarantine. This would argue against transmission by God, since the deity would not be inhibited from causing disease by the distance between persons. The three processes that eliminated the infected material—cutting out, washing, and destroying and carting off the rubble of the house—would argue for such a hypothesis.

An alternate theory, which would retain a role for God, is that an infected person was ritually impure and because of this pollution it was necessary to remove him/her from the community, lest God punish others in the community with the disease. How this would explain why the dismantling, burning, and destroying of inanimate things prevented the spread of disease is not clear. In addition, the decline of alleged ritual impurity with time, after a week or two in isolation, appears to be an event independent of God for two reasons: the deity is not mentioned as an initiator of the process, and his powers to cure are somewhat circumscribed.

Disease Transmission Without God(s)?

A story from the writings of Herodotus (Book 2, *The Histories*) suggests the transfer of the plague due to the presence of rodents carrying the infectious disease. In the war between Sennacherib, who is king of Arabia and

Syria, and the Egyptians, mentioned in 2 Kings 19, the warrior class of the Egyptians were mistreated by the king and were not willing to fight. Herodotus relates that the priest-king prayed and during these "lamentations" fell asleep and dreamed that a god would help him. The king's confidence was restored and with some men, none of the warrior group, he confronted the Egyptians and prevailed due to the "thousands of field mice" who destroyed the weapons of the Assyrians. One interpretation is that the destruction of the troops was due to rodents who were harboring fleas and the plague.

It appears inconceivable that in the century or two preceding Thucydides, who pointed out during the plague of Athens that those individuals tending the sick died more frequently, that there was not the recognition that disease could be transmitted from person to person during epidemics, with no deity involved. That is certainly one way to understand why quarantine was instituted. Nevertheless, at this time there is no general principle, no structure within which to classify a disease that appears to happen without the intervention of a god, although such possibilities may be recognized. It is all too obvious that men are wounded in battle. If the wounds are not treated, and even if treated, they may lead to death of some individuals but not others. These would be normal events, perhaps identical to the phenomena captured in the poetry of Book 6 of the "Iliad":

"Very like leaves upon this earth are the generations of men . . . old leaves, cast on the ground by the wind, young leaves the greening forest bears when spring comes in. So mortals pass; one generation flowers even as another dies away".

HOMER, *Iliad*, 1975, p. 140, trans. by Robert Fitzgerald, Anchor-Doubleday

It is recognized that the living world goes through cycles according to the seasons. This was the normal state of the world for the ancient Greeks and for all peoples. People are born, grow old, and die. These regularities that the gods have apparently set in motion appear, in Homer's narration, part of normal life, depending on antecedent events like the wind and the warmth of spring, and the aging process. Nevertheless there is no systematic writing that offers an alternate explanation for the causes of these events, including disease. Consequently, we are left with the one explicit cause, a deity.

CHAPTER 2 | Philosophers

THALES
   ANAXIMANDER
ANAXIMENES
       XENOPHENES
          HERACLITUS
              ANAXAGORAS

    PYTHAGORAS
            PARMENIDES
            EMPEDOCLES
            SOPHOCLES

     ALCMAEON   **HIPPOCRATES**

              SOCRATES
               DEMOCRITUS
              LEUCIPPUS
              PLATO
                     ARISTOTLE
               EPICURUS

                      LUCRETIUS

625 BCE   **341 BCE**   *ca.* **94–56 BCE**

General timeline of many persons discussed in this chapter

Pre-Socratic and Others; Matter Theory and the Cause of Change; the Invention of Nature; There Are Natural Causes

In Plato's last great work, "Laws", written in the form of a dialogue a few years before his death, one of the three participants in the discussion, the Athenian, the figure representing wisdom, looks back to his youth to brilliantly summarize the intellectual content of the philosophy inherited from the pre-Socratics:

> "the world are the products of nature and chance [*chance*, used by Plato. indicates a spontaneous cause rather than an intelligent one]. . . . The works of nature, they say, are grand and primary. . . . They maintain that fire, water, earth and air owe their existence to nature and to chance . . . and in no case to art (*technei*). As for the bodies that come after these—the earth, sun, moon and stars—they have been produced from these entirely inanimate substances. These substances moved at random, each impelled by virtue of its own inherent properties, which depended on various suitable amalgamations of hot and cold, dry and wet, soft and hard, and all other haphazard combinations that inevitably resulted when the opposites were mixed. This is the process to which all the heavens and everything that is in them owe their birth, and the consequent establishment of the four seasons led to the appearance of all plants and living creatures. The cause of all this, they say was neither intelligent planning, nor a deity, but—as we have explained—nature and chance".
>
> PLATO, *The Collected Dialogues*, 1961, pp. 1444–1445, ed. by Edith Hamilton and Huntington Cairns, Princeton University Press

In the *Phaedo* (96b) Plato described his early enthusiasm for this project:

> "When I was a young . . . I had an extraordinary passion for that branch of learning which is called natural science (*philosophy*). I thought it would be marvelous to know the causes for which each thing comes and ceases and continues to be".
>
> PLATO, *The Collected Dialogues*, 1961, p. 78, ed. by Edith Hamilton and Huntington Cairns, Princeton University Press

Between the time of the Homeric–Hesiod epics and the creation and compilation of the Hippocratic writings and continuing into the fourth century BCE, something revolutionary occurred. A new way of thinking about the world emerged in the sixth to fifth centuries BCE in Ionia (western Turkey) and in Italy and Sicily, initiated by a diverse group of individuals speculating about the world. They viewed it as an ordered system that is comprehensible to the mind of humans and that is composed of certain basic materials, and they believed that the processes of change are brought about by what are designated as natural causes. In these writings, a god or gods are not actors in causing phenomena in the world. Nature is endowed with powers previously attributed to a divinity. All of the phenomena take place as a result of the inherent properties of the physical objects in the world, powers previously assigned to gods. In short, they changed the discourse about the cause of phenomena in the world from gods to nature. These phenomena include the causes of disease.

The classification pre-Socratics was created in the nineteenth century to differentiate their speculations, with their emphasis on cosmology and the physical and biological world, from the interests of Socrates (469–399 BCE), whose concerns were moral issues, and to the simple fact that many, not all, preceded Socrates in time. They viewed themselves as differentiated from poets and historical chroniclers, interested in fields that include, in modern terms, physics, astronomy, cosmology, and biological sciences.

Why include pre-Socratic and later philosophical works in a book devoted to the causes of infectious/contagious disease? For the essential reason that they and others laid the philosophical base for the writers who contributed to the disease theories contained in the Hippocratic Corpus, by providing causal answers about the operation of the world that came to be designated as natural causes.

There are different varieties of pre-Socratics who contributed in different ways to their legacy. Thematically, there are philosophers before Parmenides, who speculated about the origin and composition of the world and who accepted the idea of change. There is Parmenides, who reasoned about the work of his predecessors, concluding with the principle that change does not occur; this marks the beginning of metaphysics, the beginning of philosophy. The third group, chronologically after Parmenides, are the writers who dealt with the challenge posed by Parmenides, by introducing hypotheses about the nature of matter that allows change to take place. Let us review some of their writings to get the substance of their contributions.

The New Philosophy: Milesia–Ionia–Sicily–Italy

The earliest group of philosophers came from Miletus, a small, wealthy, coastal city-state on the Aegean Sea in present-day Turkey and an important trade route where the cultures of Egypt, Babylonia, Lydia, and Phoenicia came together. Their names are Thales, Anaximander, and Anaximenes. Heraclitus came from Ephesus, some 50 km north of Miletus. From Ionia, north of Ephesus, came Xenophenes, more poet than philosopher. From Sicily and southern Italy came Empedocles and Parmenides. These individuals and the ones who followed were interested in everything about the world. They invented a new way of speculating about the world, which had an enormous impact on theology, politics and government, and natural philosophy (science), including animal and human generation and development, medicine, and theories about the cause of infectious diseases. Whereas in the Homeric–Hesiod and Israelite traditions gods were responsible for almost every aspect of the operation of the world, including the health of humans, in the world envisioned by these philosophers, deities have no such roles. The work of gods is attributed to something defined as nature. There are active principles present in matter that are responsible for origin and change.

What Is the Origin and Composition of the World?

The earliest of these pre-Socratics was Thales, born in the late sixth century BCE. We have information about him from Aristotle, living in the fourth century BCE, from Xenophenes (*ca.* 575–*ca.* 475 BCE), and from the historian Herodotus (438 BCE–post-430? BCE). Thales was a geometer and an astronomer who theorized that the earth is a flat disc that floats in water and, according to Aristotle, believed that a single source of the universe was water. Aristotle conjectured why Thales proposed this theory; among the reasons is the fact that "water is necessary for life." Thus the world and all its diversity, including all living things, came from one original substance, by a process that is not described but is uniquely important since it does not involve a deity.

Anaximander (*ca.* 610–546 BCE) wrote primarily about cosmology, astronomy, and biology. We know about him from writings by Aristotle and Theophrastus, a pupil of Aristotle. Like Thales he speculated about the origin or the principle, the Greek word *arche*, that is the first matter from which the world develops. This matter was characterized as "boundless," some kind of stuff that can give rise to air, fire, and water. In his biology, Anaximander generates life from moisture and argues that by some kind of evolutionary process humans were created, not by the intervention of gods but by the workings of the materials that compose living matter.

The third of the early Milesians is Anaximenes, who lived in the mid-sixth century BCE and was possibly associated with or a student of Anaximander. He proposed that a single substance, air, is the source of all things, not water nor some "boundless stuff." Anaximenes believed that air is a more malleable material that can be transformed by physical forces to other materials that can be organized to make the world: that provides a method to explain how air might be physically changed to other materials. By a process of rarefaction and condensation, making air more or less dense, it is possible to produce qualitatively different states of matter. Anaximenes not only proposed a theory but also suggested a way to justify, or prove, his hypothesis by stating a method to do so. This theory and method were adopted by Heraclitus (*ca.* 540–*ca.* 480 BCE) and Anaxagoras (500–430 BCE), both pre-Socratics, and rephrased by Plato (430–345 BCE) in his work:

"In the first place, we see that what we just now called water, by condensation, I suppose, becomes stone and earth, and this same element, when melted and dispersed, passes into vapor and air. Air, again, when inflamed becomes fire, and, again, fire, when condensed and extinguished, passes once more into the form of air, and once more, air, when collected and condensed, produces cloud and mist-and from these, when still more compressed, comes flowing water, and from water comes earth and stones once more-and thus generation appears to be transmitted from one to the other in a circle".

PLATO, The Collected Dialogues, 1961, ed. by Edith Hamilton
and Huntington Cairns, 1176, Princeton University Press

Air was also used as a central concept in a theory of disease in the Hippocratic treatise *On Breaths*.

The development of the complexity of the universe occurred in stages according to Anaximander, with the generation of opposites like hot and cold that balance each other and lead to a regulated system. The same laws that operate in the universe function in the human body. which leads to equating the macrocosm and the microcosm (the human).

Not only are the two systems made of the same materials but they also operate in the same way: "the first principle . . . of existing things was the boundless." It is the source of everything, and everything is dissolved back into it.

Xenophanes (575–475 BCE), possibly a student of Anaximander, directed his writings against the theology contained in the Homeric–Hesiod works. Whereas the earlier Milesians had simply eliminated gods as agents of causality in the world, Xenophanes made a direct assault on gods. Here are important examples of his attack. He considered gods to be inferior to men:

"One god, greatest among gods and men,
    In no way similar to moral men in body or in thought"

ROBERT WATERFIELD, The First Philosophers, 2000, p. 26,
Oxford University Press

Xenophanes ridiculed humans' tendency to create gods in their own image:
    "If cows and horses or lions had hands, or could draw with their hands and make things a man can, horses would have drawn horse-like gods, cows cow-like gods, and each species would have made their god's bodies just like their own"

ROBERT WATERFIELD, The First Philosophers, 2000, p. 27,
Oxford University Press

Xenophanes piled it on:

"Ethiopians say their gods are flat-nosed and black, and Thracians that theirs have blue eyes and red hair".

ROBERT WATERFIELD, The First Philosophers, 2000, p. 27,
Oxford University Press

Xenophanes did have his own version of a god, but it was without anthropomorphic characteristics and was not localized like Zeus, but rather some abstract idea of Zeus, which some commentators have suggested is a monotheistic position. He did not propose a systematic cosmogony but did comment on the presence of marine fossils. This led to speculations about various cyclical events on earth that occur without the intervention of gods, including the origin of the world:

"Xenophanes believes that the earth is becoming mixed with the sea and that it will eventually be dissolved by the moist. He adduces the following evidence: shells are found inland and in the mountains; in the quarries at Syracuse the impression of a fish and seaweeds has been found; on Paros the impression of a bay-leaf has been found buried in stone; and on Malta there are slabs of rock made up of all kinds of sea-creature. He says that these came about a long time ago, when everything was covered with mud, and that the impression became dried in the mud. He claims that the human race is wiped out whenever the earth is carried down into the sea and becomes mud, that then there is a fresh creation, and that is how all the worlds have their beginning".

ROBERT WATERFIELD, *The First Philosophers*, 2000, p. 29,
Oxford University Press

Heraclitus (*ca.* 530–485 BCE), another Ionian, wrote a systematic account of nature and held that a true explanation of all things required an understanding of its composition. His cosmogony includes a "supreme cosmic principle," which can be known by individuals like himself not only through sense experience but also "introspection," which suggests a combination of reason and revelation. The world so organized is a harmonious one.

The consensus about Heraclitus, from Plato and Aristotle into modernity, is that he believed like the early Milesians that the *kosmos* (which is everything) originates from a single substance: fire. Heraclitus also provided a scheme for the development of the other elements required for the creation of the world. The transformation begins with condensation of fire so that it ultimately becomes water, which then becomes earth. The other important basic philosophical position assigned to Heraclitus is the so-called flux doctrine, or the principle of constant change. Numerous ancient citations are available to support this contention, one of which was presented by Plato in *Cratylus*. We provide two translations of the ancient Greek, the alternate reading contained in parentheses.

"Heraclitus says somewhere (is supposed to say) that everything gives way and nothing is stable (at rest) and in likening things (he compares them) to the flowing of a river (to the stream) he says that one (you) cannot step twice into the same river (go into the same water twice)".

ROBERT WATERFIELD, *The First Philosophers*, 2000, p. 41,
Oxford University Press

Given the at times obscure language of Heraclitus, do we know that he intended such an interpretation? There appears to be one authentic "river fragment" by Heraclitus, which is subject to different interpretations, including the one offered by Plato. Whatever the reading, the concept of change is an inherent part of the different rationales. Clearly, the reading by Plato was accepted between the end of Heraclitus' life and Plato's writing, separated by more than 50 years. Later, Plutarch (*ca.* 45–120 CE) ascribed to Heraclitus the position that "It is impossible to step twice into the same river."

A fundamental theory connects pre-Parmenides thinkers. The world is made of some sort of matter, primordial stuff, or water or air, out of which everything is made. There is change in the world. This popular acceptance may have motivated Parmenides to advance an antithetical philosophy so that in the following years the representatives of contrasting metaphysics were Heraclitus and Parmenides. Constant change was Heraclitus and non-change was Parmenides.

Parmenides offered a serious objection to the idea of change. In this way he is a foundational figure. He "authored a difficult metaphysical poem that earned him the reputation as early Greek philosophy's most profound and challenging thinker" (http://parmenides.stanford.edu). His "challenge" is that there is no real change in the world, in contrast to the generally accepted position of his contemporaries. His view was contested and we will present their various answers—but first, Parmenides on metaphysics.

Parmenides' poem contains an extensive metaphysical section and cosmological speculations. He travels to a place where he obtains wisdom from a goddess who presents arguments for what is true reality. It seems that the supernatural source of learning or its role in the world has not completely disappeared among philosophers. The nature of true reality remains.

"that it is. It is . . .

It is eternal. It is a uniform substance, from which everything is composed, that did not come into existence and will not go away.

Neither will I allow you to say

Or to think that it grew from what-is-not, for that it is not

Cannot be spoken or thought.

. . . and there is no such thing as nothing" . .

In short, "it is or it is not".

ROBERT WATERFIELD, *The First Philosophers*, 2000, pp. 58, 59,
Oxford University Press

The question is, what is "it"? It is everything in the world. What is, exists, has always been present and will remain always present into the endless future:

There is being, which exists, and there is non-being that is nothing, which does not exist. What is not is nothing. Parmenides stated that to get change, something must come from nothing. However, nothing does not exist. The consequence is that nothing can come from nothing.

He started from the premise that the universe exists, that it is. What is not contained in this principle is that the universe was or will be. Why? Existence for Parmenides excludes was and will be. Existence and being are identical. Existence and becoming are incompatible. Non-being cannot come into being. The conclusion is that there is no change in the world.

Why, within the context of a discussion of disease theory, is there an emphasis on Parmenidean metaphysics that contends that change is an illusion? The reason is that this contention is disputed by a number of philosophers, including Empedocles, Anaxagoras, Leucippus, and Democritus (the last two are designated as *atomists*), who provided new matter theories to account for their contention that there is visible change. The matter theories of Anaxagoras and Empedocles became part of disease theory in the Hippocratic writings, and remarkably, a particle theory of matter, atomistic theory, is incorporated in contagious disease causation in the sixteenth and seventeenth centuries CE.

The successors to Parmenides agreed that something cannot come from nothing; however, they believed what their senses showed them, that there is change. To explain these transformations they postulated the existence of entities that indeed did not change, but by the rearrangements of these immutable entities the coming to be was achieved.

The Processes of Change in the World

Anaxagoras (*ca.* 500–430 BCE) came from Ionia, north of Ephesus. He was a pupil of one of the original Milesian philosophers, Anaximenes. Anaxagoras is classified as a pluralist, meaning he believed that the original state of the world already contained everything. Anaxagoras' metaphysical base comes from Parmenides; there is no coming-to-be nor passing away. What is, is eternal, and "everything is in everything." There is no change. Anaxagoras, nevertheless, recognized visible generation and destruction, which he explained in this way:

. . . . . "that all things, including humans, are aggregates of the stuffs that were present in the original mixtures, so that all physical change is no more than the manifestation of what was previously latent".

ROBERT WATERFIELD, *The First Philosophers*, 2000, p. 116,
Oxford University Press

What this means is that the basic physical entities remain intact, unchanged, as they form a mixture that has the visible properties characteristic of the object. The same basic entities reemerge when the visible object passes away. In this philosophy the Parmenidean principle is upheld. All of these entities and their interactions are natural occurrences, not created by gods, and do not fulfill some teleological principle, a goal-like purpose.

Anaxagoras dealt in general terms with the content of these mixtures, and it is of interest to ask why did he adopt the axiom *"everything is in everything."* He may have used such an apothegm to make comprehensible the phenomenon of nutrition. Consider that an individual will eat a food, for example bread, which then becomes part of the human body. One could argue that the grain of the bread was transformed into bone, skin, and the numerous parts of the body. However, that would suggest that change had occurred, grain into bone. But that would violate the principle that change cannot occur. Anaxagoras' contention that everything is in everything would allow the bread to contain the basic entities that are also present in the person feeding on the bread. The conversion of the bread to person results from the rearrangement of these unchangeable entities.

Another Response to Parmenides

Empedocles (*ca.* 493–435 BCE), a practicing physician, also supported Parmenides' principle that things cannot come into existence from what does not exist, but things can be generated and return to their constituent components. He also acknowledged that at the level of visibility, there is change; things can be generated. However, at a deeper level there is no change, for things revert to their constituent elements. That is, the basic elements that make up all bodies, the original entities that were there when the body was formed, remain after the dissolution of that body. He goes on to define these basic entities, introducing the tenet that the cosmos is composed of four unchangeable roots (as he calls them), earth, water, air, and fire. They are "irreducible, imperishable, and underived." In contrast to Parmenides' position that there is only one thing, what is, and the idea posited by Thales, Aximander, and Anaximene that everything comes from one substance, Empedocles says there are four. In addition to the four elements, he postulated that there are two forces in the world, love and strife. When they are balanced, when these forces interact and one is not more powerful than the other, and with the combinations of the four elements the inestimable number of things in the world can be made. The four-root hypothesis includes the biological world. which creates a philosophy in which all the processes of the world are explained by the same theory.

Empedocles also presented himself as a physician-magician and wrote about medical topics, including generation, digestion, and the motion of blood and air in the body. He theorized that the flesh and blood of humans is made of particles of the four components that underlie all the processes, like sensation, digestion, and nutrition, and these are explained by common principles. Digestion, for example, proceeds with the breakdown of food in the stomach by a process described as putrefaction that is the result of the innate heat of the body. The concept of innate heat is common among pre-Socratics and is important among Hippocratic writers and Plato and Aristotle. The digested material, the nutrients, are moved to the liver and there converted to blood and distributed to all parts of the body through vessels and become parts of all the components of the body. All of these conversions can occur because everything is made of the same four components, earth, air, fire, and water. These components in food can be converted to bodily parts by rearrangements of the four elements. It is obvious that his discussions of biological issues, like his writings on cosmology, stressed naturalistic explanations.

A major consequence of Empedocles' theorizing occurred when his four-element theory was revived by Aristotle, which influenced the development of the important four-humor theory in the Hippocratic writings, the foundation of disease theory in the Hippocratic Corpus. In Aristotle's matter theory, his physics, there are four causes. There is the material cause, the primary matter earth, air, fire, and water, which constitute a body. There is the formal cause, which makes an object what it is. In Aristotle's physics there are two more causes: the efficient cause, which elicits the effect, and the final cause, the purpose of the object. Change is the result of the conversion of one form to another, which means changing the combination of the four elements.

The Invention of Atomism: Another Answer to Parmenides

The response to Parmenides continued in the writings of Leucippus and Democritus, who proposed a theory of matter in which everything in the world is composed of indestructible, invisible, unchanging, eternal particles (atoms), present in a void (empty space) that allows for the movement of these particles; this allows for the various arrangements that make up the visible entities of the world. The atoms and the void always existed; this satisfies the Parmenidean principle that what is, is, and does not change.

The writings that prepared the ground for atomism were those of Empedocles, who stated that matter was composed of a plurality of indestructible components. The atomists adopted this axiom and defined atoms according to their shapes, sizes, and motion. Empedocles held that the processes that lead to change are the result of the rearrangements of the basic four components of matter, which do not change. The atomists appropriated the principle that there is generation and destruction, but the atoms are not altered: change occurs because of the reshuffling of atoms.

With regard to the characteristics of particles, "Leucippus and Democritus say . . . there are only three differences—in shape, arrangement, and position"
ROBERT WATERFIELD, *The First Philosophers*, 2000, p. 116,
Oxford University Press

The last of the foundational figures of atomistic philosophy was Epicurus (341–270 BCE). He developed a general philosophical system intended to lead a way to a happy, pain-free life, in body and mind, integrated by an epistemology based on sense perception, and a material world whose foundation was atomic theory. He also provided an account of the origin of the world and humans all based on natural phenomena. An important source of his work is contained in Diogenes Laertius' *Lives of Eminent Philosophers* and in the major work of a devoted Epicurean, Lucretius (98–95 or 95–55 BCE), whose work, "De rerum natura," deals extensively with his physical theories and methodology.

Epicurus took up the philosophy initiated by Leucippus and, importantly, Democritus, who contended that the matter of the world is made of atoms that are invisible and indivisible, moving in empty space. Epicurus hypothesized that atoms come in different sizes and shapes, the latter instrumental for their interactions. They do not have qualities that humans observe in visible objects such as color, smell, or temperature; they have sizes and shapes. Epicurus held that the variety of atoms is large but not infinite; however, the number of each kind is infinite. Thus, there is a limit to the types of combinations and therefore a limit to the number of things in the visible world.

An important axiom for Epicurus was that things cannot come from nothing: "nothing comes into being out of what is non-existent." Certainly,

Parmenides proposed this rule. Epicurus included another reason for accepting this principle: he recognized that if it did happen, "everything would have arisen out of anything," and therefore there would be no need for specific seeds, for example, to obtain plants and animals. The sense experience of reproduction allowed him to generalize for all objects in the world. Lucretius, more than two centuries later, reproduced the same argument:

> "Therefore when we have seen that nothing can be created out of nothing . . . For if things came to being from nothing, every kind might be born from all things, nought would need a seed".
>
> LUCRETIUS, *De rerum natura*, trans. by Cyril Baily, 1947, p. 32,
> Oxford University Press

Epicurus adopted the corollary that things that exist cannot be converted to what is not: what remains are the unalterable being of atoms, consistent with Parmenidean metaphysics. Epicurus continued to use atomic theory to explain secondary properties such as color, which is the result of different combinations of atoms, and perception at a distance. In the latter case there is the emission of thin emanations consisting of particles (Lucretius later designated it as *simulacra*) preserving the atomic pattern of the object, which impinges on the eye of the beholder without diminishing the object because other particles take their place.

The atomistic matter theory has no direct effect on medicine or infectious/contagious disease theories in the Hippocratic writings, but it returned as an important foundational component of contagious disease theory in the late sixteenth and seventeenth centuries after the recovery, during the Renaissance, of *De rerum natura* by Lucretius, the avowed Epicurean, and the revival of atomism in the seventeenth century as an integral part of the mechanical philosophy contained in the philosophy and science of that century. Therefore we shall return to that version of this theory in Chapter 8, where a particle theory of matter is appropriated to explain the cause of contagious disease.

What the natural philosophers agreed upon was the explicit principle that there are regularities in the physical world and many aspects of the biological world. Certainly uniformities had been recognized in the Homeric–Hesiod period: the changing seasons, the time for planting, the need for seeds to grow food crops. But now there was the explicit idea that the world exists as an objective reality with regularities that have general applications. This led to the important concept that the recurring processes in the world, in nature, were natural events, and their causes could be studied and understood.

This free-ranging speculation as described by Plato and initiated by the pre-Socratics included biology and medicine. The connection between philosophy and medicine proceeded logically from questions about the origin of the universe to questions about the origin of humans. The latter could actually be studied with empirical inquiries into a field such as embryology, which dealt with the manageable issue of the origin or development of a child from an apparently undifferentiated seed. In these new philosophical systems humans were part of the natural world, characterized as the macrocosm, while the human being was a microcosm, the miniature of the macrocosm, constituted of the same materials that make up the physical world and subject to the same causes. Consequently the health and diseases of humans cannot be understood unless the composition of the world is understood. The matter of the world by its own power evolves into the components of the world. The human body is composed of the same world matter already differentiated into dissimilar parts, and the health of the body is dependent on the interaction of these parts of the body. According to these new propositions disease began to be viewed as a disturbance or an alteration of the constituents of the body. The most radical consequence of this philosophical position was that the causes of disease are not imposed by gods but set off by environmental influences.

Empedocles' theory of health is based on a balance or equilibrium of four body components that are components of the cosmos. It is important to note that Empedocles' hypothesis that the body is made of four components was used later to create a scheme where the human body is made of four humors, blood, phlegm, and yellow and black bile. Health in this system is the result of the proper balance of the four humors. Such a theory appears in the Hippocratic writings that we shall discuss in the next chapter. This theory would remain regnant for the next two thousand years.

Before we discuss in some detail the contents of the Hippocratic Corpus, I would like to mention a work written in the last half of the sixth century BCE by Alcmaeon, a physician-philosopher who presented a disease theory independent of gods. He was a contemporary of the first pre-Socratics and predated Hippocrates and the Hippocratic writings by about two centuries. Alcmaeon believed that what preserves health is the equality of the powers:

> "moist and dry, cold and hot, bitter and sweet and the rest—and the supremacy of any one of them causes disease; for the supremacy . . . of either is destructive. The cause of disease is an excess of heat or cold; the occasion

of it surfeit or deficiency of nourishment; the location of it blood, marrow or the brain. Disease may come about from external causes, from the quality of water, local environment or toil or torture. Health, on the other hand, is a harmonious blending of the qualities".

JAMES LONGRIGG, Greek Rational Medicine, 1993, p. 52, Routledge

In this discourse, disease is caused by environmental effects, external factors that are not influenced or caused by a divinity. Alcmaeon's theory may provide support for the idea that diseases can pass from person to person, although such a theory was not explicitly stated during this historical period.

The interplay between philosophy and medicine and the speculations about health and disease and their connection to natural philosophy or science are seen in the writings of Alcmaeon. He was described as a pupil of Pythagoras. He "wrote chiefly about medicine . . . he touches on natural philosophy." The Pythagoreans had a catalogue of 10 principles of opposites, among them odd/even, light/darkness, and so forth. Alcmaeon extended the number that he used to construct a theory of health and disease. The famous passage, already cited, proposed that health depends on the equality among the opposite powers, or a balance among the powers of moist/dry, hot/cold, and so on. What produces disease is the *supremacy*, or in another translation of the Greek word *monarchia*, among the opposites. Put another way, an imbalance leads to disease.

Using an *a priori* principle, a doctrine of opposites, Alcmaeon proposed a series of events involving materials in the environment that can interact and interfere with processes in specific regions of the body to cause disease. To explain disease in this way is a remarkable enterprise, equivalent to explaining the production of lightning and thunder by the interaction of physical bodies. It is a natural explanation, an explanation that does not involve a deity.

A general summary of pre-Socratic thought through Epicurus would include the following principles:

1. Humans live in a world that is an orderly system that can be comprehended by the human mind.
2. The outer world and humans had their origins without the intervention of gods.
3. The world is derived from one or a small number of basic elements. About this last view there are theoretical differences.

4. The same materials that constitute the world are present in humans, and the same processes at work in the outer world are operative in the human sphere.
5. The causes of the reality of the world are due to processes inherent in the matter of the world. These are natural causes. The causes of diseases are included under the rubric of natural causes.

| Hippocratic Writings

Diseases Have Natural Causes

What to Expect in the Various Hippocratic Treatises

The Hippocratic Corpus, composed from the mid-sixth century to the fourth century BCE, reveals the theoretical and observational foundations of Greek medicine. These writings show the influence of the new philosophy initiated by pre-Socratic thinkers, whose principles were adopted to help explain the composition of humans, who are made of the same materials that constitute the physical world and are subject to diseases caused by natural events that occur in the physical world. In the pre-Socratic writings gods are not responsible for the origin and processes in the world. Among the Hippocratic authors, gods are not the originators of disease. These hypotheses are the most revolutionary consequence of a theory of natural causes.

There is general agreement that the health of humans is the result of a balance among the readily observable humors of the body (blood, phlegm, and black and yellow bile), which parallel the four-element theory of matter of Empedocles (earth, air, fire, and water) and the four qualities of Aristotle (hot, cold, moist, and dry). Any number of environmental factors, including winds, bad water, foul vapors, or faulty diet, could upset the balance of humors and cause disease.

The treatises represent somewhat different epistemologies, a preeminent issue in Greek medicine and philosophy. There are two contending positions, one a deductive system based on truths that are self-evident, which can be characterized as axioms or *a priori* philosophical principles, from which further truths are derived. A number of works are in this group. One such theoretical work is called "Fleshes." It proposes to be about medicine, but the writer holds that the proper starting point is to explain the origin of humans and animals, the nature of the soul, what constitutes health and disease. In the treatise "The Nature of Man" a disease theory is developed following the introduction of a cosmology that leads to the famous four-humor theory and the concept of balance among the humors, which is necessary for good health. This theory became emblematic of Hippocratic medicine and was elaborated in the treatises *Diseases 2* and *4*. It remained part of medicine into the nineteenth century CE.

In the treatise "Regimen" *1*, Chapter 2, there is expressed the need for a strong philosophical grounding in order to treat patients properly:

> "I maintain that he who aspires to treat correctly of human regimen must first acquire knowledge and discernment of the nature of man in general— knowledge of its primary constituents and discernment of the components by which it is controlled".
>
> HIPPOCRATES, Regimen 1, 1931, p. 227, W.H.S. Jones, Publisher/Imprint:
London, Heinemann, New York Putnam's Sons

A contrary view of Hippocratic epistemology is stated in "Ancient Medicine" (Sections 1 and 2). The Hippocratic writer states that many who write on medicine postulate the same cause for many diseases. He rejects "empty postulates" that cannot be judged either true or false "for there is no test the application of which would give certainty." There is another system of inquiry, an empirical system, where knowledge is built on experience. Such knowledge does not come about through philosophical speculation about the nature of man but comes about through the study of humans, their relation to their total environment. These ideas are made explicit in Chapter 20 of "On Ancient Medicine" (Tradition in Medicine):

> "I think I have discussed this subject sufficiently, but there are some doctors and sophists who maintain that no one can understand the science of medicine unless he knows what man is; that anyone who proposes to treat men for their illnesses must first learn of such things. Their discourse then tends to philosophy, as may be seen in the writings of Empedocles and all the others who have ever written about Nature; they discuss the origins of man and of what he is created. It is my opinion that all which has been written by doctors or sophists on Nature has more to do with painting than medicine. I do not believe that any clear knowledge of Nature can be obtained from any source other than the study of medicine and then only through a thorough mastery of this science. It is my intention to discuss what man is and

how he exists because it seems to me indispensable for a doctor to have made such studies and to be fully acquainted with Nature. He will then understand how the body functions with regard to what is eaten and drunk and what will be the effect of any given measure on any particular organ".

Hippocratic Writings, 1978, p. 83, ed. by G.E.R. Lloyd, "On Ancient Medicine", Penguin Books

Although disowning philosophy, the writer provides his own philosophical base, defining nature as the total physical environment of humans, including the food and air that the human imbibes. The interaction of the human with his surroundings then becomes the domain of the science of medicine. Thus it is obvious that there are debates between proponents of theory as a necessary foundation for medicine and those contesting this hypothesis. In summary, there is much theory contained even in those treatises hostile to philosophy.

Two important principles emerge from these writings. To know the cause of disease, is the indispensable condition of a true medical art. It is crucial to know the cause of a disease in order to treat the disease. Deities are no longer involved in causation, and the total effect of the writings is one of medicine based on experience. Natural causes are justified by observation, analysis, and even some modest experimentation.

The Hippocratic Corpus

The Corpus consists of some 70 treatises composed during the period of the fifth to the fourth century BCE. They enable us to view the development of Western medicine and science in the Periclean age following the revolutionary proposals about the material composition of the world and the causes of physical phenomena by the pre-Socratic philosophers. The Corpus is named after the premier physician of his period, Hippocrates, who was born on the island of Cos, southwest of the region of Ionia, now western Turkey, in 460 BCE. He moved to Thessaly in central Greece, a region famous in Greek mythology as the place of origin of the Asclepiads, the descendants of the Greek god of medicine Asclepius. According to Plato it was Asclepius who first established medicine as a profession.

The fame of Hippocrates is attested to by Plato in "Phaedrus," where Hippocrates is extolled as a thinker about the science of the body and as someone who in his practice of medicine is a model for the philosopher who is dealing with the "science of the soul." Plato regards Hippocrates' work as a model for a method of inquiry in philosophy.

Hippocrates' primary fame was as a teacher of medicine. He was trained as a physician according to Greek tradition, in which much of medical training took place in families. There were no medical schools, no regulations governing medical practice, nor any licensing procedures. Individuals who considered themselves physicians were competing for patients with a variety of healers, including the large number affiliated with the cult of Asclepius associated with shrines and temples in Greece and other sites scattered over the Mediterranean basin.

The Corpus is a collection of essays covering topics that in modern terms can be classified as physiology, pathology, embryology, surgery, and ethics. The treatises were composed by many authors. How do we know this?

1. Aristotle states in his book "Historia Animalium" that the author of "The Nature of Man" is Polybius, a son-in-law of Hippocrates. He also refers to a passage in the treatise "Nature of Bones" written by another follower of Hippocrates.
2. The treatises contain competing doctrines: some authors illustrate the relationship between philosophy and medicine while others argue against this connection.
3. The treatises are aimed at different audiences: some are intended as popular lectures, not for physicians, while others are directed to physicians. Some of the treatises appear not to be written by physicians.
4. By all accounts the treatises were written at different times, some during Hippocrates' lifetime but some after him.

The intellectual depth of the Corpus reveals the development of medicine and disease theory. It is a rich source of ideas, and it will be important to examine some of the treatises in detail to exhibit this fact.

Natural Causes

Four treatises in the Corpus present powerful, unambiguous support for a natural cause for all diseases: "Airs Waters Places" (AWP), "The Sacred Disease," "The Art," and "Breaths." Although the goal is the same in the writings, the forms of the presentations are distinct. In AWP it is through the content of the discourse that natural causes are revealed, although in Chapter 22 there is a definitive statement to that effect. "The Sacred Disease," at its origin and throughout, is a sustained polemic in favor of natural causes.

AWP contains a large body of observations of peoples in various lands, subject to different environmental conditions, including air and water, temperature and winds. From a detailed knowledge of these many physical factors and the geography of the region the physician will know which diseases occur in a particular place and at which time of year. These constant conditions determine the characteristics of the indigenous peoples and the diseases they acquire. From a close study of the many conditions external to humans the writer creates a philosophy of disease that can be summarized in this way:

1. Each disease has a natural cause. The cause is some element in the environment interacting with the individual, and consequently there are many natural causes. These causes are in contradistinction to causes attributed to a god.
2. The cause sets off some imbalance in bodily fluids like bile and phlegm that results in disease. Included among the humors in the Corpus is blood, a crucial component in disease theory. We shall return to this important factor in a later treatise.

Diseases come from two sources of water, foul rainwater and water derived from melted snow or ice. The writer asserts that fresh rainwater "is likely to be the best of all water," but it may become the cause of sore throat, coughs, and hoarseness when it has a "foul smell." However, if the water is boiled it is "purified" and can be consumed. This is a stunning statement without further comment. That water is a cause of disease is probably not new to the Corpus. That boiling prevents the water from spreading disease may have been discovered in some fortuitous way, and remains as a "cure." But it leaves open the challenging question: What is in the water that causes disease, and what is it that boiling does to change the character of the water? The issue is not addressed.

A theory is offered to account for the unhealthy character of water derived from melted snow or ice. When ice thaws, the "clear light sweet part . . . disappears." In support of this hypothesis the author carried out the following experiment. He took a measured amount of water and froze it outdoors. When thawed indoors, the volume was found to be less than the original. This remaining water has the unhealthy character. It is interesting that an experiment was carried out to verify the hypothesis. What this reveals is that the author considers the cause of the sore throat and coughs to be a natural event, and therefore an answer can be obtained by a methodology that manipulates the components of the system.

The explicit contention that every disease has a natural cause comes in Chapter 22. In this chapter the writer discusses the report that the Scythians (a nomadic people who lived at different times in central Asia and southern Russia) can become impotent and they "attribute this to divine visitation." Herodotus accepts this explanation and provides a reason for the incapacity. The Scythians attempted to invade Egypt but were met in Palestine by the Egyptians and persuaded to withdraw. A portion of the Scythian army robbed the temple of Aphrodite in Ascalon and were punished by the goddess with a "female disease." Their descendants still suffer from it. The writer of *AWP* objects to this explanation unequivocally,

> "Really, of course, this disease is no more of "divine" origin than any other. All diseases have a natural origin" . . . . . .
>
> *Hippocratic Writings*, 1978, p. 166, ed. by G.E.R. Lloyd,
> "The Sacred Disease", Penguin

Before we elaborate on the unusual way natural causes are portrayed in the first version, let us turn to another treatise, "The Sacred Disease," where an almost identical phrasing is used to designate its cause. The disease is epilepsy, not a contagious disease. The treatise opens with an assault on the idea that it is a sacred disease, caused by a divinity, and a criticism of the purveyors of such a theory. Their faulty theory is not just bad medicine, but it is not really medicine at all, but rather witchcraft; the practitioners are charlatans. This simultaneous attack on a divine origin of disease and the individuals who promote such a position is aimed at real persons practicing this sort of medicine. The writer intends to eliminate them as healers and as such in competition with physicians practicing Hippocratic medicine. To displace the divine theorists the Hippocratic writer provides alternate explanations for this disease and all diseases, replacing a divinity with nature. We will deal with this concept of nature later.

Here is how the writer proceeds to demolish the divine nature of the disease. He declares that there is "a definite cause" that eliminates the idea that there is a divine cause:

> "I do not believe that the "Sacred Disease" is any more divine or sacred than any other disease but, on the contrary, has specific characteristics and a definite cause. It is my opinion that those who first called this disease "sacred" were the sort of people . . . witch-doctors, faith healers, quacks and charlatans".
>
> *Hippocratic Writings*, 1978, p. 237, ed. by G.E.R. Lloyd,
> *The Sacred Disease*, Penguin

These (charlatans) provide a variety of therapies intended to counter the divine character of the disease. Imagine countering the will of a divinity! In doing so they have a win–win situation: if the patient is cured, their reputation is retained, and if the patient dies, they can ascribe it to the will of God.

Ultimately, the author of "The Sacred Disease" must provide an alternate explanation for the cause of the disease inasmuch as he has demolished sacred causes and claimed there are natural causes.

The cause is based on the complex structure of the brain connected to the nervous system linking the liver, kidney, heart, and arm, reaching into the neck and brain. What obstructs the brain is phlegm. There follows a highly speculative discourse on the relationship between phlegm and disease, including the argument that it can be a hereditary condition. A summary arbitrarily states that this disease, like all diseases, has a natural cause:

"This so-called "sacred disease" is due to the same cause as all other diseases, to the things we see come and go, the cold and the sun too, the changing and inconstant winds. These things are divine so that there is no need to regard this disease as more divine than any other; all are alike divine and all human. Each has its own nature and character and there is nothing in any disease which is unintelligible or which is insusceptible to treatment".

Hippocratic Writings, 1978, p. 251, ed. by G.E.R. Lloyd,
"The Sacred Disease", Penguin

Nature has replaced a divinity. The writer has made nature divine. Disease is now part of the regularities of the world.

The Different Forms of Natural Causality

There is no general agreement on the number of humors in the Hippocratic writings. In "The Sacred Disease" phlegm and bile are used, but in "Affections" all diseases are due to phlegm and bile. In "Diseases", "all diseases come to be from two sources, things inside the body from bile and phlegm." In "Tradition in Medicine" (also titled "Ancient Medicine") the writer uses the term forces. All ills are due to forces. A force is biliousness that leads to fever and nausea. If we get rid of yellow bile, pain and fever disappear. Health is a result of a balance between forces. An extended statement of a humoral theory is contained in the treatise "The Nature of

Man." The treatise is dated about 410–400 BCE. It is identified as a lecture in the opening sentence. It is a polemic against the use of philosophy in medicine, attacking the concept that humans are made of a single element and physicians who contend that man is made of a single humor. Although there is disapproval of physicians building their theories on the philosophical principle of one humor, the writer presents an equally speculative system influenced by Empedocles, who proposed a world composed of four roots (earth, water, air, and fire), held together by the divine opposite forces of love and strife. The author of "The Nature of Man" hypothesizes that the body is composed of four components (blood, phlegm, and yellow and black bile), and his theory of health rests on the principle of balance among the humors.

> "The human body contains blood, phlegm, yellow bile and black bile. These are the things that make up its constitution and cause its pains and health. Health is primarily that state in which these constituent substances are in the correct proportion to each other, both in strength and quantity, and are well mixed. Pain occurs when one of the substances presents either a deficiency or an excess, or is separated in the body and not mixed with the others. It is inevitable that when one of these is separated from the rest and stands by itself, not only the part from which it has come, but also that where it collects and is present in excess, should become diseased, and because it contains too much of the particular substance, cause pain and distress".
>
> *Hippocratic Writings*, 1978, p. 262, ed. by G.E.R. Lloyd,
> "The Nature of Man", Penguin

Notwithstanding his attack on philosophizing, we have a concise summary of a philosophical base for his disease theories. However, it is not a model fabricated from theoretical speculation alone. It has the author's observations on humors justifying his theory:

> "The following signs show that winter fills the body with phlegm; people spit and blow from their noses the most phlegmatic mucus in winter. . . . During the spring . . . the quantity of blood increases . . . people are particularly liable to dysentery and epistaxis [nose bleeds]. During the summer . . . bile gradually increases . . . As proof of this, it is during this season that people vomit bile spontaneously. . . . Black bile is strongest and predominates in the autumn".
>
> *Hippocratic Writings*, 1978, pp. 264–265, ed. by G.E.R. Lloyd,
> "The Nature of Man", Penguin

Another philosophical principle provides the rationale for a method of cure for diseases, and that is the theory of opposites:

> "To put it briefly: the physician should treat disease by the principle of opposition to the cause of the disease . . . This will bring the patient most relief and seems to me to be the principle of healing".
>
> *Hippocratic Writings*, 1978, p. 33, ed. by G.E.R. Lloyd,
> "The Nature of Man", Penguin

This theory becomes part of the introduction in the treatise "Breaths." "Breaths" appears to be a treatise composed by a layperson who has very firm convictions about the origin of diseases and the philosophical underpinnings of medicine. In the introduction, the "art" of medicine is extolled, as opposed to opinion, which means the understanding and the method for the cure of "disease, suffering, pain and death." In short, medicine is a science. For the Greeks in the fifth century BCE, knowing the cause of a disease was crucial for the physician, since the cure of disease depended on the principle of opposites. This is part of the philosophy that good health depends on balance. The writer states that the best physician recognizes that good medicine relies on "subtraction of what is in excess, addition of what is wanting."

The writer turns to the cause of all diseases but first explains the basis of human and animal physiology, which depends on nourishment that comes in three forms: solid food, liquid, and wind. Wind in the body is called breath and outside the body it is air. Air is cosmically active: it is the basis of weather, it causes the movements of the oceans, and it is essential for fire and all the creatures in the sea and on land. Air is responsible for life of all living things. It follows, since all living things require air, that it is the cause of disease. This presents a dilemma: since all humans take in the same air, why at the time of disease are not all exhibiting the same disease? The answer is based on the assumption that not everybody has the same nature, nor does every person take in the same nourishment or maintain the same regimen. To account for air to be the cause of disease, the author adds two additional postulates to this theory: air can be in an infected state, with an element of specificity. If the air contains "pollution" specific for man he becomes sick, but if the air is "ill adapted" to other living creatures they become sick.

Air as a conveyer of disease is presented in the treatises "Humors," "Diseases," "Aphorisms," "Affections", AWP, and "The Nature of Man." Air as a transmitter of disease is identically described in "The Nature of Man" and in "Breaths":

"When an epidemic of one particular disease is established [when a large number of people catch the same disease at the same time], it is evident that it is not the regimen [nutrition] but the air breathed which is responsible. Plainly, the air must be harmful because of some morbid secretion which it contains".

Hippocratic Writings, 1978, p. 267, ed. by G.E.R. Lloyd, "The Nature of Man", Penguin

The author of "Breaths" has taken the argument one step further: the air contains a "morbid" entity. One can envision that the writer has an ontological conception of a disease agent, a being, with, perhaps, an independent existence. However, this entity is not further defined and this concept of a disease agent is not found in other treatises. Recall in AWP there is also no comment on the fact that boiled water no longer transmits disease. The thing that was destroyed by boiling might have been characterized as a morbid object, but it did not happen. Nor was it suggested that this morbid entity may be communicated from one person to another, that the disease is contagious. It was left as an infectious disease.

The author of "Breaths" also presents a more extensive characterization of humans, hypothesizing different physiologies for different people who respond differently to the same infected air. The treatment for people with airborne infections conforms to the principle of opposites. There should be a diminished amount of air breathed! A more practical cure is recommended: to remove the person "from the infected area." This maintains the idea that it is the area that is infected and not the individual. This is consistent with the absence of the concept of contagion in the corpus.

In summary, the mode of contracting disease excludes the concept of contagion. A disease does not pass from one person to another. An epidemic disease is explained by the fact that all people breathe the same air or possibly drink the same bad water.

Aphorisms in the Hippocratic Corpus

The treatise "Aphorisms" is composed of a large collection of propositions, regarded as statements of truth, all supporting the premise that infectious diseases have natural causes. They begin with a famous saying:

"Life is short, science is long; opportunity is elusive, experiment is dangerous, judgement is difficult. It is not enough for the physician to do what is necessary, but the patient and the attendants must do their part as well, and circumstances must be favorable".

Hippocratic Writings, 1978, pp. 206, 210, 212–214, 235, ed. by G.E.R. Lloyd, "Aphorisms", Penguin

Here are a few examples. Section 3, is largely devoted to the role of the seasons in causing disease.

1. The changes of the season are especially liable to beget diseases.

5. South wind cause deafness . . . , headache . . . The north wind brings coughs sore throats, constipation . . .

8. When the weather is seasonable and the crops ripen at the regular times, diseases are regular in their appearance and easily reach their crisis.

16. The diseases usually peculiar to rainy periods are chronic fevers, diarrhea, gangrene, epilepsy, . . . and sore throats. Those peculiar to a time of drought are consumption . . . and dysentery.

Also present are some of the fundamental principles that form the basis of Hippocratic medicine, including the dangers of excess of any kind and treatment by contraries:

"Sleep and wakefulness, exceeding the average, mean disease. Disease which results from over-eating is cured by fasting; . . . So with other things; cures may be effected by opposites. It is dangerous to disturb the body violently. . . . All excesses are inimical to nature".

The Role of Hippocrates and Hippocratic Medicine

The introduction of natural causality in the Corpus did not displace conventional medicine, which was still practiced in temples and shrines. In fact, the Hippocratic physicians did not discard the idea of gods. Neither did they state that religion was incompatible with their theories, and they certainly did not call for the elimination of religious places of worship. Aristotle, an enthusiastic admirer of Hippocrates, projects such a position in the "Politics." He proscribes rules for marriage and conditions for having children. He suggests that parents should consult physicians and natural philosophers (now we would say scientists) "for both give advice, one about the human body, and the other about favorable winds." He goes on to instruct mothers to eat a nourishing diet and to exercise by taking a walk every day to some temple where they can worship the gods who preside over birth.

The respect accorded to Hippocrates is evident in the writings of Plato and Aristotle. In the "Politics" Aristotle considers the conditions for an ideal state:

"A city, too, like an individual, has work to do; and that city which is best adapted to the fulfillment of its work is to be deemed greatest, in the same sense of the word great in which Hippocrates might be called greater, not as a man, but as a physician".

ARISTOTLE, Politics, 1943, p. 287, trans. by Benjamin Jowett,
New York Modern Library

In the "Phaedrus," Socrates is commenting on the virtues of rhetoric and states that,

"Rhetoric is in the same case as medicine, don't you think?"

PHAEDRUS: How so?

SOCRATES: In both cases there is a nature that we have to determine, the nature of the body in the one, and of soul in the other, if we mean to be scientific and not content with mere empirical routine [here Plato is pushing hard for philosophy as the basis for knowledge] when we apply medicine and diet to induce health and strength, or words and rules of conduct to implant such convictions and virtues as we desire".

Socrates continues,

"Then do you think it possible to understand the nature of the soul satisfactorily without taking it as a whole?" [the whole refers to the physical body]

Phaedrus replies,

"If we are to believe Hippocrates, the Asclepiad, we can't understand even the body without such a procedure".

Socrates replies,

"No, my friend, and he is right" . . .

PLATO, The Collected Dialogues, 1971,
pp. 515–516, ed. by Edith Hamilton and Huntington Cairns
(Bollingen Series LXXI), Princeton University Press,

It appears that Hippocratic medicine, its method for obtaining knowledge, can be a model for philosophy and thus not always the other way.

Hippocrates came from a distinguished tradition and family according to folklore. In Greek medicine Asclepius is the god of medicine, and all persons practicing medicine in the classical period are considered Asclepiads. Custom relates that Hippocrates was a descendant of Polidarius, one of the sons of Asclepius mentioned in the "Iliad."

The Corpus represents the views of many writers and contains different kinds of treatises. There are practical writings, case histories, and clinical notes, and there are those with general procedures and goals of medicine. There are different views of the immediate causes of infectious disease. There are different treatments prescribed for different illnesses. These treatises reveal the contemporary state of disease theory based on natural causes.

CHAPTER 4 | Galen

The Authoritative Conveyer of Hippocratic Disease Theory

THE MOST FAMOUS PHYSICIAN after Hippocrates was Galen (130–*ca.* 216 CE), the developer, interpreter, and transmitter of Hippocratic medicine, which would dominate disease theory for the next 1,400 years.

Galen was raised in Pergamon and received a broad education in mathematics, logic, and the major philosophical schools, Aristotelian, Platonic, Stoic, and Epicurean. He began the study of medicine at age 16 and was trained in the rationalist-empiricist traditions. He spent a number of years in Egypt continuing his medical training. When he was in his early thirties he went to Rome. There he established a reputation as a physician but made some enemies and left—but returned when requested by Emperor Marcus Aurelius and Lucius Verus. For Galen the greatest medical authority was Hippocrates. He transformed Hippocrates into the absolute authority and himself as the authentic interpreter of this authority. He derived from the Hippocratic writings the theory of the four humors and four qualities (hot, cold, moist, and dry) described in the Hippocratic treatise "The Nature of Man," and the importance of diagnosis in understanding diseases. He wrote extensively on the Hippocratic writings in the form of commentaries on various treatises. Galen was philosophically sophisticated but did not subscribe to one theoretical school or one method of treatment. He was an Aristotelian subscribing to the four elements and qualities, a Platonist using "Timaeus" as a source of theory about the soul and the body, and an Empiricist with his close observation of patients and the progress of disease, a method well represented in many of the Hippocratic writings. In addition he followed the Stoics' doctrine, adopting their classification of causes in contrast to Aristotelian causality (see below). He considered knowledge of anatomy essential to understanding the operation of the human body. A precise description, however, could be gained only by dissection and vivisection. It was not possible to carry out these procedures on humans, so Galen relied on surface anatomy, skeletons of humans, and the body parts displayed on battlefields and after gladiatorial conflicts, all this supplemented with dissection and vivisection of various animals to gain understanding of human anatomy. His elegant dissection of the nerves from the brain to the spinal cord allowed him to propose the interactions of these systems and provided a basis for explaining the relationship between physiology and behavior.

He based his disease theory on the four humors and contended that individuals were susceptible to disease when their unique balance was disturbed. Galen reworked Aristotle's principles of causality to a form that he considered more appropriate for diagnosis of the causes of human diseases. We will initiate our discussion with Galen's theory of the causes of contagious diseases. A fundamental characteristic of these diseases is fever, which Galen designates as "unnatural heat":

"Everybody knows that fatigue, anger, sorrow, a chill and many other things will start a fever. Again, persons not entirely bereft of their wits are aware that a pestilential state of the atmosphere will give rise to a fever, and they also know that it is risky to come in contact with the plague-stricken for there is a danger of contagion, just as there is from scabies or ophthalmia or indeed from phthisis or any who expire bad air so that their living places reek".

There are a specific number of diseases named and phrases used containing concepts relating to the causes of disease. The diseases that were selected are those that could be contracted from individuals who had the disease. They are plague, scabies (a skin disease), ophthalmia (an eye disease), and phthisis (consumption or tuberculosis). Galen was not the first to recognize the possibility that a disease may be contagious (that is, transmitted from person to person). Recall that Thucydides wrote that physicians died in great numbers during the plague in Athens for they attended the sick, while Aristotle attempted to explain the rapid spread of the identical diseases listed by Galen. Aristotle had asked the question.

"why are those in contact with phthisis, ophthalmia, scabies (psora) infected by them, but are not infected by dropsy, fever and apoplexy nor by many other diseases?"

Aristotle provided a number of answers according to the disease:

"In ophthalmia it is due to the fact that the eye is most easily moved, and is more inclined than the other senses to assimilate itself to what it sees. On phthisis [the person has] decay of breath, such as plague in any form. The one who is in contact breathes this breath. He becomes diseased because the breath is diseased. Scabies is more infectious than other diseases, like leprosy and similar complaints, because the discharge is superficial and viscous; for itching diseases are of this kind. This disease is therefore infectious, because the discharge is superficial and viscous".

ARISTOTLE, "Problems," Book VII.8, 1961, 887a, Harvard

Aristotle's account is in agreement with explanations in Hippocratic writings where a disease is the result of infected air or contact with diseased materials. In Aristotle's narration there is the added proviso that the bad air or diseased material could come from another person.

Galen stressed another factor necessary for acquiring a disease; he stated

"that no cause can be efficient without aptitude of the body."

The term *efficient cause* is part of Aristotle's complex four-cause theory, which we will return to shortly. But first, Galen is addressing the important question: Why doesn't everyone contract a disease when exposed to a disease? He begins by briefly reviewing the external origins of disease:

"The beginning of the putrescence may be a multitude of unburned corpses, as may happen in war; or the exhalations of marshes or ponds in the summer; sometimes it is the immoderate heat of the air itself as in that pestilence of which Thucydides writes: "a corruption seized the bodies of men in their close and ill-ventilated hovels." The starting point of the pestilence was the preparation for putrefaction of the humours of the body due to bad food. It may be, too, that atmospheric continuity was a factor, that products of putrefaction came from which were destined to be a cause of fever in persons whose bodies were prepared to sustain this injury".

These conditions are necessary but not sufficient:

"Always this is to be remembered, that no one cause can be efficient without an aptitude of the body; otherwise all who are exposed to the summer sun, move about more than they should, drink wine, get angry, grieve, would fall into a fever.

Or, again, all would fall sick at the rising of the dog star and in a pestilence all would die. But, as I have said, the chief factor in the production of disease is the preparation of the body which will suffer it".

Galen provides an extended description of the susceptibility of certain individuals to contagious disease:

"Let us imagine, for instance, that the atmosphere is carrying divers seeds of pestilence and that, of the bodies exposed to it, some are choked with excrementous matters apt in themselves to putrify, that others are void of excrement and pure. Let us further suppose an obstruction of orifices and resultant plethora in the former, likewise a life of luxury, much junketing, drinking, sexual excess and the crudities which must attend on such habits; in the latter let us suppose cleanliness, freedom from excrementous matters, orifices unobstructed and uncompressed, desirable conditions, as we may say, free transpiration, moderate exercise, temperance in diet. All this being supposed, judge thou, which class of body is the likelier to be injured by the inspiration of putrid air. Is it not plain that the former class from the first inspiration will receive a beginning of putrefaction and that bad will go to worse, while those which are pure and void of excrement will either escape all together or suffer so little damage as easily to return into the way of nature? Hence then, although, so often as the [quality] of the atmosphere departs from its proper nature into the hot and humid, pestilential diseases must needs arise, yet will those chiefly be affected who were beforehand saturated with excrementous moisture while those who labour moderately and are temperate in diet remain refractory to such diseases. This has been said of one example but it is an universal truth".

These excerpts are from a paper by M. GREENWOOD, "Galen as an Epidemiologist," Proc. Roy. Soc. Med., 1921, XIV (Sec. Hist. Med. pp. 3–16).

Galen proposed a theory of causality different from Aristotle's. In Aristotle's theory of causality there are four components. There is a formal cause, the essence of a thing that makes it what it is. There is the material cause, the substance that constitutes it. There is the efficient cause, which signifies its action. The fourth is the final cause, namely the goal. The concept of teleology has been included. Galen provides a theoretical structure more useful to the practicing physician, and in assigning the origin of a disease to a "manifest" or original cause, equivalent to Aristotle's efficient cause, he introduces a scenario in which the "atmosphere is carrying divers seeds of pestilence" as one possible initial or "efficient" cause. We will discuss shortly what meaning Galen gives to the term *seeds*.

Galen has adopted the efficient cause as the most useful for the physician, for to know the event that caused the disease is to begin to know how to effect a cure. Galen proposed a complex efficient cause theory and subdivided it into three parts: initial, antecedent, and cohesive causes:

1. The initial causes are external events, such as contact with individuals with diseases, inhalation of bad air from patients with certain diseases, or exposure to the vapors of unburned corpses or air emanating from marshes or ponds, or eating bad food, which causes the humors to undergo putrefaction.
2. Antecedent causes are preconditions of the individual so that the person is affected by the initial cause.

The background for this causal theory has its origin in Stoic philosophy, which can be summarized as follows. A cause is a body (not only a physical body) that does something to affect another body, which brings about an effect. The formal, the material, and the final causes of Aristotle do not fit this depiction, although the efficient cause of Aristotle fits this description. The external cause and the antecedent cause, working together, appear to be the ideal notion of causality with regard to contagious disease.

Bad air is an important external cause of disease. If the bad air "is carrying divers seeds of pestilence," it offers a possible explanation for the fact that not everyone who comes in contact with air would contract the disease because they might not inhale the "seeds" of disease. This brings us to the question: What does Galen intend to convey using the term *seeds*? Over a two- to four-year period Galen wrote about seeds of disease as part of a discussion of contagious or communicable diseases. He recognized that diseases could roam the world, as Hesiod had written, and could be contracted by contact with persons having certain infections such as scabies, plague, or tuberculosis, and by contact with a drop of saliva of a rabid dog. For Galen the latter phenomenon was an example not only of transfer of disease by direct contact with material from a source of disease but also of an event that involved something small or invisible that nevertheless resulted in a lethal effect.

There are in the historical record many references to seeds. Such an entity was described by Anaxagoras, one of the pre-Socratics, who stated that from seeds "watered by rain, plants are produced." Aristotle, in *Generation of Animals* and *History of Animals*, described the role of seeds in reproduction. He stated a general principle, that animals and plants spring from seed. This mode of reproduction is common to animals and plants. It is clear that, for Aristotle, a seed is something with the potential to give rise to a unique, living organism.

Theophrastus (*ca.* 372–*ca.* 288 BCE), head of the Lyceum after Aristotle, wrote two works, *Enquiry into Plants* and *On the Causes of Plants*, dealing with the question of plant generation. He included growth from a seed or a cutting and also the possibility of spontaneous growth.

The discourse about seeds is considerably enriched by the philosophy of Epicurus that all things are made of invisible atoms. Such a position was laid out in some detail by Lucretius in the poem *De rerum natura* (The nature of things), which was infused with the philosophy of Epicurus, a materialist philosophy that envisioned the world as constructed of an infinite number of atoms. Marcus Terentius Varro (116–27 BCE), a Roman writer who was a contemporary of Lucretius, wrote in an agricultural treatise, *De Re Rustica*:

"Seed, which is the principle of vegetation is of two kinds; one which is not perceptible to our senses, the other is that which lies open to our perception."

The invisible seeds may be the atoms of Lucretius. Lucretius used a theory of atoms to explain the origin and development of living beings, plants, animals, and humans. Book VI ends with a description of the plague of Athens very similar to the description provided by Thucydides. It was here that seeds are described as causing disease, but this was not the only place where Lucretius used the term *seeds*; there are numerous occasions where seeds are invoked to give rise to life:

"Therefore, we must confess that nothing can be brought to being out of nothing, in as much as it needs a seed for things from which each may be produced.

All things are born of fixed seeds and a fixed parent and can, as they grow, preserve their kind".

LUCRETIUS, *De rerum natura*, 1947, p. 274, trans. by Cyril Baily, Oxford)

We return to the basic question: In employing the term *seed*, did Galen intend it to indicate something living? Does Galen think that the seed is the thing that initiates the disease? Is it a living thing that is itself the disease agent, just as a plant seed gives rise to a plant? Initially the answer is an ambiguous one, but within a short time it appears clear that the answer is "no." In one instance, in a commentary on the Hippocratic treatise *Epidemics*, he warned of relapses in certain fevers that may be due to the survival of "seeds of disease." To be consistent he provided an antecedent cause (that is, not observing a good regimen). However, in a commentary on the treatise *Aphorisms,* written prior to his work on epidemics, he ascribed the recurrence of disease to bad humors. Putting it all together one cannot make a strong argument that Galen thinks of seeds as things giving life to disease. After this brief period of time where he wrote of seeds, he abandoned the use of this term.

For Galen contagious entities (not more precisely defined), those objects that affect the humors, arise from the environment, the atmosphere, food, and decaying organic materials. These causes would not have surprised the Hippocratic writers. An in depth discussion of these issues is found in, "The Seeds of Disease: An Explanation of Contagion and Infection from the Greeks to the Renaissance" by V. Nutton in Med. Hist. v. 27, pp. 1–34, 1983.

CHAPTER 5 | # After Galen

*The Fracture of the Roman Empire; Greek
Learning Survives in the Islamic East;
Recovery of Greek Sources in the Twelfth
Century CE in Christian Western Europe*

IN THE CENTURIES FOLLOWING Galen a new religious, political, and linguistic landscape developed in the Roman Empire. In the east a new center emerged with Constantinople as its center and under the Emperor Constantine adopted a new religion, Christianity. The empire was physically split in the European west by internal cultural differences and by a number of destructive invasions by Visigoths and Vandals in the fourth century CE. What may have contributed to the decline in the west was the outbreak of a persistent plague in 165 CE, followed by various epidemics in the following decades that killed perhaps half the population, and another outbreak of plague in the third century (252–270 CE). In the latter part of the fourth century the empire was converted to Christianity and was ruled by two emperors, but after the invasions of the west it was broken into various independent entities. The separation of "east" and "west" was complete with the various Muslim conquests in the east, culminating with the Arab conquest at the end of the seventh century. The separation of the two cultures had a profound effect on the transmission of science and medicine in the western region, which retained as its religion Christianity while its writings were in Latin. In the east there survived Greek medical learning in the form of encyclopedias created by a number of scholars located in cities of Asia Minor, now Turkey, and Alexandria, Egypt. These encyclopedic works became part of the authoritative medical teaching in the east, and the commanding figure in these works was Galen. Galen had died in

Rome and left volumes of writings in the Greek language on medicine and philosophy, including many commentaries on the Hippocratic writings. This was the learning that Oribasius (*ca.* 320–400 CE) of Pergamon, in Asia Minor, and others included in the encyclopedias in the eastern regions.

Sources of ancient learning in the west included Pliny's "Natural History" (23–79 CE), a large collection of information (and misinformation) about the world. A work by Boethius in the sixth century CE contained mathematics of Euclid (325–265 BCE), astronomy by Ptolemy (90–168 CE), and some Latin treatises of Aristotle. A work by Isidore of Seville (560–636 CE) contained material on medicine and astronomy. After the conversion of much of the east to Islam the west was completely separated from the east by the ever-present hostility between the two religious traditions and by scholarly language. The knowledge of Greek medicine and science became inaccessible to the Latin west when the Greek texts were translated into Arabic. In that language there was a development of medicine, science, and philosophy that was unknown in the west.

By the end of the eighth century Muslim rule extended from the area now known as Afghanistan to the Tigris and Euphrates region, including Baghdad, to the Mediterranean, all of Arabia, the north African coast from Egypt to the area inhabited by the Berber tribes, present-day Libya and Tunisia, and a good deal of Spain. An important center of science and medicine developed in the eighth and ninth centuries CE in Baghdad, and later in Andalusia in Spain.

In Western Europe the study of nature was encouraged; however, the intent was to reveal the power and creativity of God. The view of Vincent of Beauvais "might equally well have come from the pen of Albert Magnus" (1193/1206–1280 CE), expressed this way in the thirteenth century CE:

"I am moved with spiritual sweetness towards the creator and ruler of the world, because I follow Him with greater veneration and reverence, when I behold the magnitude and beauty and permanence of His creation".

A.C. CROMBIE, *Augustine to Galileo*, 1969, p. 174, Penguin, Harvard University Press, 1979

The Greek impulse to understand the causes of physical and biological phenomena was condemned: the causes of these phenomena were already known, and curiosity was dangerous. St. Augustine (354–430 CE) assured the faithful that God is the creator of all things and that all good will come from faith in Christ. He continued with the injunction not to search for the secrets of nature, which are unknowable:

"From this, then, one can the more clearly distinguish whether it is pleasure or curiosity that is being pursued by the senses. For pleasure pursues objects that are beautiful, melodious, fragrant, savory, soft. But curiosity, seeking new experiences, will even seek out the contrary of these, not with the purpose of experiencing the discomfort that often accompanies them, but out of a passion for experimenting and knowledge . . .

This is also the case with the other senses; it would be tedious to pursue a complete analysis of it. This malady of curiosity is the reason for all those strange sights exhibited in the theater. It is also the reason why we proceed to search out the secret powers of nature—those which have nothing to do with our destiny—which do not profit us to know about, and concerning which men desire to know only for the sake of knowing. And it is with this same motive of perverted curiosity for knowledge that we consult the magical arts. Even in religion itself, this prompting drives us to make trial of God when signs and wonders are eagerly asked of him—not desired for any saving end, but only to make trial of him".

AUGUSTINE, *Confessions, Book 10:55*, 1955, Tr. by Albert C. Outler

Such a position would be incomprehensible to the Greek philosophical tradition and the medicine so intimately tied to it. The heritage of Greek inquiry into nature was carried on in the Muslim world. The philosophical divide between Latin, Christian "science" and that of Arab-Muslim learning is vividly illuminated by a dialogue between Adelard of Bath (*ca.* 1080–1152) and his nephew, contained in a book titled *Questiones naturales*. This book, covering topics of biological and physical sciences, was written in the twelfth century CE by Adelard on his return to England from studies in southern Italy and Sicily, where he had access to Arab science and medical works. Adelard located his return to England in the reign of Henry 1 (*ca.* 1068–1135 CE). His nephew was interested in natural science, although not an expert, and was curious about what Adelard learned in his "Arab studies." He asked Adelard:

"Why do plants spring from the earth? What is the cause and how can it be explained? When at first the surface of the earth is smooth and still, what is it that is then moved, pushes up, grows and puts out branches? If you collect dry dust and put it finely sieved in an earthenware or bronze pot, after a while when you see plants springing up to what else do you attribute this but to the marvelous effect of the wonderful divine will"?

Here is Adelard's answer:

"I do not detract from God. Everything that is, is from him and because of him. But nature is not confused and without system and so far as human knowledge has progressed it should be given a hearing. Only when it fails utterly should there be recourse to God".

A.C. Crombie, Augustine to Galileo, p. 45, 1979
Harvard University Press

This is a significant transformation. The contrast between the two views represents the renewal of the divide in Western Europe between supernatural and natural explanations of phenomena that characterized the period before and after the pre-Socratic revolution.

Adelard did not learn this view in the west but in a center of learning where Arabic writings of ancient Greek texts were translated into Latin by Muslim scholars, among them Constantine the African (*ca.* 1020–1087). He was associated with the first center for medical education at the University of Salerno, Italy, part of the Kingdom of Sicily. This was where Adelard had studied. The city of Salerno had a distinguished history of medical practice and medical education. Constantine powerfully contributed to the new curriculum by translating the works of Hippocrates and Galen from Arabic to Latin. Additional works by Islamic scholars were also translated. The most famous of the Arabic texts translated into Latin was the *Canon of Medicine* by Avicenna. These translations provided the first introduction of Hippocratic–Galenic medicine to Western Europe. This is the learning that Adelard obtained during his sojourn in Salerno. These translations were rapidly disseminated throughout Europe.

The influence of Hippocratic–Galenic writings was apparent in Avicenna's sections on disease theory. Avicenna acknowledged the Greek philosophical position that to understand a phenomenon it is essential to know its causes and antecedents. The Aristotelian four causes, modified by Avicenna, are the basis of disease theory. The human body is made of four primary components (elements), fire, air, earth, and water, which cannot be further divided. The primary qualities are hot, cold, moist, and dry. There are four humors, blood, phlegm, and yellow and black bile. The causes of disease include air, influence of the seasons, climate, psychological factors, and water, particularly stagnant water and water obtained from melted snow or ice. This last category of disease-causing materials was explained in essentially identical language in the Hippocratic work "Airs Waters Places."

Avicenna's version is

"Snow water and water from melted ice are coarse in texture. . . . When pure and free from admixture with deleterious substances, such water is good and healthy. . . . Water from snow water or melted ice is harmful for persons suffering from neuritis. Boiling renders such water wholesome".

AVICENNA, *The Canon of Medicine*, 1999, p. 228, Adapted
by Laleh Bakhtiar, trans. by O. Cameron Gruner and Mazar H. Shah,
Great Books of the Islamic World, Inc. Kazi Publications

The Hippocratic version in "Airs Waters Places" is

"Water from snow and ice is always harmful because, once it has been frozen, it never regains its previous quality".

To prevent this,

"it needs to be boiled and purified".

Hippocratic Writings, 1978, p. 155, ed. by G.E.R. Lloyd,
"Airs Waters Places", Penguin

A close comparison of Galen's and Avicenna's causes of disease demonstrates great similarities. Galen blamed corrupted air due to putrid exhalations from heated swamps or ponds in summer and from unburned decaying corpses. In Avicenna's theory, pestilential fevers come from air that is damp, turbid, and noxious, produced from marshy, stagnant places or from unburned decaying corpses.

Air is to be made free of disease by almost identical means. Galen recommended building fires; Hippocrates advised this procedure during the plague of Athens. He also recommended spreading flower-scented air. Avicenna advised burning herbs or incense.

Western European medicine, suffused with Christian belief, had recovered the contents of Hippocratic–Galenic medicine with its naturalistic, philosophical base. It now had a way of providing proximal natural causes of contagious disease, while maintaining that the deity is the ultimate conveyor of disease.

CHAPTER 6 | The Causes of Plague (The Black Death) in Europe 1348–1350 CE

"The great mortality appeared at Avignon in January, 1348, when I was in the service of Pope Clement VI. It was of two kinds. The first lasted two months, with continued fever and spitting of blood, and people died of it in three days. The second was all the rest of the time, also with continuous fever and with tumors in the external parts, chiefly the armpits and groin; and people died in five days. It was so contagious, especially that accompanied by spitting of blood, that not only by staying together, but even by looking at one another, people caught it, with the result that men died without attendants and were buried without priests. The father did not visit the son, nor the son his father. Charity was dead and hope crushed".

Description by the physician Guy de Chauliac, condensed by
Anne M. Campbell in *The Black Death and Men of Learning*,
1931, pp. 2–3, Columbia University Press, New York

THE AUTHENTIC PLAGUE OF the mid-fourteenth century was the most devastating pandemic in recorded history—although it is impossible to know the number of deaths with any certainty, since fatalities varied in different geographical regions, and the numbers of fatalities were estimates by contemporary chroniclers who experienced the disease. From the many reports it appears that Europe may have lost one-third to half the population.

There is general agreement that the disease came from Asia because of a significant trade route established with the Mongol empire, which included China and a good part of Russia, during the period 1279–1350 CE. These trade routes linked cities such as Astrakhan on the Volga and Caffa in the Crimea with posts in the Far East. Caffa, now known as Feodosiya, is a port city in Crimea, Ukraine, on the Black Sea. Caffa, under the control of the Genovese Republic, carried out vigorous trade with Mediterranean Europe and was certainly one of the sources of the plague. In the spring of 1347 the disease was present in Constantinople, where it remained for a year. In October 1347 it arrived in Cyprus, Malta, Greece, Sardinia, Corsica, and Sicily, and by the end of the year it had reached Genoa, Livorno, Marseilles, and the Dalmatian coast. In early 1348 it spread throughout southern France, Italy, and Spain, and by mid-year it was in Paris and within a few months in all of France. It crossed the Channel into England and Ireland in 1348–49 and moved north and west into the Germanic lands, Netherland, Iceland, and Russia in succeeding years.

The descriptions of the effects of the disease, the inability of medicine or prayer to alleviate suffering, and the apathy that swept over the populations in the face of such a disaster are reminiscent of the events described by Thucydides during the "plague" of Athens some 1,780 years earlier, although they were not the same diseases.

In response to this overwhelming catastrophe there was produced a large literature on the disease, many works by individual physicians and an authoritative treatise by the Medical Faculty of the University of Paris. These reports reveal the state of contagious disease theory in mid-fourteenth-century Europe. There was general agreement that the disease was sent by God as punishment for the sins of the population. At the same time there were proximal causes that, it could be argued, were used by God to inflict punishment. These causes can be characterized as natural causes contained in the available writings of the Hippocratic Corpus, in Galen, and in Avicenna (980–1037 CE). In addition, there was created a human cause, blaming Jews, foreigners, and the poor for the disease.

God Is the Original Cause of Disease

All contemporary physicians agreed that the plague was an act of God, although the behavior that was so unacceptable that it elicited such a calamity was not specified. However, one Gabriele de' Mussis, a lawyer in the city of Piacenza, about 90 miles northeast of Genoa, was not so reluctant to describe the human failings that brought on the disease. In his hands the overwhelming tragedy became "an expression of divine anger." Humanity was to blame: its actions brought this judgment on everyone, the guilty as well as the innocent. He wrote:

"Because those I appointed to be shepherds of the world have behaved towards their flocks like ravening wolves, and did not preach the word of God, but neglect all the Lord's business and have barely even urged repentance, I shall take a savage vengeance on them. I shall wipe them from the face of the earth".

ROSEMARY HORROX, *The Black Death*, pp. 15–16

"I am overwhelmed, I can't go on. Everywhere one turns there is death and bitterness. the hand of the almighty strikes repeatedly . . . The terrible judgment gains in power as time goes on . . . We know what we suffer is the just reward of our sins".

ROSEMARY HORROX, *The Black Death*, 1994, p. 23, trans. and ed. by Rosemary Horrox, Manchester Medieval Sources Series, Manchester Univ. Press, Manchester

De' Mussis described the progress of the disease, which struck men, women, and children equally. Although the cause of the disease was God, the Medical Faculty of the University of Paris urged the population to rely on physicians, who are instruments of God's power to heal, to provide immediate causes and cures:

"But this does not mean forsaking doctors. For he created earthly medicine, and although God alone cures the sick, he does so through the medicine which in his generosity He provided".

ROSEMARY HORROX, *The Black Death*, p. 163.

Natural Causes

Numerous contemporary treatises provided natural causes. We will refer to four of them. The most important is the summary written in 1348 by members of the Medical Faculty of the University of Paris, a center of Christian orthodoxy; excerpts from the treatise by Gentile de Foligno (d. 1348) written at the request of the University of Perugia; and two works by Islamic physicians living in Spain and writing in Arabic. First, we will present an abstract of the Paris Medical Faculty statement:

"the prudent soul . . . strives with all its might to discover the causes of the amazing events. To attain this end we have listened to the opinions of many modern experts on astrology and medicine about the causes of the epidemic . . . we, the masters of the faculty of medicine at Paris . . . have decided to compile, with God's help, a brief compendium of the distant and immediate causes . . . (as far as these can be understood by the human intellect) . . . drawing on the opinions of the most brilliant **ancient philosophers** (my emphasis) and modern experts, **astronomers** (my emphasis) as well as doctors of medicine. And if we cannot explain everything . . . it is open to any diligent reader to make good the deficiency. . . . we shall investigate the causes of this pestilence and whence they come, for without knowledge of the causes no one can prescribe cures".

R. HORROX, *The Black Death*, 1994, pp. 158-159, trans. and ed. by Rosemary Horrox, Manchester Medieval Sources Series, Manchester Univ. Press, Manchester

It is obvious that the Paris Medical Faculty returned to antiquity for guidance as to the cause of the disease. For example, the last few lines are similar to the wording found in the Hippocratic treatise *Breaths*.

The report of the Paris Medical Faculty continued:

"We say that the distant and first cause of this pestilence was and is the configuration of the heavens. In 1345, at one hour after noon on 20 March, there was a major conjunction of three planets in Aquarius. This conjunction . . . by causing a deadly corruption of the air around us, signifies mortality and famine . . . Aristotle testifies that this is the case . . . And this is found in ancient philosophers, and Albertus Magnus [who states] that the conjunction of Mars and Jupiter causes a great pestilence in the air".

R. HORROX, *The Black Death*, p. 159

The influence of astronomical events on the cause of disease had already appeared in Homeric literature before the development of philosophical speculations about the connections between the macrocosm (celestial bodies) and the microcosm (humans). In the *Iliad*, Book 22:21, at the time of the battle between Achilles and Hector during the siege of Troy, Homer includes a description of the sky:

"Then toward the town with might and main he ran, magnificent, like a racing chariot horse that holds its form at full stretch on the plain. So light-footed Akhilleus held the pace. And aging Priam was the first to see him sparkling on the plain, bright as that star in autumn rising, whose unclouded rays shine out amid a throng of stars at dusk—the one they call Orion's dog, most brilliant, yes, but baleful as a sign: it brings great fever to frail men".

HOMER, *The Iliad*, 1975, trans. by Robert Fitzgerald, p. 515, Anchor-Doubleday

A direct connection between the health of man and the stars and the injunction to the physician to know astronomy is made in the Hippocratic treatise "Airs Waters Places."

"Being familiar with the progress of the seasons and the dates of rising and setting of the stars, he could foretell the progress of the year. Thus he would know what changes to expect in the weather and not only would he enjoy good health himself for the most part but he would be very successful in the practice of medicine. If it should be thought that this is more the business of the meteorologist, then learn that astronomy plays a very important part in medicine since the changes of the seasons produce changes in diseases".

Hippocratic Writings, 1978, p. 149, ed. by G.E.R. Lloyd, Penguin

In the Hippocratic treatise *Fleshes* there is a philosophical discourse on the origin of the cosmos and humans that leads to the conclusion that all the components of the physical world are present in the body of humans and that everything in the world, the macrocosm and the microcosm, has a common origin. Since human beings are created from the same components as celestial bodies there is a belief that there are correspondences between man (the microcosm) and the macrocosm. There was a general belief that there are astral influences on humans.

To continue with the views of the Paris Faculty, their claim was that the astral influence generates corrupt air that is more potent than noxious food or drink as a disease cause for the obvious reason that everyone is exposed to the same air that rapidly enters the body. The air is corrupted because it receives pestilential vapors from the earth and various waters at the time of the conjunction of the planets. When this vitiated air (corrupted air from the earth or water) enters the body it penetrates the heart and

"corrupts the substance of the spirit there and rots the surrounding moisture, and the heat thus caused destroys the life force, and this is the immediate cause of the present epidemic".

R. HORROX, *The Black Death*, p. 161

The contents of this paragraph have their roots in a theory that states that the heart creates and distributes a spirit, a life force, throughout the body that if destroyed by the noxious air it will cause death. The treatise goes on to assert that contracting a disease depends on the susceptibility of the individual:

> "We must therefore emphasise that although, because everyone has to breathe, everyone will be at risk from the corrupted air, not everyone will be made ill by it but only those . . . who have a susceptibility to it".
>
> R. HORROX, *The Black Death*, p. 163

Who are these predisposed individuals most likely to contract the disease? They include bodies with "evil humors," persons leading a bad lifestyle, individuals who are too thin or very heavy, young children, and women.

The Paris Faculty acknowledged that there are other causes of pestilential diseases, such as bad food or water, and that the air may be corrupted by sources other than the configuration of the planets; diseased air can emanate from "swamps, lakes and chasms" and is even more dangerous when it comes from decaying corpses. There are additional near (proximal) causes. Unseasonable weather is a special cause of disease; Hippocrates is quoted to that effect. When seasons deviate from the norm, then plagues will occur that year. Again, the Hippocratic writings are a resource; an aphorism is quoted that when there are fogs and damp weather, there are illnesses. Obviously, there is nothing new here regarding causality. Furthermore, there continues to be a strong belief that there are multiple causes for a specific disease.

Gentile of Foligno (d. 1648) was a physician who practiced medicine in Perugia, lectured at Padua, and wrote commentaries on Galen and Avicenna. He was in Perugia when the plague struck Genoa and spread to neighboring cities. He prepared a document intended to provide instructions to protect the Perugian population, but nothing helped. He attended the sick, contracted the disease, and died in June 1648.

Gentile acknowledged astrological events that cause the air to become putrid and thus cause disease but stressed more immediate causes. They are poisonous materials that affect the heart and lungs brought to the person by air. Where do these pestilential vapors come from? There are two sources, one geographically far away (south winds) and the other one nearer ("wells and caverns shut up too long" or confined air in walls or roofs). Another source is stagnant water or decaying corpses. The common carrier is air, which transmits disease. The air enters the body by way of breath or through "wide pores" and passes to the heart. This is similar to the process described by the Paris Faculty. Gentile concurred that the bad air displaces the vital spirit and added that the "pestilential air multiplies itself while feeding upon the moist humors of the body."

Gentile also presented arguments for contagion, disease transmitted from one person to another, recognizing views of scholars including Avicenna and Galen:

> "But the greater strength of this epidemic and, as it were, instantaneous death is when the aerial spirit going out of the eyes of the sick strikes the eyes of the well person standing near and looking at the sick, especially when they are in agony; for then the poisonous nature of that member passes from one to the other, killing the other".
>
> ANNE M. CAMPBELL, *The Black Death and Men of Learning* 1931,
> p. 61, Columbia University Press New York

What it is that passes from one person to another is not further described.

A similar view, the contagious nature of the disease, is expressed by the distinguished surgeon Guy de Chauliac (1300–1368), who was quoted at the head of the chapter. In the face of this great epidemic, de Chauliac stated, doctors were useless:

> "It was so contagious, . . . that not only by staying together, but even by looking at one another, people caught it, . . . the more so as they [doctors] dared not visit the sick, for fear of being infected. And when they did visit them, they did hardly anything for them".
>
> ANNE M. CAMPBELL, *The Black Death and Men of Learning*, p. 3

Plague Writings by Ibn Khatima and Ibn el-Khatib

Both of these Muslim scholars were from the Kingdom of Granada in Spain. They wrote Arabic tracts at the time of the disease and disagreed on whether the disease could go from person to person.

Ibn Khatima was a theologian, physician, and poet living in the city of Almeria. The plague had been going on for eight months when, in February 1349, he wrote his work in the form of answers to questions about the disease. The disease had started in a small section of Almeria and spread slowly to the rest of the city:

> "This is an example of the wonderful deeds and power of God, because never before has a catastrophe of such extent and duration occurred. No satisfactory

reports have been given about it because the disease is new, . . . God only
knows when it will leave the earth".

ANNE M. CAMPBELL, *The Black Death and Men of Learning*, pp. 51–52

He referred to two categories of causes of the plague, remote and nearer.
The remote causes are various astronomical influences and God's decree, but
he did not stress these preconditions. He attributed the disease to pestilential
air as a result of irregularities in the weather, such as temperatures, winds,
and rain. His preoccupation with the role of air in disease led him to discount
transmission of disease among individuals, but only after he produced an
excellent case for infection, or what is generally referred to as contagion:

"The best which long experience has taught me is that when anyone comes
in contact with a sick man, forthwith the same disease seizes him, with the
same symptoms; . . . if with the first buboes appeared on the groin, so do like
buboes appear in the same place with the other; . . . The family suffers in the
same way, shows the same symptom; if the sickness of one member, runs a
fatal cause, the others suffer the same fate; . . . In this way on the whole,
with slight differences, has it gone in our city".

ANNE M. CAMPBELL, *The Black Death and Men of Learning*, p. 57

Ibn Khatima went on to deny infection (contagion) "in which the Arabs
in their ignorance [before Islam] believed." His concentration on the corruption of air led him to postulate that the constitution of the air, its constituent parts, is altered. Although air is regarded as a principal element,
and it does not undergo change, it does contain other components that do
change. These confer disease. In the case of the plague the factors that lead
to the production of foul components in the air are fumes from decaying
matter, manure, swamps containing stagnant air, decaying plants, unburied
corpses, and plague-stricken cattle.

Ibn el-Khatib's positions were in contrast to those of Ibn Khatima. He
contended that air is responsible for infection, not that the components are
altered. Infection, as used by Ibn el-Khatib, is contagion—that is, disease
that passes from person to person. He forcefully proposed a theory of
infection:

"But it belongs among evident principles that a demonstration evolved from
tradition [Islamic law], if it is opposed to the perception of the mind and the
evidence of the eyes [the existence of infection stands firm through experience, el-Khatib wrote], must necessarily be subjected to explanation or

interpretation. And this, in the present instance, is exactly the idea of many who defend infection".

ANNE M. CAMPBELL, The Black Death and Men of Learning, p. 59

Anne Campbell noted that this disagreement between the two Arab authors indicates that in the western Muslim tradition (both authors were writing in Spain), the "Black Death stirred up somewhat of a revolt of medical science against theology." Theology, according to Ibn Khatima, argued against contagion. A belief in contagion was held before the adoption of Islam. Ibn el-Khatib differed, relying on experience to justify his belief in contagion.

Other treatises proposed "divine wrath" as the ultimate cause of disease. The manifestation of this wrath may be earthquakes or astrological events.

In summary, there are supernatural explanations and naturalistic explanations for disease that can be reconciled in the Western European culture by differentiating ultimate and proximal causes. The sources of proximal causes during the plague are found in the Hippocratic writings, in Galen and Avicenna.

Human Cause

There is a new, human cause of the plague not derived from Greek or Muslim sources but created in Western Christian society. Rosemary Horrox explained its origin:

> "In the panic caused by the epidemic feelings ran so high that accusations might be leveled against almost anyone perceived as an outsider, including foreigners, the poor . . . travellers. But the most frequent scapegoats were the Jews. Local officials exchanged details of alleged Jewish enormities in their districts, and of confessions extorted under torture" . . .

ROSEMARY HORROX, The Black Death, 1994, p. 109, trans.
and ed. by Rosemary Horrox, Manchester Medieval Sources Series,
Manchester Univ. Press, Manchester

Horrox reproduced "a contemporary account of a Franciscan friar from Franconia" of the assault on Jews, an event repeated in numerous localities in Germany and France:

> "In 1347 there was such a great pestilence and mortality throughout almost the whole world that in the opinion of well-informed men scarcely a tenth

of mankind survived. The victims did not linger long, but died on the second or third day. The plague raged so fiercely that many cities and towns were entirely emptied of people. Some say that it was brought about by corruption of the air; others that the Jews planned to wipe out all the Christians with poison and had poisoned wells and springs everywhere. And many Jews confessed as much under torture: that they bred spiders and toads in pots and pans, and had obtained poison from overseas; . . . God, the lord of vengeance, has not suffered the malice of the Jews to go unpunished. Throughout Germany, . . . they were burnt".

R. HORROX, *The Black Death*, p. 207

Almost 600 years later Jews were again burned by Germans. God played no role in the twentieth-century *Shoah* (Holocaust).

CHAPTER 7 | The Late Renaissance Period
Paracelsus and Fracastoro, Sources of Their Contagious Disease Theories

"In the pre-modern period theories in physics came from philosophical speculations. No period between remote antiquity and the end of the seventeenth century exhibited a single generally accepted view about the nature of light. Instead, there were a number of competing schools and sub-schools, most of them espousing one variant or another of Epicurean, Aristotelian, or Platonic theory. Each of the corresponding schools derived strength from its relation to some particular metaphysic".

THOMAS S. KUHN, *The Structure of Scientific Revolutions*, 1964, p. 12, University of Chicago Press

SCHOLARS IN THE RENAISSANCE criticized existing systems such as Scholasticism, with its source in the philosophy of Aristotle, and in its place proposed Platonism, Hermeticism, and so forth. There were consequently many "true" systems. How was one to know which was the right one?

Theophrastus von Hohenheim (1493–1541), known as Paracelsus, and Giralamo Fracastoro (1478–1556) are the two individuals we have selected to represent different contagious disease theories at mid-sixteenth century. The sources of their speculations came from sharply contrasting philosophies. Paracelsus constructed a contagious disease theory from a number of philosophical and religious strands that included neo-Platonism, Hermeticism, and alchemy, which emerged as a complex system imbued with religion, the occult, and chemistry and an epistemology, a way of knowing, that depended on a mystical union of the human with objects in the world. Fracastoro produced a contagious disease theory that relied on the matter theory of Epicurus, a fourth-century BCE philosopher, as transmitted in the work of Lucretius, *De rerum natura* (On the nature of things). At the heart of this philosophy is a particle theory of matter, atomism, which is fundamentally a materialist philosophy.

The Cultural Environment: Criticism of Scholastic Philosophy

In the fifteenth and sixteenth centuries there emerged a number of critiques of Aristotelian–Scholastic philosophy allied with various Christian concepts that had the effect of undermining Galenic–humoral medicine. Both the philosophy and the medicine were solidly entrenched in the great universities of Western Europe:

> "Because this [Scholasticism] formed the basis of education of every literate person in early modern Europe, the works of Aristotle . . . offered a common vocabulary and conceptual framework with which to view the natural world".
>
> S. GARBER, *Cambridge History of Science*, 2006, vol. 3, p. 26,
> Cambridge University Press

The process of dismantling such a formidable edifice took some two centuries and led to new philosophical bases for medicine and disease theories. Various criticisms came from Renaissance humanists, primarily men of letters, a cosmopolitan group associated with cities, close to governmental officials, some close to the Roman Catholic church hierarchy. They admired classical antiquity, particularly the writings from Rome. They were expert Latinists. The contents of antiquity that the humanists esteemed were the literary styles, the philosophy, the legal arguments, and the active role these individuals played in society.

The Scholastic philosophy as a basis for the study of nature and medicine was criticized by Rabelais (1494–1553). He encouraged the study of nature through observing the world. Studying the world for Rabelais also involved observing craftspersons, alchemists, dyers, and instrument makers to learn directly the composition of things, and herbalists and apothecaries to gain understanding of natural products. In addition, one should study the great variety of medical writings of different cultures:

> . . . "so that there be no sea, river or lake of which thou knowest not the fish; so that all the birds of the air, and all the plants and fruit of the forest, all the flowers of the soil, all the metal hid in the bowels of the earth, all the gems

of the East and the South, none shall be foreign to thee. Most carefully pursue the writings of physicians, Greek, Arab, Latin, despising not even the Talmudists and Cabalists; and by frequent searching gain perfect knowledge of the microcosm, man".

ALLEN G. DEBUS, *The Chemical Philosophy*, 1977, vol. 1, p. 30, Science History Publications, Neale Watson Academic Publications, New York

An influential Renaissance humanist was Marsilio Ficino (1433–1499), born in Florence and selected as a young man to head the Florentine Academy by Cosimo de Medici. He was indeed a Renaissance figure: a philosopher, a priest, a physician, and a musician, providing the first translation from Greek into Latin of the Corpus Hermeticum in 1463. This was a collection of writings by Hermes Trismegistus, alleged to be an author from ancient Egypt (not true) comparable to Moses and Zoroaster and thus providing a foundation for religious thought comparable to the Hebrew Bible. The Corpus Hermeticum contained two classes of writings. One was technical, including works on alchemy, astrology, astronomy, botany, magic, medicine, and pharmacy. On the theoretical side the major theme was spirituality; it contained speculations about God, the cosmos, and the human condition. The study of nature is declared a divine activity since the world was created by God. The physician could be viewed as a natural magician who understood the human being as a microcosm containing the same materials as the macrocosm. The Corpus stressed the role of alchemy, which dealt with the creation of the world and the production of the elements, ultimately leading to the production of inanimate and animate matter, plants and animals. It was also held that the methods of alchemy were able to extract from objects their true essences, the signatures placed there by God. One of the important consequences of Ficino's efforts was that alchemy acquired cosmological and religious associations that transformed it into a philosophy, and it became the origin of ideas for medicine and disease. This happened when Ficino made the association between "the spirit of the world" and the "alchemical quintessence." Quintessence or elixir was transformed and characterized in the alchemical literature as a substance that could be obtained by distillation and functioned as the philosopher's stone:

"Between the tangible and partly transient body of the world and its very soul, whose nature is very far from its body, there exists everywhere a spirit, just as there is between the soul and the body in us, assuming that life everywhere is always communicated by a soul to a grosser body. . . . When this spirit is rightly separated and, once separated, is conserved, it is able like the power of seed to generate a thing like itself, if only it is employed on a material of the same kind. Diligent natural philosophers, when they separate this sort of spirit of gold by sublimation over fire, will employ it on any of the metals and will make it gold. This spirit rightly drawn from gold or something else and preserved, the Arab astrologers call Elikir. But let us return to the spirit of the world. The world generates everything through it (since, indeed, all things generate through their own spirit); and we can call it both "the heavens" and "quintessence."

WILLIAM R. NEWMAN, "From Alchemy to "Chymistry": The Cambridge History of Science, 2006, vol. 3, p. 497, Cambridge University Press

The total Hermetic package gained acceptance in intellectual circles for a number of reasons. The Renaissance neo-Platonists in particular were drawn to the occult philosophy and magic. In an era of religious ferment following the Reformation, this Corpus provided different materials for religious speculation.

Another foundational philosophical component in the disease theory of Paracelsus is neo-Platonism. Neo-Platonism is a term used to describe the versions of Plato's philosophy that evolved over the centuries to include Christian theology. Plato initially wrestled with two philosophical traditions about change in the world: either it is unchanging according to Parmenides or it is a dynamic entity. Plato "solved" the dilemma by assuming the rabbinical position, postulating that both are right. To do this he proposed that there is a real world, eternal and unchanging, and a visible world that is active and consequently changing. How do we know this is happening? The world of appearances, the dynamic world, is perceived by the senses. The real world, the immutable world, is composed of abstract ideas and understood by the mind. These abstract entities are designated forms. They exist independently of any visible object. They are perfect manifestations of an object, immaterial, eternal, and unchangeable. In this Platonic world the forms are the thoughts of a divine being, designated the *Demiurge* in Plato's creation story, *Timaeus*. Beginning in the third century CE there were elaborations of Plato's philosophy labeled neo-Platonism. Among the developments was the principle of *Emanations*. Everything in the world comes from God by necessity but not directly, rather through various entities from more perfect ones to less perfect ones. The result is that everything in the world, the macrocosm and the microcosm, has a common origin from the One. The world can be characterized as an *anima mundi*, a living world. There is a soul in humans and animals and all the heavenly spheres according to Ficino. Everything is subsumed in a single world soul that is the ultimate source of all activity in all bodies. Since human beings are created from the same components as celestial bodies, there is a belief that there are correspondences between man (the microcosm) and the macrocosm. This leads to the belief that there are astral influences on humans. Such ideas were already present in Greek sources such as the Hippocratic Corpus. In the Hermetic texts there are also links between the stars and humans, sympathy or attraction between humans and parts of the universe. Interactions, influences, may occur in both directions.

Contagion is closely related to beliefs in sympathy and antipathy in the cosmos, the direct effect of the *astrum* on humans that led to the contraction of disease:

"Infection is transmutation of like into like comparable to the resonance given by one of two guitars attuned to each other when the twin instrument is played. Hence the more two persons are related to each other, by birth, complexion or **constellation**, the greater danger of one being infected by the other".

W. PAGEL, *Paracelsus, An Introduction to Philosophical Medicine in the Era of the Renaissance*, 1958, p. 183, Karger

The concept of the physician as a magician who has the power of magic was endorsed by Agrippa von Nettesheim (1486–1535), a physician and philosopher. His contribution was a manuscript titled *On Occult Philosophy* in which he described natural magic. A major theme was the way causality circulates, from ideas in God's mind to heavenly bodies to all bodies on earth. Natural magic is

... "the pinnacle of natural philosophy and its most complete achievement ... it produces works of incomparable wonder ... Observing the powers of all things natural and celestial, probing the sympathy of these same powers ... it brings powers stored away and lying hidden in nature into the open. Using lower things as a kind of bait, it links the resources of higher things to them ... so that astonishing wonders often occur, not so much by art as by nature".

BRIAN COPENHAVER, "Magic", "*The Cambridge History of Science*, 2006, vol. 3, p. 519, Cambridge University Press

The interaction of astral bodies and humans, what passes for astrology, was a common belief in the society of the period, although later it was not treated so kindly. A critique of its reality is offered by Shakespeare in King Lear, composed about 1600–03 CE:

"GLOUCESTER: These late eclipses in the sun and moon portend no good to us. Though the wisdom of nature [Shakespeare's reference to natural philosophy or science] can reason it thus and thus, yet nature finds itself scourge [afflicted] by the sequent effects [disasters that follow].

EDMUND: This is excellent foppery [foolishness] of the world, that when we are sick in fortune often the surfeit [excess] of our own behavior we make guilty of our disaster the sun, the moon, and stars, as if we were villains on necessity; fools by heavenly compulsion; knaves, thieves, treachers [traitors] by spherical predominance [according to Ptolemaic astronomy the stars moved about the earth in circles]; drunkards, liars and adulterers by an enforced obedience of planetary influence . . . An admirable evasion of whoremaster man".

SHAKESPEARE, *King Lear*, 1993, Act 1, Scene 2 Folger
Shakespeare Library, Washington Square Press

In 1417 CE a manuscript was found in the monastery of St. Gall, near Constance (southern Germany), by an indefatigable manuscript hunter, Poggio Bracciolini (1380–1459), who had been a secretary in the Papal Curia. The manuscript is titled De rerum natura. It was written in the first century BCE by the poet Lucretius and revealed to the scholarly world of the Renaissance an extended treatment of the philosophy of Epicurus, who wrote in the fourth century BCE. This work containing an atom theory of matter had a profound effect on the contagious disease theory of Girolamo Fracastoro more than 125 years later in the sixteenth century. The Epicurean version transmitted by Lucretius in De rerum natura contended that all bodies were physically constituted of particles; this led to the generalization that the visible characteristics of bodies are the emergent properties of the interactions and combinations of these invisible, indestructible particles designated as atoms.

Lucretius provided a range of analogies from processes in the visible world to justify a particle theory of matter.

"For look closely, when ever rays are let in and pour the sun's light through the dark places in the hours: for you will see many tiny bodies mingle in many ways all through the empty space right in the light of the rays . . . so that you may guess from this what it means that the first-beginnings of things (primordia rerum) are forever tossing in the great void. . . . . . . and for this reason it is more right for you to give heed to these bodies, which you see jostling in the sun's rays, because such jostlings hint that there are movements of matter too beneath them secret and unseen"

LUCRETIUS, De rerum natura, Book 2, p. 243, 1947, Oxford

In addition, Lucretius describes a sequence of events such as a food chain, where plant material is ultimately converted to the components of the human body. These interconversions can take place because all matter is composed of the same basic particles:

"Why, we may see worms come forth alive from noisome dung, when the soaked earth has gotten muddiness from immoderate rains; moreover we may see all things in like manner change themselves. Streams, leaves, and glad pastures change themselves into cattle, cattle change their nature into our [human] bodies . . . and so nature changes all foods into living bodies".

LUCRETIUS, De rerum natura, p. 283, 1947, trans. and commentary by
Cyril Bailey, Oxford

The same particles make up the matter of leaves, cattle, and humans. They are rearranged in various forms to yield the great variety of structures of living things. Lucretius also explained the cause of contagious disease via *semina*, which are physical particles.

Some Personal Histories: Paracelsus and Fracastoro

In the breadth of their interests and writings these two are representative of Renaissance men. Both were physicians. Paracelsus also wrote extensively on theological matters and was knowledgeable in chemistry, while Fracastoro wrote poetry and philosophy. They were contemporaries but appear to be unaware of each other's work. Paracelsus came to maturity in the Swiss-German region of Europe and wandered over much of Eastern Europe and the Netherlands. Fracastoro lived all of his life in the Verona-Padua region of the Venetian Republic.

Paracelsus

Paracelsus was born in Einsiedeln, Switzerland, a town that was the site of pilgrimages. His father was a physician with an interest in chemistry. Paracelsus lost his mother at an early age. He received an extensive religious education, which included elements of mysticism and magic. His father moved his practice to Vellach, a mining city in Austria. What happened to Paracelsus after that is only revealed by some of his later writings. He appears as a physician, although there is no evidence that he graduated from any university medical school. By the early 1520s he claimed he traveled in Eastern Europe and the Netherlands as a physician and from each place was driven out due to the hostility of local physicians, whom he declared to be incompetent. His contempt for the conventional practice of Galenic medicine is paralleled by his theological critiques of the Church of Rome. Both of these radical stances undermined his position in various cities, and he was generally on the run.

In 1525 he was in Salzburg, where he wrote theological critiques of the Roman church, at the time of the Peasants' Revolt, which brought him into conflict with the local theologians. He left Salzburg and eventually went to Basel, where his successful treatment of a humanist publisher led to his appointment as town physician. Such a position would have given him access to the university medical facilities, where he could give lectures; however, the medical faculty prevented him from doing this. In response he issued a manifesto expressing his contempt for the medical authorities, contending that their ideas were inferior to his own. In addition, he is alleged to have burned a medical textbook, which presumably contained notes by Avicenna, the great Islamic scholar of the eleventh century CE.

The flavor of Paracelsus' personality is revealed in this polemical writing:

"I am Theophrastus, and greater than those to whom you liken me; I am Theophrastus, and in addition I am monarcha medicorum [king of physicians] and I can prove to you what you cannot prove. I will let Luther defend his cause and I will defend my cause, and I will defeat those of my colleagues who turn against me; . . . It was not the constellations that made me a physician; God made me. . . . I need not don a coat of mail or a buckler against you, for you are not learned or experienced enough to refute even one word of mine. I wish I could protect my bald head against the flies as effectively as I can defend my monarchy . . . I will not defend my monarchy with empty talk, but with arcana. And I do not take my medicines from the apothecaries; their shops are but foul sculleries, from which comes nothing but foul broths. As for you, you defend your kingdom with belly-crawling and flattery. How long do you think this will last? . . . Let me tell you this: every little hair on my neck knows more than you and all your scribes, and my shoe-buckles are more learned than your Galen and Avicenna, and my beard has more experience than all your high colleges".

ALLEN G. DEBUS, *The Chemical Philosophy*, 1977, vol.1, p. 52, Science History publications, Neale Watson Academic Publications, New York

In summary, he declares, he is the king of medicine. He is more learned than his contemporaries. But more important, he is more learned than Galen and Avicenna and consequently more learned in philosophy than the Aristotelianism upon which Galenic medicine is based. He is a revolutionary like Luther. He does not require university degrees to qualify as a physician; he is a physician made by God. His methods include natural magic and chemistry. With his chemistry he can produce medicines more effective than those made by apothecaries. One can easily comprehend that these declarations did not make him very popular with the medical establishment.

Fracastoro

Fracastoro was born in Verona (the site of Shakespeare's play *Romeo and Juliet*) in northeast Italy, which was then under the control of Venice. He came from a prominent family, some of whose members were on the city council and in the medical profession. He attended the University of Padua, a remarkable institution where Copernicus studied in 1501. At Padua Fracastoro studied medicine, taking the standard curriculum for a physician: medicine, botany, geology, astronomy, and philosophy. In addition he studied mathematics. At a young age he became a lecturer in logic at Padua for a number of years. Later he settled in Verona. However, life was not to remain tranquil. France invaded Italy in 1494, and later Maximilian, Emperor of the Austrian-ruled confederation the League of Cambria, formed a coalition with Pope Julius II, Louis XII of France, and Ferdinand V of Aragon (Spain) in 1508 and entered Italy. Verona was occupied and destroyed. A plague occurred and Fracastoro left with his family to a country home, where he wrote. Among the works he composed was a poem on syphilis. Verona later returned to control by Venice, and thereafter Fracastoro practiced medicine, studied medicinal plants, and continued to write in Verona.

He became famous as a poet. In 1530 CE the poem *Syphilis or the French Disease* was published in Rome and in Paris. In later years it was published in Basel, Antwerp, Lyons, and Venice. He was learned in astronomy and in 1538 wrote a work on astronomy dedicated to Pope Paul

III, a move that gave him some cover since writing astronomical works after Copernicus was a dangerous activity without clerical approval. In addition he was cited as an authority on medicine equal to Giambattista da Monte, who in the early 1540s became chair of medical theory at Padua. While Paracelsus presented himself as a radical reformer at a time of religious and political conflict, Fracastoro was a substantial member of the establishment, expressed conventional religious views, and in his science and medicine opposed metaphysical explanations while supporting naturalistic explanations of phenomena.

He wrote:

"That ancient theology known indifferently as divine or first philosophy examines truly noble and imperishable things. **However, so few of them can be known to us, and then our knowledge of them is so uncertain, that we may as well confess that no knowledge, or next to none, has come our way in such matters** (my emphasis). On the other hand, we can achieve no small certainty with regard to those things which belong to the realm of nature, and nevertheless have some nobility, such as substances and bodies, the heavens, and living things; and what can be known is practically limitless, inasmuch as nature is everywhere, wherever we turn, whatever we see or hear. On this account, the philosophy that deals with these things is to be considered, among all others, the greatest and most worthwhile".

SPENCER PEARCE, *The Cultural Heritage of the Italian Renaissance: Essays in Honour of T. G. Griffith*, ed. by C. Griffiths and R. Hastings, 1993, p. 236, Edwin Mellen Press

Disease Theory of Paracelsus

"How long is it that medicine has been in the world? They say that a newcomer, whom they call Paracelsus, is changing and overthrowing the whole order of the ancient rules, and maintaining up to this moment it has been good for nothing but killing men. I think that he will easily prove that; but as for putting my life to the test of his new experience, I think that would not be great wisdom".

MONTAIGNE, *The Complete Essays of Montaigne*, trans. by Donald M. Frame, 1958, pp. 429–430, Stanford University Press

Paracelsus rejected much of Galenic medicine not because it had suddenly failed to explain the cause of contagious diseases in the sixteenth century CE, while it had successfully done so in previous centuries, but because he adopted religious and philosophical principles different from those underlying Galenic disease theory:

"The distinguishing feature of Paracelsus' own philosophy is the consequential view of cosmology, theology, natural philosophy [science] and medicine in the light of analogies and correspondences between macrocosm and microcosm . . . Paracelsus was the first to apply such speculation to the knowledge of Nature systematically".

WALTER PAGEL, *Paracelsus: An Introduction to Philosophical Medicine in the Era of the Renaissance*, 1958, p. 50, Karger, Basel

"All this you should know exists in man and realize that the firmament is within man, the firmament with its great movements of bodily planets and stars which result in exaltations, conjunctions, oppositions and the like . . . for, none among you who is devoid of astronomical knowledge may be filled with medical knowledge".

ALLEN G. DEBUS, *The Chemical Philosophy*, 1977, vol. 1, p. 53, Science History Publications, a division of Neale Watson Academic Publications, New York

Humans are linked with elements of the "firmament" by mutual concordances or, expressed another way, by correspondences or sympathy. Why is it important that there is something in the human that is an analogue, "an inner representation of a thing in the external world?" For Paracelsus it is decisive, since it is one of two ways for humans to gain knowledge about themselves and the world. By some magical, mystical process, union between a body in the world (macrocosm) and the human (microcosm) is achieved. This is a "sovereign (supreme) means of acquiring ultimate and total knowledge." Rather than gaining knowledge of nature by the rational/ Scholastic method, an obvious allusion to Aristotelian–Galenic practice, the process of union proposes another method employed by various non-rational means such as dreams, trances, clairvoyance, or imagination.

Paracelsus indicated a second way to understand the *virtues* (processes) of objects: by experimentation, which is characterized as a divine activity:

. . . "by means of unprejudiced experiment inspired by divine revelation, the adept may attain his end. Thus, knowledge is a divine favour, science and research divine service, the connecting link with divinity. Grace from above meets human aspiration from below. Natural research is the search for God".

ALLEN G. DEBUS, *The Chemical Philosophy*, 1977, vol. 1, p. 54, Science History Publications, a division of Neale Watson Academic Publications, New York

The practical tools of experimentation are provided by alchemy, such as fire and the methods of separation.

Paracelsus rejected Galenic medicine and its philosophical base and provided a new base that includes chemistry with a new elementary system to replace the four-element system (earth, air, fire, water) used by Galenists. It is a system that consists of three principles, or essences: salt, sulfur, and mercury. They function in a spiritual way, with sulfur conferring structure and combustibility, salt giving solidity, and mercury designating a vaporous quality.

The processes of the human body and those of the macrocosm are defined by Paracelsus as chemical events. This principle transforms the contagious disease process into a chemical event that can be studied using the tools developed by alchemy.

Practical methods of alchemy were firmly established in Western Europe after they were imported during the eleventh century CE from Arabic sources. Their theoretical roots go back to antiquity, from the matter theory of Empedocles and Aristotle, who proposed that the world is composed of earth, air, fire, and water paired with primary qualities where hot/dry is equivalent to fire, hot/wet equals air, cold/wet equals water, and cold/dry equals earth. Aristotle introduced the idea that by changing the primary contrary qualities, one element might be transformed into another. Alchemists adopted this philosophy and searched for the *philosopher's stone*, a material that would facilitate these changes. Roger Bacon (1214–1294) expressed respect for and the utility of alchemy:

"There is another alchemy, operative and practical which teaches how to make . . . things . . . better . . . than . . . made in nature. And science of this kind is greater than all preceding because it produces greater utilities . . . It also teaches how to discover things as are capable of prolonging human life".

A.C. CROMBIE, *Augustine to Galileo*, 1979, vol. 1, p. 54,
Harvard University Press

From the thirteenth to the fifteenth century the utility of alchemy was demonstrated. Metallurgy made great advances with the development of blast furnaces for melting iron, which was used to prepare alloys, and this led to techniques needed to control the metal mixtures. Measuring devices such as balances were introduced, marking the beginning of quantitative chemistry. Other industries that required chemistry and chemical techniques included pottery, tile, and brick manufacture, tanning and soap production, and wine and beer making. In medicine, the most important field was the production of herbal and chemical remedies. Alchemy was firmly established as a base for an extensive variety of commercial activities.

For Paracelsus alchemy was essential for understanding the processes of nature and the cause of disease by revealing the hidden activities that are responsible for these phenomena:

"The great virtues [Paracelsus means processes] that lie hidden in nature would never have been revealed if alchemy had not uncovered them and made them visible. Take a tree, for example; a man sees it in the winter, but he does not know what it is, he does not know what it conceals within itself, until summer comes and discloses the bud, the flowers, the fruit . . . Similarly the virtues in things remain concealed to man, unless the alchemists disclose them, as the summer reveals the nature of the tree".

ALLEN G. DEBUS, *The Chemical Philosophy*, 1977, vol. 1, p. 1, Science History Publications, a division of Neale Watson Academic Publications, New York

Paracelsus believed that a component of the disease process is analogous to a chemical process:

"What a disease is, its cause and its cure is learnt from inorganic nature outside man, i.e., from the growth and transmutation of minerals and metals . . . He who is ignorant of what makes copper and gives birth to Vitriolata [a class of corrosive materials such as sulfuric acid] does not know what makes leprosy; he who is ignorant of what makes rust on iron does not know what makes ulcers or what makes earthquakes" . . .

W. PAGEL, *Paracelsus, An Introduction to Philosophical Medicine in the Era of the Renaissance*, 1958, p. 272, Karger, Basel

This philosophy of disease was derived from his thinking in analogies. Paracelsus postulated that the disease cause was analogous to a parasitic invader, in contrast to the principle in humoral medicine, which regarded disease as a physiological imbalance of the fluids of the body (blood, phlegm, and the two biles). In humoral theory the various remedies, sweating, bloodletting, and vomiting, were intended to remove morbid matter from the body and restore the balance of the humors. This humoral theory of contagious disease did not recognize specific disease-causing entities that could have required different treatments. Paracelsus contended that there was a specific effect on one part of the body introduced by the conventional means, air, food, or liquids. However, an additional entity accompanied these sources: "seed-like factors, seeds of disease."Diseases grow like grass.

These seeds are analogous to the metallic seeds in the earth that are the source of veins of metals. Their origin is astral. They affect a particular organ in the body to which they are related "by a kind of predestined sympathy." The air transmits this astral "poison." This theory is in conformity with Hermetic tradition, where plague is an astral disease, "fiery and contagious." It appears not to be a material entity and certainly not a living thing.

What Paracelsus proposed is that there is a disease agent, present in air, water, or food, that is a specific thing that has a specific effect. The entire disease scenario is explained in his description of the acquisition of plague. The philosophical basis of this theory may be summarized in this way. The contraction of disease involves a complicated interaction of the two domains, the macrocosm and the microcosm. Plague originates as a noncorporeal event, as a passion or imaginative occurrence in the portion of man connected to the different planets, such as Saturn and Mars. This is a psychic process, where a part of the body that is volatile travels to the appropriate planet, where it stays as a *semina* (seed), and incurs the wrath of God. Subsequently, it returns to the earth, to food, water, or air or directly to the human body, where it causes a liberation of a poisonous substance, leading to disease.

Fracastoro's Disease Theory

In his writings on disease he used two modes, one poetic and the other scholarly and analytical. The poem published in 1530 CE in the classic Latin hexameter style used by Vergil and Lucretius is titled "Syphilis Sive Morbus Gallicus" (Syphilis or the French Disease), which, remarkably, provides the name for the disease to this day and describes its arrival in Europe at the end of the fifteenth century CE. The poem comprises three books. Book One discusses the origin and causes of the disease and provides us with an earlier view of his disease theory before the publication of his major treatise On Contagion in 1546 CE.

Fracastoro began writing the poem during his stay in his country retreat after the capture and destruction of Verona. He dedicated the poem to Cardinal Bembo, whose contribution to the work was to recommend that Fracastoro change the treatment of the disease from mercury to guaicum, a wood product. Fracastoro did not alter the story but added a third book containing a fable concerning a king and his shepherd, named Syphilis, who contracted a disease as a punishment for not regularly worshipping the god Apollo. He was cured of the disease by using guaicum after he admitted his impiety and promised to reform.

Cardinal Bembo was able to read a first draft of the poem in 1525 and wrote to Fracastoro that

"you write with more charm than Lucretius often does . . . that the soul of Vergil has passed into you".

FRACASTORO, Syphilis or the French Disease, Tr. H. Wynne-Finch 1935, p. 46,
Heinemann [Elsevier]

In the poem there are a number of philosophical principles regarding causes of disease, not only syphilis, which represent Fracastoro's views in the early decades of the sixteenth century. Causes are natural although nature is mysterious because it operates in invisible ways. The disease agents are invisible semina. This Latin word is generally translated as seeds, although the word "germ" is sometimes used by translators. These semina are conveyed by air. Syphilis is conveyed by air. In his later work he recognizes that the disease is passed from person to person and not by air. There are planetary influences on disease causation. Syphilis is one kind of disease among many specific diseases that infect plants, animals, and humans.

To appreciate Fracastoro's contemporary disease views (1525–1530 CE) expressed in verse we present the following passages from H. Wynne-Finch's English version. Book One opens with the following verses on the origin of the disease (p. 53):

"What various chances in Life, what semina conveyed this strange disease, unknown of any through long centuries, which, in our own day, has raged throughout Europe . . .
Tell me, Goddess, what causes after so many ages brought forth for us this unaccustomed disease?
Was it borne by the Western Sea [the Atlantic Ocean] and so came to our world at the time when a chosen band [Columbus and crew] set sail from the shores of Spain" . . . (p. 55)

Fracastoro is skeptical of this explanation that Columbus returned with the disease (p. 61)

"since Time flows on forever, we must hold that this disease has appeared in the world not once only, but many times" . . .

To find the origin of this disease:

"look to the conjunctions of the planets for the signs they have given and the fate heaven has presaged for our years. Because here, perchance, the origin of this novel pestilence Was it that, when so many stars came into conjunction with the blazing sun, the fiery energy drew many vapours from sea and land; and that these, mingled with thin air and seized by the new taint, conveyed a corruption too fine to be seen"? (p. 69)

On the cause of the disease, the "seat of evil must exist in the air itself." And thus it is air that brings disease to all living things; however, something in the air confers a high degree of specificity:

"air becomes noxious to trees alone . . .
sometimes it seizes upon the cornfields and the fertile crops . . .
and consumes the straw with a scabby blight . . . at other times it is only
living creatures, few or many that have been punished . . .
In the spring too, and in the following summer (strange to relate) a
dread pestilence, accompanied by an evil fever, carried off the
suffering cattle . . .
So various are the semina of the infected sky, and so various the
appearance of things
and yet there is a fixed order, both in the motions and the forces
that move.
Do you not see that when the chest is laboring, although the eyes
are exposed and more tender, the disease does not, nevertheless,
seize them, but plunges into the depths of the lungs?
And although grapes are softer than apples, one does not infect the
other; but grape taints grape".

FRACASTOR, *Syphilis or the French Disease*, (pp. 75 and 77) 1935,
Tr. H. Wynne-Finch, Heinemann, Elsevier

Once again we encounter the term "seed" as the cause of disease. We shall provide an extensive analysis of Fracastoro's meaning when we discuss his mature work, *On Contagion*.

In 1546 he wrote a medical text, *Contagion, Contagious Diseases, and their Treatment*, on the cause and treatment of many of the contemporary contagious diseases. *Contagion* was the second part of a two-volume treatise, the first titled *Concerning the Sympathy and Antipathy of Things*, which contains little medical information but provides a philosophical grounding for his contagious disease theories. Therefore, before we examine the mature product of his speculations about the causes of contagious disease we will consider his philosophical writings. His intention, he states, is to discuss:

> "the hidden agreement and disagreement of things, called Sympathy and Antipathy . . . There is nothing more marvelous in the whole of nature, nothing that is more worth knowing".
>
> Hand written English translation obtained from Deakin University
Library, Australia

The phenomenon of Sympathy, which so elicits his admiration and curiosity, may be characterized as action-at-a-distance. Among the examples of Sympathy are the navigator's box and the herculean stone (lodestone). Both are instances of magnetic phenomena, the first where the compass points to the poles of the earth and the second where iron is moved to the magnet. These magnets reveal a power that acts over small or large distances without direct contact between the magnet and the object that is acted upon. Fracastoro makes the important conjecture that "the nature of contagion" is included under the category of action-at-a-distance. He is reaffirming the often-observed fact that one can contract a contagious disease without physical contact. Linking magnetism and contagion suggests that they operate by the same process. If that is the case, let us examine Fracastoro's treatment of action-at-a-distance. We see that it is embedded in the philosophical discussions of the composition of matter, the origin of motion, and various emanation theories that have been part of philosophical discourse since antiquity, but most importantly the philosophy of Epicurus as transmitted by Lucretius in his great poem *De rerum natura,* where he adopts a particle theory of matter and *semina* as the cause of disease.

Lucretius' poem is presented in six books. Books One and Two discuss the material that makes up the world, "atoms," an infinite number of solid, indivisible, eternal particles moving in a void of limitless empty space. Books Five and Six deal with the phenomena of the world, its creation, the growth of all life on earth, the phenomenon of weather, the magnet, and an extended discussion of epidemic disease, concluding with the plague in Athens, the disease described by Thucydides.

Lucretius provided an extended Epicurean rationale for a natural world where humans and all other creatures are similarly created not by gods, who exist in some distant place but are not involved in the lives of people. Everything in the world is made of the same basic particles; the variety of things in the world is due to the rearrangement of these basic "atoms." There are no miracles. There is no reason for humans to consider themselves privileged creatures. According to Epicurean philosophy humans should overcome their fears, accept that everything is transitory, and enjoy the emotional and physical pleasures of the world.

After the poem's recovery numerous editions were produced in various cities and it was attacked as a how-to book of atheism by M. Ficino; in 1516 its reading was prohibited by the Florentine Synod. Nevertheless, it remained a seminal source of Fracastoro's philosophy and disease theory and remained in circulation as a prime example of Latin poetry from antiquity. Recall Cardinal Bembo comparing favorably Fracastoro's poem in Latin on syphilis to the work of Lucretius.

Lucretius' Model for Action-at-a-Distance

Lucretius specified a physical theory for phenomena such as the propagation of light, sound, smell, magnetism, and disease. Lucretius began with the statement that there are simulacra, images or representations of surfaces, a principle applicable to sight:

> . . . "which like films stripped from the outermost body of things, fly forward and backward through the air . . . a thin image from things too must needs be given off from the outermost body of things".
>
> LUCRETIUS, *De rerum natura, Book 4*,1947, trans. and commentary by Cyril Bailey, pp. 363 and 365, Oxford

What is the content of simulacra?

> "First of all from all things whatsoever we can see, it must needs be that there stream off, shot out and scattered abroad, bodies such as to strike the eyes and wake our vision".
>
> LUCRETIUS, *De rerum natura, Book 6*, 1947, pp. 561–562, Oxford

With regard to scents, "often moisture of a salt savour comes into our mouth when we walk by the sea" Book 6, p. 563). What are these bodies that stream off the surface of things, that strike the eyes, that transfer the taste of salt to our mouths, that are the cause of magnetism?

Although the actual physical events envisioned by Epicurus and Lucretius differ, they both state that the emanations from the magnet are atoms. In Epicurean philosophy these atoms, the ultimate, irreducible, immutable components of matter, attach to the iron through hooks, while in the Lucretian description the atoms fill a vacuum. Lucretius' theory is contained in De rerum natura.

> "For what follows, I will essay to tell by what law of nature it comes to pass that iron can be attracted by the stone which the Greeks call the magnet, from the name of its native place . . . the land of the Magnets. At this stone men marvel.
>
> First of all it must needs be that there stream off this stone very many semina or an effluence which with its blows parts asunder all the air which has its place between the stone and the iron. When this space is emptied and much room in the middle becomes void, straightway first-beginnings of the iron start forward and fall into the void".
>
> LUCRETIUS, De rerum natura Book 6, pp. 561–562, Oxford

In this theory there are no spiritual or occult emanations. The word *semina* is introduced, which can be translated as seeds and in this text is equivalent to atoms.

Lucretius' Model for Contagion

Immediately following the Book 6 section on magnetism there is a discussion of the cause of plague, and what is interesting and not surprising is that Lucretius ascribes the cause to *semina*, those particles (atoms) that are the material basis of things in the world and of necessity are the source of disease agents.

> "Now what is the cause of plague, and whence on a sudden the force of disease can arise and gather deadly destruction for the race of men . . . of cattle . . . First I have shown before that there are seminaria [seeds] of many things which are helpful to our life, and on the other hand it must needs be that many fly about which cause disease and death".
>
> LUCRETIUS, De rerum natura, Book 6, p. 571

It is reasonable to conclude that Lucretius believed that the same particles are responsible for magnetism and disease.

What did Fracastoro believe? In Sympathy and Antipathy he subscribed to natural, material causes justified in the following way:

"When we are in search, not of the first and universal cause (of some natural phenomenon) but of one that is particular and proper to it, this cannot be anything immaterial for thus would nature be destroyed".

This leads him to the important practical operating principle:

"The philosopher may not be able to explain in detail what any given natural process involves, but it is his duty to seek for some cause intermediate between the complete explanation that may forever elude him and the illegitimate appeal to the supreme explanation [God], which is scientifically inadequate because it does nothing to further understanding".

SPENCER PEARCE, The Dialogues of Girolamo Fracastoro, The Sixteenth Century Journal, 1996, vol. 27, No. 1, p. 119

Natural phenomena, including contagious disease, occur in compliance with natural laws. He rejected occult causes and at this time generally celestial influences.

"Nature tolerates nothing, admits nothing that is purposeless and that frustrates the order and laws of the universe . . . a universe full, continuous and harmonious. The harmony that governs the activity of all its parts guarantees the cohesion of the entire organism and the fulfillment of the Creator's purpose"

SPENCER PEARCE, The Dialogues of Girolamo Fracastoro, The Sixteenth Century Journal, 1996, vol. 27, No. 1, p. 120

This beautiful rhetorical position leaves open the question of what is his theory of matter and is it used to explain contagion. In "Concerning the Sympathy and Antipathy of Things" he designated motion induced by a magnet as a form of movement that Aristotle characterized as forced movement, different from movement such as that of a falling body. For forced motion to occur, Fracastoro adopted the classical emanation theory. Something is sent out from one body to another. The "something will be either a body" (a substantial physical thing) "or some simple form, material or spiritual." There is ambiguity here. In an attempt to provide clarity Fracastoro cites ancients such as Empedocles, Epicurus, and Lucretius (the latter two are atomists), who:

"posited effluxions of bodies which they called atoms . . . these effluences should by no means be negated [denied] . . . but the manner which these men taught [treated them] was crude [rough] and inept [unsuitable]. We must perhaps seek another principle than that of atoms . . . If it is not a body . . . it will be called spiritual".

Hand written English translation obtained from Deakin University Library, Australia

But he is not satisfied with this position. It is not an either/or position, material or spiritual, for "nothing can be moved per se which is not either a body or at least nature and substance in a body." Furthermore, there can be no effect—that is, one object having an impact on another—barring contact. If that is true, that only a material entity can move another body through interaction, how can something *spiritual* be a principle of motion? Fracastoro provided an explanation, although not a precise one, in *Concerning the Sympathy and Antipathy of Things*. There are entities called spiritual species (spiritual manifestations), a concept with its source in Ficino's neo-Platonic philosophy. These species emanate from objects. In Fracastoro's scheme they are not conventional bodies; they are manifestations of a process characterized as "tenuous and superficial," an image that may reflect the surface of the object and is produced instantaneously. It is a subtle, film-like part of the surface. "It shares an extremely fine grade of materiality." They are substantially identical with the material forms from which they proceed, differing only in their grade of existence.

Let us conclude that Fracastoro accepts that it is a material entity that is propagated from a magnet and it is an example of action-at-a-distance. Contagion is also an example of action-at-a-distance, as Fracastoro contended in "Concerning the Sympathy and Antipathy of Things." Does he provide the same emanation theory for disease transmission? To obtain an answer we turn to the treatise On Contagion.

In On Contagion, diseases can be caused by *semina*. In the poem syphilis and in *Contagion* there also are varieties of infectious agents with different specificities. In the poem on Syphilis, he wrote:

"Nature is, above all incomprehensible and infinitely variable in matters of disease . . . (p. 75) air becomes noxious to trees alone . . . sometimes it seizes upon the cornfields and the fertile crops . . . and consumes the straw with a scabby blight . . . at other times it is only living creatures, few or many, that have been punished" . . .

There is a lot going on in these lines. There are entities designated as *semina* that cause diseases. There are specific *semina* that cause specific diseases. Diseases of plants, animals, and humans are caused by the same category of things called "semina." This is "the fixed order." This is the natural order of the world. Fracastoro arrived at an ontological view of disease at a time when Galenic medicine considered disease an imbalance of humors set off by various environmental conditions. Now, the question is what Fracastoro intends by the use of the word "semina," which occurs countless times in On Contagion. Is it an emanation comparable to the object emitted by the magnet? For answers we turn to Fracastoro's descriptions of the entities that cause contagious diseases. There are three kinds of contagions based on the mode of transfer: the first infects by direct contact; the second infects via fomites; the third infects at a distance.

To explain infection by direct contact he relies on an analogy. It is often observed that one fruit undergoes decay. If that fruit is in contact with another, the second one will show the identical decomposition that destroyed the first fruit. This decay or putrefaction is equivalent to an infection. What passes from one fruit to the other are unseen particles that evaporate from the first fruit and bring heat to the second fruit, causing decay. These imperceptible particles are designated *seminaria* of contagion, or seeds of disease, in later chapters. Willmer Cave Wright, Professor of Classics at Bryn Mawr College, who translated Fracastoro's Latin text in 1930, rendered the word "seminaria" as germs. She claimed that it is the nearest English equivalent, although that word can be translated as "seeds," which it was by the translator of Fracastoro's poem on syphilis. The contemporary (1930) use of the word germ almost always refers to a living agent, and by its use Wright suggested that Fracastoro implied that characterization. However, Fracastoro did not claim that the carriers of disease are living agents, although there are instances where his description of such semina have inspired modern commentators to interpret him that way (see the Appendix).

It is not surprising that semina could be translated to refer to a living agent, since in the historical literature, in the writings of Aristotle and Theophrastus, "semina" (seeds) give rise to living things. Galen used the term briefly, and Lucretius employed the term numerous times in De rerum natura, alone and in combination—for example, semina rerum (seeds of things), and that seeds are the cause of plague. Lucretius generalized this doctrine that all living things come from a fixed seed: "cattle, fowl, fruit trees, corn, roses, vines, and humans."

Thus, in these writings, except in Galen, seeds can be the source of living plants, animals, and humans. In De rerum natura disease has the same source: "there are seminaria [seeds] . . . many fly about which cause disease and death."But Lucretius does not classify these semina as living entities.

The Nature of Fracastoro's Agent of Contagion

Fracastoro asks where these particles come from. There are two sources: some come from without, an extrinsic origin, and "some arise from us" (that is, generated in human bodies).

Air is the most efficient vehicle for spreading disease, since everyone requires it to live. Bodily humors, when confronted with "*semina*," become "fouled, putrified, occluded" and generate more "*semina*." From then on a train of events is initiated; another person receives seeds from the first infected individual, which reach an analogous humor, starting another round of putrefaction due to the propagation of the "semina."

The second mode of contagion is by "fomites." The term appears to be used to indicate intermediaries such as clothing or other inanimate objects. This mode of transfer obligates Fracastoro to explain how the disease-causing agent can survive after a period of time on some object. He stated that they are the same particles that can infect on contact, but they have some additional characteristics that make them strong. These are hardness, like iron, and increased viscosity that allows them to adhere to fomites. Fracastoro asserted that they must have these qualities, since the particle–fomite combination may not immediately encounter the proper object to infect, the object for which the particle has a "selective affinity" or specificity.

These *semina* have characteristics, solidity and viscosity, indicating a physical entity.

The third form of contagion occurs at a distance—that is, these diseases can be acquired without contact with individuals or objects carrying the putative disease agent. Fracastoro stated that the particles that can infect at a distance can continue to infect by direct contact or by their presence on fomites. This postulate established in Fracastoro's theory that there is only one kind of contagious particle. Three examples of diseases acquired at a distance are ophthalmia, certain pestiferous fevers, and phthisis (tuberculosis).

Fracastoro presents in detail the properties of "*seminaria*" that work over distances. *Seminaria* are material and not spiritual. Spiritual factors cannot produce effects like corruption, the term used to describe the result of the disease process, and generation, the propagation (multiplication) of the disease agent. The second argument for the material nature of the disease agent is that the seed that is contagious persists after traveling a long distance and can cause disease when the primary infected object is no longer present:

> "This is proof that this same thing [the agent] is a body that can be carried and endures even when far from its place of origin. For if the contagion has been correctly defined, it follows that what is developed in the second thing must be of the same sort as was developed in the first thing and that the principle is the same in the fourth and fifth and so on . . . from the original semenarus other semina must be . . . propagated . . . similar to those former semina both in their nature and combination. But nothing spiritual can effect this per se—the spiritual cannot generate in a second body the same sort of thing that was in the first body, for all generation occurs through primary qualities (size, shape, motion, quantity)".
>
> FRACASTORO, *On Contagion, Contagious Diseases, and their Treatment,*
> 1930, p. 25, Putnam

It is categorically clear that these four qualities are all material properties. They are the characteristics ascribed to atoms by Epicurus and Lucretius.

The important question is whether Fracastoro's seeds of disease have these same characteristics. In On *Contagion* they are physical bodies because, according to Fracastoro, they can persist and be transported over a long distance over an extended period of time. He claimed that no spiritual entity has these properties. The *semina* of contagion are propagated in the appropriate host and give rise to the same *semina*. Again, Fracastoro stated, no spiritual entity can do this. Fracastoro endowed seeds with the physical properties of viscosity, hardness, and sharpness. They are destroyed by burning and by great cold. What about the ability to propagate or generate more of their identical kind? This suggests a living agent, but Fracastoro never claimed they are.

Fracastoro's matter theory, following Epicurus and Lucretius, stated that everything comes from seeds. They are the origin of all things, including agents that cause disease. In Lucretian physics they are physical particles, they are atoms. In *On Contagion* we appear to have an early example of a physical model of contagious disease. This model can be viewed as representing Epicurean–Lucretian atomism; it remained an isolated event in the mid-sixteenth century, competing with disease theories that involved humors and entities like morbid matter or poisonous vapors.

Neither Lucretius nor Fracastoro provided a process whereby physical particles are reproduced or reproduce themselves.

There are elements of Fracastoro's disease theory that are similar to mainstream Galenic medicine. Air was considered the prime conveyer of disease. There is agreement that there is something harmful in air that brings disease. There is recognition that the physical condition of the individual influenced the contraction of disease. It is clear that in an epidemic not everyone was infected. There is general agreement that contagion can be transferred from one individual to another. This was acknowledged by the implementation of procedures that segregated ill people. In Venice *lazarettos* (quarantine areas) were set up during the plague, clothing was burned, and the sale of food and drink was regulated.

An immediate controversy arose concerning Fracastoro's use of the word *semina*, seeds of disease. It was raised by Giambattista Da Monte (1489–1561), a contemporary of Fracastoro from Verona, who was once personal physician to Cardinal de Medici, became Professor of Medical Practice in 1540 BCE, and three years later was named Chair of Medical Theory at Padua. He was a leading Galenist. He attacked *On Contagion* on two grounds: it rejected Galen and accepted the philosophy of Epicurus and Democritus, and it introduced a new and unnecessary agent as the cause of disease. Da Monte already knew there was something in the air that causes disease, and that is putrid matter and foul vapors. By claiming that he invented "absurd seedlets" and accusing Fracastoro of Epicureanism, he would turn Fracastoro into an atomist, equating *semina* and atoms. Fracastoro had indeed invented, or rather reinvented, seeds to accomplish two interrelated objectives. The first was to account for the specificity of the disease agent. The second was to explain the obvious fact that there were unique diseases. The introduction of seeds into the discourse about contagion added a new dimension to disease theory. There was now a distinction between seeds and "miasms," the latter preferred by Da Monte.

There is, obviously, no fixed meaning for the term "seeds of disease." In Paracelsian disease theory there are seeds of disease whose origin is the result of a complex interaction between spirit and an astral body, whose nature is determined by a chemical process. Whatever its final identity, it is not called a physical particle according to Epicurus and Lucretius.

The differences in general disease theory include many different names for the thing present in the air that causes disease. There are morbid secretions, poisonous vapors labeled miasms, seeds of disease, and poisonous occult properties. It was clearly recognized by Fracastoro that poisons are not contagious: a person who had ingested poison never infected anyone else.

In conformity with Galenic medicine, discussions about disease included environmental influences such as the time of year and atmospheric conditions that influenced the contraction of diseases. In "On Contagion" there are humors that became obstructed as a precondition for the effect of a seed of disease. There are seeds that had analogy, selective affinity, for "spirits and the more volatile humours." There are celestial events that precede disease occurrences. These ideas were part of the discourse not only in Galenic medicine but also in the writings of Paracelsus.

Summary: Comparing Paracelsus and Fracastoro

Paracelsus and Fracastoro were contemporaries in the sixteenth century who did not know each other and were not aware of each other's writings. They were equally devoted to understanding the causes of contagious diseases and equally committed to the idea that Galenic– humoral medicine cannot explain the cause of diseases such as plague, tuberculosis, syphilis, and a variety of fevers. They both concluded that there is something specific carried by air, for example, that is responsible for disease. However, from that point on they produced conflicting scenarios. The process of acquiring the disease and the nature of the entity that causes disease is different because Paracelsus had included vitalistic and occult components in the process and a matter theory incompatible with the monist particle theory of matter of Fracastoro. In brief, they each based their theory on dissimilar philosophical principles.

After Paracelsus' death a large number of physicians in Europe could be classified as Paracelsians since they vigorously supported the use of chemistry to understand bodily functions and malfunctions, although many did not necessarily accept Paracelsian philosophy. For some writers support for the study of chemistry and its useful role in medicine was justified in religious terms.

R. Bostocke's book, "Difference between the Ancient Phisicke . . . and the latter Phisicke," written in 1585, was a defense of Paracelsian medicine in England. Bostocke's book expressed a deep faith in the ability of Paracelsian physicians to use chemistry to understand disease. He claimed chemical philosophy was founded on "God's books" and referred to Paracelsian physicians as chemical philosophers and compared them to "heathenish" Aristotelians and Galenists. The three forms of prime matter that make up the world in Paracelsian thought, the semina, salt, sulfur, and mercury, reflect the "divine unity in the Trinity." The general principle was adopted among the chemical philosophers that diseases are not the result of imbalance of humors but "arise from seeds which grow within the body as the earth brings forth fruit from seed."

In the early seventeenth century the theoretical and chemical practitioner successor to Paracelsus was Joan Baptista Van Helmont (1578–1644):

> "Pessimism, skepticism, and criticism are the outstanding key-notes of all of Van Helmont's works and researches. He rejected the world into which he was born because he felt rejected by it . . . His country had been occupied by Spain with all its attendant cultural and doctrinal convulsions; among these could be included the Jesuits, whom Van Helmont regarded as pseudo-scientific doctrinaires . . .
>
> Complacent human reason had ousted the only true source of knowledge of the physical world: the spiritual union of man and nature. Man could now attain knowledge only by patiently knocking on nature's door, observing, weighing, and measuring in a way that was informed by imagination and vision. This search had to be crowned by personal illumination through the divine mind that should inhabit the ground of the soul, untainted by dogma".
>
> WALTER PAGEL, *Joan Baptiste Van Helmont*, 1982, p. 1,
> Cambridge University Press

He was born in Belgium and received his university training in classics and philosophy. He rejected this teaching "as nothing that was sound, nothing that was true." He read ancient writings on herbal medicines. He admired them because they revealed the power of the creator, but the writings themselves dissatisfied him because they lacked rigor; there were no "maxims and rules." In these intellectual wanderings he retained an interest in natural philosophy, including medicine and Hippocratic writings, Galen, and the translated work of Islamic scholars. But again his persistent searching for answers about the causes of phenomena in nature, of ways to explain the operation of the human body, led him to conclude that this learning was of little value.

Nevertheless, he went on to receive a medical degree, at the age of 22, via the study of Galenic medicine, which was the only kind of training that was available. He continued to study and travel, and most important of all he read Paracelsus and converted to the utility of chemistry:

> "In turning to chemistry Van Helmont felt that he was obeying a divine call. I praise, he said, the bounteous God who called me to the art of fire [pyrotechnia], away from the "dregs"—the so-called sciences and professions;

[Van Helmont followed Paracelsus in his contempt for the academy and their inferior standards and bad scholarship]—

"for [the sciences] its principles do not rest with syllogism, [symbolic of scholastic philosophy] but are made known by nature and manifest by fire [one of the methods of chemistry]. It enables the mind to penetrate to nature's secrets thus to ultimate truth. It admits the worker to the first roots of things through separating . . . nature's deed's [processes] . . . developing the virtues [activities] of the semina". . . . . .

WALTER PAGEL, *Joan Baptiste Van Helmont*, 1982, p. 6–7, Cambridge University Press

Van Helmont's belief in acquiring knowledge through unification with the object reflects Paracelsus' position and has its source in neo-Platonism, where the whole cosmos represents a coherent body and all its parts are connected by a cosmic sympathy.

An early example of Van Helmont's philosophical writings, *Magnetic Cure of Wounds,* brought him into serious conflict with the medical authorities at Louvain. In the early seventeenth century there was a dispute about an issue once engaged by Paracelsus concerning the treatment of wounds inflicted by a weapon. One method was to apply the healing salve to the weapon that caused the wound and not to the wound. This action was intended to act by sympathy regardless of the distance between the patient and the doctor treating the weapon. The dispute was between a Protestant professor who believed in natural magic that operated through sympathy and antipathy between natural objects and a Jesuit professor who considered any form of magic as the work of the devil. Van Helmont entered this controversy and engaged the issue seriously, labeling his views as Christian philosophy. He incurred the opposition of the Jesuit authorities and the magic elements in his writings were declared heresy. He recanted his views in 1630 but was convicted by the Medical Faculty of Louvain for adhering "to the monstrous superstitions of the school of Paracelsus since that school distorted nature by ascribing its activities to magic." There is no doubt that Van Helmont was a Paracelsian.

Van Helmont rejected Galenic medicine with its four humors. His philosophy included homage to the Trinity and alchemy because it is the key to understanding medicine and nature. Natural phenomena could be revealed by an analytical technique like fire. It was one of the various methodologies of chemistry that is the premier tool for "revealing the secrets of nature." What are these secrets of nature? They are the *root* of things; they are the processes that underlie the visible manifestations of living beings such as growth and development, and the cause of disease. Using fire, for example, he revealed the essence of an object, when a gas was released upon burning. The gas was not a separate entity but represented the object itself, which in alchemical terms had been transmuted by fire.

There are two terms in Van Helmont's philosophy of disease: "semina" and "archei," both present in Paracelsian disease theory. His philosophy contained the concepts of idea, image, and imagination. The morbid entity is the idea or image contained in a specific disease-causing *semina*, a spiritual entity. There are many different diseases, each with its own image. Each "*semina*" is endowed with its own "*archeus*," its vital principle, which gives it its identity and is responsible for its activity. How, then, does infectious disease come about? The host has its own "*archeus*," which is affected by an external agent, for example the agent of plague with its "*archeus*." The "*archeus*" of the external agent gains superiority over the "*archeus*" of the host. At this point it is the interaction of two spiritual factors. The image of the external agent is conceived of by the host through some complex "psycho-physical" process. In the case of plague the host vital principle is converted to the vital principle of the plague agent. For Van Helmont this process was a transmutation equivalent to that carried out by the chemist in the laboratory.

Van Helmont applied this process to the contraction of rabies. The vital principle (*archeus*) of the invader, contained in the saliva of the dog, overcomes the analogous receptor in the patient, converting it to the form of the invader. This is also an act of transmutation that the chemist achieves in the laboratory. From this theoretical base each disease has its own identity. The interaction represents the conversion of spirit into *body*. What is generated becomes an external thing, a *parasite*, that has its own *archeus* and can impose its image on another host. From this perspective there are specific disease agents and consequently specific diseases. Walter Pagel characterized this as the *ontological* view of disease. "Disease is not an abstraction. It is like any object in the world. It is from an external efficient seminal cause." A specific disease-semen suggests the idea of a biological entity, but that is not the case: the entity was defined in ideal terms, *image* and *vital principle*.

There is another part of the disease process that involves a ferment. Where did this fit in with Van Helmont's theory of disease? Fermentation is an essential component of bodily functions. The process is responsible for digestion, nutrition, and actions of the body. Ferment is also a component of the *psycho-physical* process that leads to the production of a disease. Ferments were historically associated with effervescence, which was clearly visible in wine and beer production using yeasts. The ferment was thus able to convert matter, for example the pulp of the grape or the components in the brewing vat, to a gas. The ferment in some mystical way is involved in production of the *archeus*.

Van Helmont, like Paracelsus, sought to change the practice of medicine by demolishing its philosophical base, now 2,000 old, and by replacing it with a theoretical foundation derived from religious mystical traditions and the works of Paracelsus. His disease theory contained concepts and terms used by Paracelsus with some differences, such as the absence of macrocosmic–microcosmic interactions so that astral influences are absent in Van Helmont's theory.

A large number of writings during the last half of the seventeenth century in Europe illustrate how the discourse about contagious disease was influenced by Helmontian chemistry, without the religious-philosophical elements, but instead conjoined to the *mechanical philosophy*. We will discuss this issue in the following chapter.

CHAPTER 8 | Mechanical Philosophy, the Revival
of Atomism, and Contagious Disease
Theory in the Seventeenth Century

IN 1989 A book entitled "The Medical Revolution of the Seventeenth Century" was published. The opening line of the Introduction asked, "How did medicine fare in the age of revolution?" The revolution is, of course, the Scientific Revolution, spanning the century, including the works of Copernicus, Kepler, Galileo, and Isaac Newton, with a host of distinguished natural philosophers and physicians as participants in these dramatic scientific developments. It was the intent of the editors to show that medicine "was open to external forces of change," and from the title of the volume it is obvious that they had concluded that medicine was profoundly affected. A major external force driving this change was a new philosophy represented by macro- and micro-mechanical ideas, the latter containing a "new" material base for the composition of matter with the revival of atomism.

A significant number of theorists in the early decades of the seventeenth century who for philosophical reasons proposed different forms of a particle, atomist–corpuscularian, theory of matter. The number and varieties of philosophers involved in this enterprise revealed the shift away from Aristotelianism, the rejection of forms and essences as physical explanations, and the collective acceptance among natural philosophers, whatever their differences, of explanations of the properties of all things and the cause of all phenomena by the mechanics of particles. This view was expressed by Robert Boyle:

. . . ."notwithstanding these things, wherein the Atomists and Cartesians differed, they might be thought to agree in the main, and their hypotheses might by a person of a reconciling disposition be looked on as, upon the matter, one philosophy. Which because it explicates things by corpuscles, or by minute bodies, may (not very unfitly) be called corpuscular".

ROBERT BOYLE, The Works of the Honorable Robert Boyle in Six Volumes. Some Specimens of an Attempt to Make Chymical Experiments Ufeful to Illuftrate the Notions of the Corpuscular Philosophy, v. 1 1772, p. 356

In this century a particle theory of matter accounted for natural phenomena. The mechanical interactions of these invisible particles are solely responsible or are crucial for all the phenomena of the world. The seventeenth-century proponents of a physical particle theory of matter may be characterized as mechanical philosophers and thus conveyers of the mechanical philosophy. This philosophy holds that the visible properties of objects, the macro-properties, are due to the actions and interactions of the micro-composition, invisible particles, that constitute bodies, large and small. The mechanical philosophy explains the real composition of objects in the world and purports to explain change. An important consequence of this philosophy is that it provides a new theoretical base for explaining the causes of contagious diseases. There are many contributors to this philosophy, and we will cite a few who were influential on contagious disease theory in the latter half of the seventeenth century and the early decades of the eighteenth century. To support this contention it is necessary to demonstrate the general acceptance of such a philosophy and to show that a new contagious disease theory depended on a particle theory of matter. It is interesting to note that Fracastoro does not appear in the discourse about a particle theory cause of contagious disease in the seventeenth century.

Micro-mechanism: Early Atomism in England

For a significant portion of the century there was a continuing debate between the atomists and the corpuscularians concerning the basic building blocks of matter; however, there was no disagreement about the physicality of these particles.

The philosophical choice for Thomas Hariot (1560–1621) was one of two contradictory systems offered by Aristotle and Democritus/Epicurus. Aristotle had contended that the universe is finite while the matter composing the universe was infinitely divisible. Democritus/Epicurus contended that the universe was infinite while matter is divisible up to a limit. These "limit" particles were characterized as atoms.

There was nothing of evidence for either theory; thus, a choice was made on the basis of reason. Hariot's choice emerged from a complex mathematical argument, a logical argument. One might suggest that Hariot had already decided on the Democritus/Epicurus model and constructed a reason to support it.

Hariot was one of a group of intellectuals in the Northumberland Circle in England, a physicist and mathematician, a supporter of the Copernican heliocentric model, a critic of Aristotelian philosophy, and a proponent of Epicurean atomic theory. In 1606 he wrote a letter to Johannes Kepler (1571–1630) questioning him on the subject of optical phenomena. He asked why, when a light ray strikes a surface of a transparent medium, it is "partially reflected and partially refracted." Reflected light is displaced at some angle, while refracted light is transmitted but bent to some degree. Hariot provided an answer:

"Since by the principle of uniformity, a single point cannot both reflect and transmit light, the answer must lie in the supposition that the ray is resisted by some points and not others".

Furthermore,

"A dense diaphanous body, therefore, which to the sense appears to be continuous in all parts, is not actually continuous. But it has corporeal parts which resist the rays, and incorporeal parts vacua which the rays penetrate".

ROBERT H. KARGON, Atomism in England from Hariot to Newton, 1966, p. 26, Oxford

Kepler disagreed and provided an Aristotelian answer. The phenomenon was due to the property of two opposing (contrary) qualities, transparence and opacity—transparence where light passes completely through and opacity when no or a limited amount of light passes through the medium. According to Heriot, the atomist, Kepler's answer is a verbal explanation and not a structural explanation for the cause, while Heriot has provided a physical cause based on the hypothesis that the medium is composed of particles.

At the beginning of the seventeenth century the writings of Hariot concerning the composition of matter are essentially the theories of the atomists of antiquity, particularly Epicurus and the writings of Lucretius. Heriot's world is made of atoms moving in a void. The physical properties of visible bodies are the result of the physical properties of these particles, their size, shape, and motion. Heriot wrote:

"Nothing is done without motion. There is no motion without a cause. Out of nothing comes nothing".

ROBERT H. KARGON, *Atomism in England from Hariot to Newton*,
1966, p. 13, p. 26, Oxford

"For the last line Hariot has revived the words of Lucretius as well as his philosophy contained in Book 1:155 of *De rerum natura*. The contention that **all** bodies are physically constituted of particles led to an all-encompassing generalization, that the visible characteristics of all bodies are the emergent properties of the interactions and combinations of these indivisible, indestructible particles designated as atoms.

Lucretius provided a range of analogies from the visible world to justify a particle theory of matter:

"For look closely, whenever rays are let in and pour the sun's light through the dark places in houses: for you will see many tiny bodies mingle in many ways all through the empty space right in the light of the rays . . . so that you may guess from this what it means that the first-beginnings of things (*primordia rerum*) are forever tossing in the great void . . . And for this reason it is more right for you to give heed to these bodies, which you see jostling in the sun's rays, because such jostlings hint that there are movements of matter too beneath them secret and unseen".

LUCRETIUS, *De rerum natura*, trans. and commentary by Cyril Bailey,
1947, p. 243, Oxford

In addition, Lucretius described a food chain where plant material is ultimately converted to the human body. These transformations can take place because all matter is composed of the same basic particles:

"Why, we may see worms come forth alive from noisome dung, when the soaked earth has gotten muddiness from immoderate rains; moreover we may see all things in like manner change themselves. Streams, leaves, and glad pastures change themselves into cattle, cattle change their nature into our [human] bodies . . . And so nature changes all foods into living bodies".

LUCRETIUS, *De rerum natura*, 1947, p. 283, Oxford

The adherents of a particle theory of matter stressed structural explanations for the causes of phenomena. This form of explanation appeared superior to causes based on qualities that did not explain the changes of forms. Robert Boyle described the alleged weakness of Aristotelian explanation in 1666:

"The schools [have] made it thought needless or hopeless for man to employ their industry in searching into the nature of particular qualities and their effects. As if [for instance] it be demanded how snow comes to dazzle the eyes, they will answer that it is by a quality of whiteness that is in it, which makes all very white bodies produce the same effect; and if you ask what this whiteness is, they will tell you no more substance than it is a real entity which denominates the parcel of matter".

STEVEN NADLER, *Doctrines of Explanation in Late Scholasticism and in the Mechanical Philosophy*, The Cambridge History of Seventeenth Century Philosophy, 1998, vol. 1,p. 513, Cambridge

Boyle's critique reveals that the answer to the question "What is white?" or "Why is something white?"—"Because it has the quality of whiteness"—is not an explanation. It is a tautology, a simple repetition of the statement. Compare this explanation to that given by Epicurus and Lucretius to the question of why something is white. For Epicurus, atoms have the properties of shape, size, weight, and motion but lack, among other characteristics, color. Only the former properties are required to make physical bodies. Epicurus has therefore excluded color as a property of atoms, which leads to the idea that color comes about through the arrangement of atoms in a body.

He wrote

. . . "every quality changes (i.e., color) but the atoms do not change at all, since there must needs be something which remains solid and indissoluble at the dissolution of compounds, which can cause changes";

ANDREW PYLE, *Atomism and its Critics*, 1997, p. 135, Thoemmes

Lucretius concurred when he wrote that white bodies are not composed of white atoms, which he characterizes, at different times, as *first-beginnings* and *seeds:*

"all things . . . each is made of first-beginnings of a different shape . . . lest by chance you should think that these white things, . . . are made of white first-beginnings, . . . For the bodies of matter have no color at all . . . I will now teach you that [the first-beginnings] are [deprived of all color]. For any color, whatever it be, changes . . . ; but the first-beginnings ought in no wise to do this. For it must needs be that something abides unchangeable".

LUCRETIUS, De rerum natura, trans. and commentary by Cyril Bailey,
1947, p. 275, Oxford

Hariot and his followers explained the operation of bodies by mechanical causes, properties of visible bodies. In England a particle theory of matter as an essential part of the mechanical philosophy was initially confined to a small group of Heriot's students and followers. In the 1630s another small group, the Cavendish Circle, with one of its members tutored by Thomas Hobbes, is committed to the mechanical philosophy. Hobbes' world contained invisible and indivisible atoms that differ in "hardness" and in motion. Atoms come together to form visible bodies. Hobbes rejected the idea of spirit, immateriality. For this he was attacked as an atheist, an Epicurean, an atomist, and a variety of other epithets. The connection of heresy and atomism, particularly in the person of Hobbes, undercut the acceptance of atomism in England.

This particle philosophy offered a mechanical model at the micro-level of events. Shortly thereafter a particle theory became part of a macro-mechanical model for the operation of human bodies whose activities are analogous to the function of machines, such as mechanical clocks, with their intricate parts.

Macro-mechanism: William Harvey and René Descartes

The important experimental component in the construction of macro-mechanism is the great work of William Harvey (1578–1657) on the circulation of blood in humans.

Harvey's work owed much to the developments in anatomy in the sixteenth century, which came from two sources. The first source was artists, who needed to know body structure for their work; they included Leonardo da Vinci (1452–1519), Raphael (1483–1521), Durer (1471–1528), and Michelangelo (1474–1564). Their art became the model for the representation of the body in anatomical texts. The other indispensable source was the anatomist, Andreas Vesalius (1514–1564), who revitalized the teaching of anatomy. In 1543 he published his great work, *On the Fabric of the Human Body*. At that time he was lecturing at the University of Padua, where he had received a medical degree. His concern with human anatomy and his knowledge of Galenic writings on anatomy led him to realize that Galen had not dissected humans. He concluded that reading Galen was not the way to learn anatomy; this knowledge could be obtained only by working on the human body. Vesalius carried out his own dissections and presented lectures on this new anatomy. In the later edition of his work in 1555 he made a number of corrections of Galen's work. An important revision concerned the pathway of blood in the heart. He showed that there was no permeable septum, as Galen held, to allow blood to pass from one side of the heart to the other.

William Harvey received his doctorate in Padua in 1602 and was practicing medicine in London and lecturing on anatomy and physiology. In 1628 he published his landmark of experimental biology, *Movement of the Heart and Blood in Animals*. His description of the circulation of the blood in humans differed from the Galenic model, now more than 1,400 years old. Harvey derived his model by a number of techniques, including dissection, vivisection, and observing the comparative anatomy of a variety of adult species of animals and studying embryological material. In particular, he used the technique of ligature to demonstrate that circulation was linked to the generation of the pulse and that blood is carried away from the heart by arteries and returns from the periphery of the body to the heart via veins. Harvey calculated the volume of blood pumped by the heart during a certain period to show that this great volume of blood could not be contained in the human body but could be explained by a model of circulation. From this variety of experiments and observations he developed a reasoned scheme to describe the flow of blood, since he could never actually see this continuous flow. The description of an organ of the body as a pump suggested an analogy to a visible piece of machinery in the writings of René Descartes.

Descartes read Harvey's book in 1632 in the midst of his studies of animal anatomy, which was to be the basis for understanding the physiology of the human body and was an integral part of his grand project to understand the operation of the world from celestial bodies to human bodies. In a letter written to Marin Mersenne (1588–1648) in the summer of 1632 he related his interest in human physiology and wrote that he was "dissecting the heads of different animals in order to explain what imagination, memory, etc consist of." A reasonable interpretation of these remarks is that Descartes had adopted the important principle that knowledge of a structure will reveal its function. This is a precept that becomes an integral part of the mechanical philosophy. He informed Mersenne that he had seen Harvey's book after having finished writing about this matter. The "matter" was the circulation of the blood. Descartes had gained enough experience from his own studies to say he differed little from Harvey's work.

Descartes' 1637 work, *Discourse on Method*, and the *Treatise on Man*, published 12 years after his death in 1664, continue to reflect Harvey's influence. In the *Discourse* he acknowledged that "an English physician" (obviously Harvey) had the credit of "having broken the ice" concerning the circulation of the blood, which he has proved by "experience." Descartes, already committed to a macro-mechanical philosophy, used Harvey's description of circulation to designate the body as a machine. In the *Treatise on Man* he continued to compare the body to a machine: "Man's body . . . is equivalent to some mechanical device fabricated by humans like clocks or systems of pulleys and levers." At this point Descartes introduced another dimension to this mechanism. The mechanical characteristics of the visual bodies are the emergent properties of the mechanics of invisible particles that have various shapes and motions. This is Descartes' particle theory of matter that we discuss below.

The summation in the *Treatise* leaves no doubt that Descartes is committed to explaining every aspect of human physiology by a mechanical model.

"I desire you to consider, further, that all the functions that I have attributed to this machine, such as (a) digestion of food; (b) the beating of the heart and arteries; (c) the nourishment and growth of the members; (d) respiration; (e) waking and sleeping; (f) the reception by the external sense organs of light, sounds, smells, tastes, heat, and all other such qualities; (g) the imprinting of the ideas of these qualities in the organ of common sense and imagination; (h)the retention or imprint of these ideas in the memory; . . . I say, . . . these functions . . . follow naturally . . . entirely from the disposition of the organs . . . Wherefore it is not necessary, on their account, to conceive of any vegetative or sensitive soul or any other principle of movement and life than its blood and its spirits [air], agitated by the heat of the fire which burns continually in the heart and which is in no other nature than all those fires that occur in inanimate bodies".

RENÉ DESCARTES, *Treatise on Man*, 1972, p.113, trans. by Thomas Steele Hall,
Harvard University Press

Continental Philosophy: Cartesian Mechanics and Pierre Gassendi's Atomism

On the Continent a particle theory of matter became a dominant philosophical force in the persons of Descartes and Pierre Gassendi (1592–1655). Descartes' matter theory is an essential part of his philosophy that explains all natural phenomena as a consequence of the mechanical inter-actions among invisible corpuscles that have configurations and move-ments. Such Cartesian mechanics emerged epistemologically from a met-aphysical foundation that, according to Descartes, is the source of all knowledge. This claim must be read in Descartes' words to appreciate the ambitious goal he proposed for his method. First, Descartes stated that the study of philosophy is important for it leads to wisdom, which leads to "a perfect knowledge of all things that man can know" that is essential for proper conduct and the preservation of health. All knowledge depends on "first causes," which are true without further justification. And so true phi-losophy begins with metaphysics, the principles of knowledge. These in-clude the attributes of God, the immateriality of the soul, and the clear concepts that are intrinsic to us. Then we have:

. . . "physics in which, after having found the true principles of material things, we examine generally how the whole universe is composed, and then in particular what is the nature of this earth and of all the bodies which are most commonly found in connection with it, like air, water and fire, the loadstone and other minerals. It is thereafter necessary to inquire individu-ally into the nature of plants, animals, and above all of man, so that we may afterwards be able to discover the other sciences which are useful to man. Thus philosophy as a whole is like a tree whose roots are metaphysics, whose trunk is physics, and whose branches, which issue from the trunk, are all the other sciences. These reduce themselves to three principal ones, viz. medicine, mechanics and morals". [**Fig. 8.1**]

RENÉ DESCARTES, *The Philosophical Works of René Descartes*, 1911,
vol. 1, p. 211, trans. by Elizabeth S. Haldane and G. R. T. Ross,
Cambridge University Press

Cartesian mechanics is a consequence of a generally applicable particle theory of matter. All phenomena are the result of the interactions among particles assumed to have shapes and movement. The particles are differ-entiated from atoms and classified as corpuscles by Descartes since he envisaged that they are divisible beyond the range of atoms.

The qualities of visible objects are the result of the size, shapes, exten-sion (length, breadth, width), and motion of particles, the latter initiated by God. Earth, air, and fire are composed of corpuscles, each with different shapes and associations that give them their different realities.

Descartes did compare human bodies to machines and attributed their operation to the mechanics of very small particles. The physiology he was

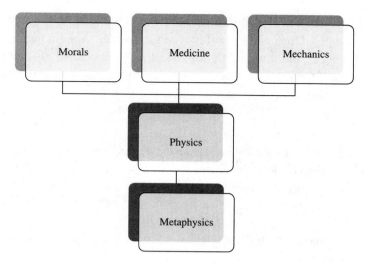

FIGURE 8.1

replacing derived from the Greek tradition already under attack prior to the seventeenth century. For example, the idea that the body contains four humors, which in the Hippocratic and Galenic tradition was involved in disease, was discarded by Descartes; in its place he postulated that all bodily processes depended on corpuscular physics. In the Galenic system there were three "spirits" that worked somehow as intermediaries between the body and soul, where the soul was the cause of physiological activities. Descartes only retained the animal spirit. He also rejected Galen's theory of "faculties," which are presumably processes, not defined, that served as functions in various organs. For these alleged functions he substituted mechanical processes. Most physiological functions, including the heat of the body, he ascribed to the motion of corpuscles.

Pierre Gassendi

Pierre Gassendi provided an important alternative to Cartesian physics. He was born in Provence to a peasant family and emerged as an exceptional student, studying Latin from the age of seven and going on to study philosophy and theology, receiving a doctorate in theology at age 24. He was ordained one year later as a priest. In 1617 he assumed a chair in philosophy at the University of Aix. He left Aix in 1622–23 and began his philosophical writing. In 1623 he arrived in Paris, where he was introduced to Marin Mersenne and other intellectuals, including Hobbes and members of the Cavendish family, all intimately involved with developing mechanical philosophy. During the next few years he began work on the philosophy of Epicurus, but it was not until 1647 that he published his first work on Epicurus, "On the Life and Character of Epicurus." It was the first of four publications on Epicureanism and was a significant offering by Gassendi because it:

> . . . "was intended both to demonstrate the impeccable personal morals of Epicurus and to render his doctrines consistent with the Catholic faith".
>
> ANTONIA LOLORDO, *Pierre Gassendi and the Birth of Early Modern Philosophy*, 2007, p. 14, Cambridge University Press

Gassendi started with the matter theory of Epicurus, providing the following justification:

> . . . . "this theory of matter has the advantage that it does not do a bad job of explaining how composition and resolution into the primary elementary particles is accomplished, and for what reason a thing is solid, or corporeal, how it becomes large or small . . . and so forth . . . these questions and others like them are not so clearly resolved in other theories where matter is considered as both infinitely divisible and either pure potentiality . . . or endowed with a certain shape from among a very small range of possibilities, or endowed with primary and secondary qualities, which either do not suffice to explain the variety in objects or are useless".
>
> SAUL FISHER, *Pierre Gassendi's Philosophy and Science: Atomism for Empiricists*, 2005, pp. 213–214, Brill

He went on to make clear that the classical view, the Epicurean view, may not be sufficient as it exists. It can be improved by removing what is "recognizably false." Here is an example of Gassendi's improvements:

> . . . "in order to recommend the theory, we declare first that the idea that atoms are eternal and uncreated is to be rejected and also the idea that they are infinite in number and occur in any sort of shape; once this is done, it can be admitted that atoms are the primary form of matter, which God created finite from the beginning, which he formed into this visible world, which finally he ordained and permitted to undergo transformations out of which, in short, all the bodies which exist in the universe are composed".
>
> SAUL FISHER, *Pierre Gassendi's Philosophy and Science: Atomism for Empiricists*, 2005, p. 214, Brill

Gassendi eliminated the reason for the charge of heresy against the philosophy of atomism by hypothesizing that atoms and their activities are the creation of God.

Gassendi's atomistic philosophy is fundamentally Epicurean atomism with the aforementioned additions. Matter is composed of particles with defined sizes, shapes, and motion moving in a void (vacuum), a space empty of particles. In contrast to the metaphysical foundations of Descartes' corpuscularian theory, Gassendi's theory can be epistemologically characterized as derived from empirical knowledge, relying as much as he can on experimental evidence. The existence of a void is an essential element in Epicurean atomism to allow for the movement of atoms. For Gassendi, the reality of the void was established in the barometer experiment that Gassendi described, carried out by Toricelli in 1644. In this setup a three-foot-long tube is filled with mercury and placed with its closed end up and its open end placed in a bowl of mercury open to the environment. At ground level the mercury in the tube falls to 29 inches, leaving a space above the mercury. When the experiment is carried out at higher altitudes, the height of the column changes, inversely proportional to the altitude, so that the space above the enclosed column of mercury varies in height. Gassendi explained that the height of mercury is determined by the external atmosphere. He made the argument, simply stated, that the space above the column of mercury contains no air, for there is no way for it to enter when the column of mercury falls to a position determined by the external weight of the air. For Gassendi, this is experimental evidence for the presence of a vacuum.

It is clear that Gassendi considered empirical knowledge, obtained by sense experience, crucial for understanding phenomena, and he believed he could construct hypotheses based on these data. However, he was also willing to propose hypotheses about things and events that could not be observed, for example the presence of unseen moving atoms, which are the foundation of a philosophical system that hypothesizes that all objects in the world are composed of these particles. However, he was not satisfied with speculations alone and continued to search for evidence for such structures. He believed he found them in two instances when he employed the microscope to examine a crystal structure and a living animal, the almost invisible mite, to infer that there are a range of small structures that ultimately exist at the atomic level in both objects. This was made clear in his own words starting with comments on the mite:

"We imagine all these parts without the whole with which nothing that nourishes itself, lives, senses, imagines, or moves, can subsist. And we understand that nature must be able to distinguish, put in order, and organize into a whole, innumerably tens of thousands of particles to form a small organism which to the naked is like a point. Nevertheless, since nature cannot go to infinity in its dissolution, but stops finally at something non-decomposable which is the [physical] minimum, this is what philosophers have generally called the atom. In this sense, it is convenient to call as atoms those extremely small and non-decomposable particles—of which we can conceive that many tens of thousands exist in one mite. If nevertheless we admit that fine membranes are formed from only one layer of atoms, what a large number—what an innumerable quantity—must be brought together before we arrive at the thickness of a mite or of a spider's web".

SAUL FISHER, Pierre Gassendi's Philosophy and Science: Atomism
for Empiricists, 2005, p. 231, Brill

It is fascinating to observe Gassendi confronting the issue of the number of atoms that make up the body of the mite without a measure of the dimensions of the animal or the size of the atom. He estimated that the number of the atoms ranges between tens of thousands and innumerable. Certainly in the mid-seventeenth century not much more could be said about the size of near-invisible objects, although in England Christopher Wren was trying to develop the technology to measure microscopic materials.

In his microscopic examination of crystals Gassendi followed the breakdown of crystals as they are being dissolved, noted that the structure is preserved as they pass into solution, and inferred that they continue to do so, retaining their shape, as they arrive at the atomic level. As they grow larger they again regain their structure:

"I recognize in particular that these large solid forms—whether cubes, octahedrons, or others—are composed of other lesser ones of the same shape, and that those [forms] of the lesser ones—until they are resolved into very minute ones—are almost insensible and remain shaped in the same way, from which I conclude that they can be resolved until their [constituent] atoms, which by some sort of necessity must be of the same shape. The composition demonstrates this, by the manner in which I observe that they grow larger from the moment that they become like [the size of] mites".

SAUL FISHER, Pierre Gassendi's Philosophy and Science: Atomism
for Empiricists, 2005, p. 213, Brill

Gassendi was an important figure in the revival of Epicurean atomism. He offered a philosophical position different from Descartes. For Gassendi there was no epistemic foundation that can produce certain knowledge. Knowledge can be gained generally by sense experience, but as a Catholic priest he excluded this methodology in matters of theology. He recognized that atoms are not available to the senses, yet he relied on this hypothesis as the best among possible explanations. Gassendian atoms are solid, hard particles, have different sizes and shapes, are indivisible, and have movement. The issue of movement and origin of atoms was a crucial element differentiating Gassendian from Epicurean theory. According to Gassendi, atoms are created by God and have motion because He placed in them the intrinsic ability to move:

"It should be granted that atoms are mobile and active because of a force of moving and acting that God gave to them in his creation of them".

ANTONIA LOLORDO, *Pierre Gassendi and the Birth of Modern Philosophy*, 2007, pp. 142–143, Cambridge University Press

Motion is the essential characteristic of atoms, responsible for the formation of composite bodies. In short, atoms constitute everything in the world.

If atoms have only the qualities designated by Gassendi (size, shape, weight, or motion), how can there be so many different objects in the world? Gassendi argued that the named qualities do not change, but there are "accidental" qualities that account for the diversity. The diversity of objects is the result of "position and arrangement." He provided an analogy to represent his hypothesis. Let atoms be regarded as letters so that it is consequential whether a sequence is XY or YX:

"Just as the same letters are different in respect of seeing and hearing when they are positioned differently . . . so the same atom positioned differently affects sense in different ways".

ANTONIA LOLORDO, *Pierre Gassendi and the Birth of Modern Philosophy*, 2007, p. 156, Cambridge University Press

He continued with the analogy to explain arrangement:

"As the same two or many letters, when they proceed or follow in different ways, suggest different words to the eye and ear and the mind . . . so the same atoms, transposed in different ways, can demonstrate very different qualities or species to the senses".

ANTONIA LOLORDO, *Pierre Gassendi and the Birth of Modern Philosophy*, 2007, p. 156, Cambridge University Press

It is not surprising that Gassendi reproduced the Lucretian analogy of letters and atoms contained in Book1 of De rerum natura, starting at line 817:

"And after it is of great matter with what others those first-beginnings [atoms] are bound up, and in what position, and what movements they mutually give and receive; for the same build up sky, sea, earth, rivers, sun . . . crops, trees, living creatures, but only when mingled and moving with different things in different ways. Indeed scattered in my verses you see many letters common to many words, and yet you must needs grant that verses and word are unlike both in sense and in the ring of their sound. So great is the power of letters by a mere change of order. But the first-beginnings of things can bring more means to bear, by which all diverse things may be created."

LUCRETIUS, De rerum natura, 1947, pp. 217–218, trans. and
commentary by Cyril Bailey, Oxford

Motion is responsible for the origin and arrangement of objects in the world, and God has played the indispensable role giving motion to atoms. What about the issue of the generation of plants and animals? Certainly they come from seeds. Where do seeds come from? What is their origin? Gassendi provided two versions, one designated direct divine creation, a conventional account in the religious context of the seventeenth century. The other possibility is, remarkably, a scenario that is part Lucretian cosmology and part Epicurean physics, where nature is involved in the formation of seeds:

"Nature, having gradually become accustomed [to propagation], learned to procure the propagation of animals similar in kind, so that from the perpetual and ordering of atoms [nature] acquired a certain necessity to continually operate in this way".

LUCRETIUS, De rerum natura, 1947, p. 198, trans. and commentary
by Cyril Bailey, Oxford

But there is one necessary proviso in Gassendi's formulation, and that continues to be motion placed into atoms by God, a cause external to the particle. Gassendi argued that once motion is conferred on atoms, seeds will come into existence due to their motion. Gassendi reconciled these apparently discordant explanations by contending that both were reasonable hypotheses and neither could be demonstrated to the detriment to the other. The willingness of Gassendi to accept a level of uncertainty, and to argue on the basis of reasonableness, suggests a contemporary position in the philosophy of science that accepts the idea of "inference to the best explanation."

Epicurean–Lucretian atomism is a materialist philosophy. All motion is inherent in matter and is responsible for everything in the world. God has been eliminated from the world. From this perspective Gassendi's adoption of atomism appears incomprehensible since he is a Catholic priest. Nevertheless, he accepted the philosophy of atomism as the best explanation but allayed the materialism by introducing God as the creator of atoms and their motion. This tactic was characterized in the following way:

> ... "it is entirely correct that by inserting God's role in the cause of atomic
> ... [movement] Gassendi is trying to mitigate worries about the atheism
> associated with Epicurean theories".
>
> ANTONIA LOLORDO, *Pierre Gassendi and the Birth of Modern Philosophy,*
> 2007, p. 144, Cambridge University Press

Atomism in its Epicurean–Lucretian form appeared unacceptable in England while Gassendian atomism was received enthusiastically by the English physician Walter Charleton (1629–1707). He helped to introduce it into England in the mid-seventeenth century, where it received a less hostile reception among the proponents of the mechanical philosophy, such as Robert Boyle, who relied on a physical particle theory as the cause of contagious disease. In the next chapter we will take up this narrative.

The differences between the matter theories of Descartes and Gassendi, which include their different epistemological positions, their forms of argument, their views on the nature of atoms, the existence of a void (vacuum), and the generation of the visible properties of objects from the activity of particles, are representative of the general discourse in the seventeenth century among the many contributors to the atomistic–corpuscularian theory of matter. Although there were differences among the adherents of such a philosophy, there was enough agreement so that Aristotelian and Paracelsian matter theory was replaced and the development of the mechanical philosophy was reenergized in England, when the micro- and macro-mechanical ideas were imported from the Continent.

The New Learning in England: The Acceptance of a Particle Theory of Matter

In England a number of important factors contributed to the rise of this new philosophy in the first half of the seventeenth century. The major philosophical stimulus was the writings of Francis Bacon (1561–1626), stressing the inductive process and using experiments to gain knowledge. Bacon published "The Advancement of Learning" in 1605 and the "Novum Organum" in 1620, designated as a new instrument for obtaining knowledge. This treatise was intended to replace the old *Organum,* a compilation of Aristotle's writing on logic. Bacon's methodology, based on observation and experiments, would be the way of understanding the natural world. Knowledge gained from experiments is reproducible and true and by a process of logical induction "generalizations" and "universal laws" about nature would be revealed. Knowledge gained in this way would reveal the underlying mechanisms, the causes of things. The third book, the "New Atlantis," was published in 1627. In the "New Atlantis," Bacon laid out a plan for the creation of a society built on a foundation of science. It was not just a theoretical science, but one that provided for all the needs of society by its ability to understand and to control nature toward practical ends. To carry out this program there needed to be a collective effort of individuals who practiced and promoted this new philosophy. In time this view inspired some of the founders of the Royal Society in London.

The idea that the processes of the world could now be understood because the atomistic–corpuscularian theories of matter provided a structural base captured the allegiance of a remarkable group of men in London and subsequently at Oxford University. They were individuals with different backgrounds but with common interests, investigating natural phenomena. These researchers transcended their political and religious differences at a time of civil war in England. In 1645 such a group met at Gresham College in London. Among the members of this group were Robert Boyle (1627–1691) (then only 18 years old) who was to become the premier chemist of the seventeenth century, was awarded an honorary Doctor of Medicine, and was a founding member of the Royal Society; Dr. John Wilkins (1614–1672), a Doctor of Divinity who became Warden of Wadham College at Oxford and was responsible for gathering an outstanding group of experimentalists in Oxford; Christopher Wren (1632–1723), mathematician, astronomer, and architect; William Petty (1623–1687), who became a Doctor of Medicine; and John Wallis (1616–1703), a Doctor of Divinity and a Professor of Mathematics.

Around the year 1645 the group began to meet to discuss what was labeled the New Philosophy or Experimental Philosophy. Boyle described the issues they discussed, including natural philosophy, and "mechanics and husbandry" based "on the principles of the Philosophical College." Boyle also commented that they sometimes referred to themselves as the "Invisible College." In the year 1648–49 Wilkins left the group to go to Oxford University and was followed by Wallis and Jonathan Goddard (1617–1675) and others in the group.

Another factor enhancing the development of the new science was the growth of Oxford University. What initiated these institutional changes was the recognition by

... "the merchant and gentry classes ... that a university education, previously reserved for the poorer sort intent upon a church career, was desirable background for personal improvement and public advancement".

ROBERT G. FRANK, Harvey and the Oxford Physiologists, 1980, p. 45,
University of California Press

Significant sums of money were supplied to Oxford, which resulted in an exceptional increase in the number of students, an increase in land holdings, and an increase in the number of colleges. Science and library holdings were expanded. In the 50 years between 1620 and 1670 friends enabled the library to triple its holdings in books. It now contained "almost every book that a physician or natural philosopher might wish to consult." The library, now called the Bodleian, contained 30 works by Francis Bacon, 15 by Pierre Gassendi, 10 by Descartes, "almost all of the Continental writers on anatomy and chemistry" and those by English writers including Robert Boyle, the physician Thomas Willis, and Robert Hooke. Additional professorships were created that survived the period of the civil war. The institution continued to flourish under a committee appointed by Parliament to oversee the university. In time the governance of the institution was returned to the faculty, who selected new fellows. Among the new appointments were adherents of the new philosophy.

Medical training was profoundly changed by mid-century. An important example of the innovation in training was the appointment of Thomas Willis (1621–1675) as the Sedleian Professor of Natural Philosophy in 1660. Willis was an outstanding physician-researcher and one of the founding members of the Royal Society in 1660; his medical lectures did not include Aristotle but did contain the corpuscular philosophy.

One can imagine the excitement engendered by the new philosophy that claimed it provided the physical composition of everything in the world, which for believers was the true description of matter. In England and on the Continent the followers of the new philosophy in the first half of the seventeenth century may have agreed with Annaliese Maier, who wrote in 1942,

> "One could even say that the question [of the structure of matter] has been the crossroads of the concept of nature. After all this has been the problematic point where, at the turn of the seventeenth century, the decisive change in the world view of natural philosophy took place, when atomism stepped at the place of the medieval doctrine of the place of forms and qualities in the structure of matter and thereby ushered in the natural science of the modern period".
>
> Quoted in CHRISTOPH HERBERT LUTHY, *Matter and Microscopes in the Seventeenth Century*, Ph.D. Thesis, 1995, Microform Edition, UMI

For medical practitioners the expectations were equally high, since it was now possible to comprehend, by mechanics and a number of emerging chemical techniques, the composition and physiology of the human body.

Atomism, the Mechanical Philosophy, Medicine in England

> "An alternate theory of disease had been proposed and gained adherents in the seventeenth century. Medicine was affected by the development at that time of a new natural philosophy or science which replaced Greek and especially Aristotelian natural philosophy. This meant that the theoretical foundations of Galenic learned medicine were undermined. The plague poison and its transmission were re-interpreted by the language of the "new science" in chemical corpuscular terms".
>
> ANDREW WEAR, *Knowledge and Practice in English Medicine, 1550–1680*, 2000, pp. 275–276, Cambridge University Press

The mechanical philosophy gained powerful support in mid-seventeenth-century England from Walter Charleton, an early student of Wilkins, who obtained a Doctor of Medicine degree at Oxford in 1643 and became one of the earliest fellows of the Royal Society. His fame and influence rest on two works, one a theological treatise published in 1652,

The Darkness of Atheism Dispelled by the Light of Nature, where he praises the writings of Mersenne, Descartes, and Gassendi. In 1654 he published a treatise whose title reveals its contents, "Physiologia-Epicuro-Gassendo-Charltoniana: A Fabric of Science Natural Upon the Hypothesis of Atoms, Founded by Epicurus, Repaired by Petrus Gassendus, Augmented by Walter Charleton". 1654, London (hereafter designated PEGC).

Charleton gravitated to the philosophy of Gassendi. Why to Gassendi? Gassendi had retained in his philosophy the immateriality of the soul and that God is the creator of and the source of motion of atoms, a theological position that did not compromise Charleton's religious beliefs.

Support for Gassendi's position is contained in Charleton's 1654 book, which combined his choice of atomism and his theological beliefs. Charleton invented the term "physico-theology" and subsequently applied it to writings intended to counter the philosophical arguments of particle theorists and mechanical philosophers who had eliminated divine creation and continuing Providence in the world. Physico-theology contained the "argument from design" that was already implicit in the writings of Gassendi and explicit in the efforts of Charleton. Later in the century the argument from design was vigorously supported by Robert Boyle in his *Theological Works* and explained by John Ray (1627–1705), a distinguished botanist, who wrote *Wisdom of God in the Works of Creation* in 1691, and William Derham, a physicist and mathematician, who wrote *Physico-theology* in 1714. Adherents of physico-theology established a middle ground to reconcile faith in divine creation and science, the physics of the mechanical philosophers.

Charleton's commitment to Gassendi's physical theory was evident in the 1654 book *Physiologia*. His mechanical philosophy was summarized at the end of an extended passage where he disposed of those who explain "Secret Sympathies and Antipathies" in "Windy terms" that are "a refuge for the Idle and Ignorant." He continued, in ungentlemanly terms,

"To lance and cleanse this Cacoethical [evil] . . . Ulcer" one must invoke "the General Laws of Nature . . . [which] . . . produceth All Effects." These are

1. Every effect must have a cause.
2. That no cause can act but by motion.

"That all attraction, referred to Secret Sympathy; and all repulsion, ascribed to Secret Antipathy, betwikt the Agent and Patient, is effected by Corporeal

Instruments, and such as resemble those, wherby one body Attracteth or Repelleth another, in sensible and mechanique operations".

WALTER CHARLETON, PEGC, p. 343, 1654

The complete chapter where this quote appears is devoted to the properties of atoms, which include magnitude, figures, and motion, essentially using Gassendi's terminology and descriptions. In describing this atomism he made clear that he had rid it of its atheism. First, Charleton stated the Epicurean version that is to be eliminated:

"Atoms were eternally existent in the infinite space . . . their Motive Force was eternally inherent in them, and not derived from any External Principle."

Charleton provided the substitutes.

"That atoms were produced ex nihilo, or created by God . . . that at their creation, God invigorated or impregnated them with an Internal Energy, or Faculty Motive, which may be conceived the First Cause of All Natural Actions, or Motions performed in the world".

WALTER CHARLETON, PEGC, p. 126, 1654

This is the essence of physico-theology. What Charleton did, or rather what he and Gassendi accomplished, was to provide an acceptable learning, since it was now possible for natural scientists to operate using this new philosophy because it had been shorn of its atheistic component:

"For, by virtue of these corrections the poisonous part of Epicurus' opinion, may be converted into one of the most potent Antidotes against our Ignorance".

WALTER CHARLETON, PEGC, p. 126, 1654

Chemical Philosophy

The chemical philosophy that had been promoted in the sixteenth century by Paracelsus and successfully in the seventeenth by Van Helmont with its unique theory of matter was continued and changed by the inclusion of a particle theory of matter.

By the 1650s Helmontian medicine was a force in England. Chemistry was viewed by English physicians in the Oxford group as having explanatory power for theories of medicine due to the teaching of chemistry by

Peter Stahl, who was brought to England through the efforts of Robert Boyle, and the availability of the writings of Peter Sennert (1572–1637). Sennert produced a number of texts that included Greek atomism into his chemical theory. The four Aristotelian elements were each composed of physical particles whose interactions could give rise to new compounds. Chemistry would reveal the components of natural compounds and refine them so that they could be used as medicines: "Chymistry . . . belongs to Physick [medicine] and is the perfection of it."

Another influential text was produced by Nicholas Lefevre (*ca.* 1610–1669) in 1660. He too had been invited to England, where he established a chemistry laboratory and was elected to the Royal Society. His view was that "Chymistry is the key to Nature."

This work was published in Europe and went through many editions.

Robert Boyle

Robert Boyle (1627–1691) was born into the most auspicious circumstances, in a castle in Ireland, the seventh son of the Earl of Cork. As a child he was educated by tutors and at the age of eight was sent to Eton, where he resided until the age of 11. One year later he began a five-year tour of Europe, returning to England in 1644. Within a year he joined the 1645 Group, as Boyle described it the "invisible college," which included John Wilkins (1614–1672) and Samuel Hartlib (1660–1662), a group devoted to the discussion of the emerging new natural philosophy. At this time Boyle was 18 years old and Wilkins was 31. In 1647 he established residence at his country seat in Stalbridge, Dorsetshire, which he inherited in 1642, and there founded a laboratory and hired assistants to perform experiments. He did not attend university, was self-taught, and maintained written contact with the most distinguished natural philosophers in England and on the Continent. Included among his friends and correspondents was Samuel Hartlib, who was the coordinating member of a network of individuals whose goal was to bring about the "Great Instauration," called for by Francis Bacon, which was intended to put knowledge to use to improve education, trade, husbandry, and medicine. This group was organized in the late 1620s and lasted until 1660, at the time of the founding of the Royal Society. Hartlib maintained a large correspondence with natural philosophers in England and France, which included Boyle, who wrote to Hartlib about his scientific activities, and from whom Boyle learned about the work of many others. In 1654 Boyle moved to Oxford, where he again established a laboratory with Robert Hooke (1635–1703), then 19 years old, as his assistant. He remained at Oxford until 1668, although traveling on many occasions.

He came to Oxford already committed to a particle theory of matter. He came to such a theory in the following way. Boyle had traveled in Europe from 1639 to 1644 and was knowledgeable of the writings of Descartes and Gassendi. He wrote to Hartlib in 1647 that he was reading a mechanical treatise of Mersenne. At the same time (1647) he settled in Stalbridge, where he set up his laboratory. Again he wrote Hartlib praising Gassendi: "a great favorite of mine." In 1650 he composed an essay, *Of Atoms,* in which he praised the writings of "Gassendus and DesCartes" and others who reintroduced the atomism of Democritus, Leucippus, and Epicurus, which had been overshadowed by Aristotelian philosophy. Boyle believed in the reality of particles and interpreted the evidence in chemistry and biology from this perspective.

At Oxford he was part of a club of remarkable individuals who met in John Wilkins' rooms in Wadham College. When Wilkins left to marry Oliver Cromwell's sister, the group continued to meet in Boyle's residence. He was introduced in the Oxford environment to a distinguished group of individuals working on physiology and anatomy, related to the circulation of the blood and respiration, a continuation of the work initiated by the studies of William Harvey. He came to appreciate the relationship between the macro-structure of the human body as necessary to carry out various physiological processes. Underlying all these processes was a micro-structure based on a particle theory of matter.

Boyle, through four works, helped spread the influence and acceptance of the corpuscular or mechanical philosophy: "The Sceptical Chymist" (1661), "Usefulness of Experimental Philosophy" (1663), "Origins of Forms and Qualities" (1666), and "The Excellence and Ground of the Mechanical Hypothesis "(1674). Boyle also demonstrated a deep concern about the improvement of medicine and wrote an essay in 1649 under the auspices of Hartlib in which he expressed the view that chemistry would contribute to medicine.

In 1674 Boyle provided a compelling opening to his work "The Excellence and Grounds of the Mechanical Philosophy" that demonstrated a commitment to a particle theory of matter and the physico-theology of Gassendi and Charleton:

"By embracing the corpuscular, or mechanical philosophy, I am far from supposing, with the Epicureans, that, atoms accidentally meeting in an infinite vacuum, were able, of themselves, to produce a world, and all its phenomena: nor do I suppose, when God had put into the whole mass of matter, an invariable quantity of motion, he needed do no more to make the universe; the material parts being able, by their own unguided motions, to throw themselves into a regular system. The philosophy I plead for, reaches but to things purely corporeal; and distinguishing between the first origin of things, and the subsequent course of nature, teaches, that God, indeed, gave motion to matter; but that in the beginning, he so guided the various parts of it, as to contrive them into the world he design'd they should compose; and establish'd those rules of motion, and that order amongst things corporeal, which we call the laws of nature. Thus, the universe being once fram'd by God, and the laws of motion settled, and all upheld by his perpetual concourse, and general providence; the same philosophy teaches, that the phenomena of the world, are physically produced by the mechanical properties of the parts of matter; and, that they operate upon one another according to mechanical laws".

ROBERT BOYLE, The philosophical works of the Honourable Robert Boyle,
1725, p.187, vol. 1, London

The specifics of Boyle's matter theory were presented in *Origin of Forms and Qualities*

. . . "there is one . . . universal matter common to all bodies, a substance extended, divisible, and impenetrable....... to discriminate the catholick matter into variety of natural bodies, it must have motion . . . matter and motion, being thus established . . . matter must be divided into parts . . . must have two attributes, its own magnitude . . . and its own figure or shape. It is from the size, shape and motion of the small parts of matter . . . that the colour, odour, taste, and other qualities . . . are to be derived. . . . . The various manner of the coalition of several corpuscles into one viable body is enough to give them a peculiar texture, and . . . exhibit divers sensible qualities . . . it will naturally follow, that from . . . innumerable swarms of little bodies that are moved to and fro in the world, there will be many fitted to stick to one another, and so compose concretions."

ROBERT BOYLE, The philosophical works of the Honourable Robert Boyle,
1725, pp. 197 and 213, vol. 1, London

The importance of Robert Boyle to the mechanical philosophy rests importantly on his range of experiments intended to provide support for a corpuscular theory of matter. We will discuss one that illustrates how his interpretation of results was determined by an *a priori* principle, a philosophical position, he accepted.

The experiment was similar to one performed by Van Helmont, called the willow tree experiment. Van Helmont's goal was to demonstrate that the matter of the tree was derived from water. In the context of Aristotelian philosophy water is, of course, one of the basic components of matter. Van Helmont placed 200 pounds of earth, dried in a furnace, in a container, moistened the earth with water, and planted in the earth a stem of a willow tree that weighed five pounds. He covered the soil so that no extraneous matter would fall on the earth and watered the tree regularly for five years. At the end of that period the tree weighed about 169 pounds, which included the leaves that fell off during the years. The soil continued to weigh 200 pounds, and thus Van Helmont concluded the 164 pounds of tree was derived from water.

Boyle confronted the same problem: Does the tree originate from water? He began by stating that water is a homogeneous, colorless, odorless, volatile, and transparent material. However, a plant grown in water does not have any of the characteristics enumerated for water but does contain innumerable parts that are solid, are colored, have odor, and so on. To turn the conjecture that a plant is derived from water into a general principle, that all plants are derived from water, Boyle placed a number of different plants in vials of water and maintained them for many months. During that time some produced roots and green buds and gained weight. These were the expected results. What did he make of these data? Plant matter does indeed come from water. What is happening is that there is assimilation and transmutation of water into plant matter. The question is, how did this happen? Whatever are the underlying processes that led to the conversion of water to plant components, they are invisible. Nevertheless, Boyle offered an explanation that should be apparent from his philosophical position that all matter is made of corpuscles. Therefore, it must be the case that the same corpuscles present in the water, which imbued water with its physical characteristics, by their unique *concretions* (combinations), in the plant are arranged in different combinations to give the plant its qualities. It is clear that once Boyle had accepted the corpuscular philosophy, his interpretation of experimental results remained consistent with this theory.

Founding of the Royal Society

A mechanical philosophy had great support in mid-century England before and after the founding of the Royal Society and the appearance of the publication *Philosophical Transactions of the Royal Society*, with Henry Oldenburg as the editor. Many individuals from the Oxford group had come to London by 1660 and were meeting at Gresham College. On November 28 of that year Christopher Wren presented a lecture on astronomy, and following his talk there was a discussion about forming a permanent institution to promote "physico-mathematical experimental learning." In 1662 this group received a charter from Charles II that created the Royal Society. In that year Oldenburg was appointed Secretary and Robert Hooke was made Curator of Experiments.

In 1665 Hooke published "Micrographia" as "ordered" (Hooke's description; "commissioned" might be another way to designate it) by the Royal Society. Both Hooke's preface and the review of the book in the Philosophical Transactions expressed the view that true knowledge can be gained only by the mechanical, experimental philosophy. The book contained the microscopic observations of a large number of biological entities and a few inanimate objects. Prior to the publication Hooke had presented to the society, at many meetings, material that became chapters in the book.

Hooke presented numerous images of the constituent parts of gnat larvae, the head of a fly, a mite, and small parts of larger creatures, among many other observations. He hoped his studies "may be in some measure useful to the main design of a reformation in Philosophy." He wrote that the activities of bodies that have been attributed to Qualities, and some to occult qualities,

> . . . "are perform'd by the small Machines of Nature . . . seeming the meer (sic) products of Motion, Figure, Magnitude . . . which a greater perfection of Optics [Hooke refers to better microscopes] may make discernable by these Glasses".

They (microscopes) will help reveal the subtle structure of whole bodies that will uncover the structural basis for the processes of the body:

> "It seems not improbable but that by these helps [microscopes] the subtilty of the composition of Bodies, the structure of their parts, the various texture of their matter, the instruments and manner of their inward motions . . . may come to be more fully discovered".
>
> ROBERT HOOKE, *Micrographia*, The Preface, (1960, Dover)

There was an extended and enthusiastic review (no surprise) in the first volume of the Philosophical Transactions of the Royal Society:

"The ingenious and knowing author of this treatise, Mr. Robert Hook, considering with himself, of what importance a faithful History of Nature is to establishing of a solid System of natural Philosophy, and what advantage Experimental and Mechanical knowledge hath over the Philosophy of discourse and disputation (scholasticism) . . . hath lately published . . . which is very welcome . . . for the new discoveries in Nature".

From these studies there

. . . "may emerge, many admirable advantages towards the enlargement of the Active and Mechanick part of knowledge, **because we may perhaps be enabled to discern the secret workings of Nature** (my emphasis), almost in the same manner, as we do those that are the productions of Art, and are managed by Wheels, and Engines, and Springs, that were devised by Humane wit".

Philosophical Transactions of the Royal Society, 1665–1666, vol. 1, pp. 27–28

The conclusion is clear. It may be possible to visualize the workings of the particles, atoms or corpuscles, that constitute the material basis of all things.

The New Philosophy and Contagious Disease Causation

A particle theory of matter had an important impact on the disease theories of Thomas Willis and Robert Boyle.

Thomas Willis

An important figure in seventeenth-century medicine in England was Thomas Willis (1621–1675). He was a Professor of Medicine at Oxford for seven years. While at Oxford in the 1650s Willis and others, including Robert Boyle, Robert Hooke, and Christopher Wren, carried out "chymical experiments" on the blood and the nature of respiration. Thereafter he moved to London to practice medicine and conduct chemical research.

Willis' major treatises related to disease were "A Medical Philosophical Discourse of Fermentation" (p. 1) and a Treatise on Fevers (p. 47)

(THOMAS WILLIS, *A Medical-Philosophical Discourse of Fermentation, or, of the intestine motion of particles in every body*, 1684, London)

Willis' writings reflect the influence of the new learning, the Baconian experimental philosophy, provide alternatives to the classical and new matter theory, and integrate a particle matter theory and chemistry to explain the cause of contagious disease. Willis' chemical theory is different from that of Van Helmont. He, like Paracelsus and Van Helmont, replaced the philosophical base of medicine with his own version of the chemical philosophy. First, he recognized two theories that attempt to explain the makeup of matter. There was the "Empedoclean–Aristotelian" hypothesis that matter is composed of four elements, earth, air, fire, and water. The second was the "Democritus–Epicurean" theory that matter is made of atoms. Willis rejected the first and modified the second. In its place he adopted a chemical theory in which all bodies are made of particles having the chemical characteristics of spirit, sulfur, salt, water, and earth. Bodies are composed of particles of different classes. Willis' theory retained the atomistic view that there are particles; however, they have chemical characteristics. The other major base for his medicine was the circulation of blood as worked out by William Harvey. He recognized that the movement of the blood can be discussed in mechanical terms and that now the composition of the blood could be described in chemical terms.

In Willis' philosophy all matter, including blood, contained five active principles, "water and spirit . . . salt, and sulphur and . . . earth. Spirits are Substances . . . subtle and Aetherial Particles" imparting "Life . . . Soul . . . Motion . . . Sense" to "every thing." The structure of everything is determined by spirits. Spirits have a divine origin. Other principles confer physical characteristics on bodies; for example, sulphur particles regulate temperature, while salt gives solidity to objects. This description employed the language of the chemical and mechanical philosophy, a materialist phraseology, that differed from, for example, Van Helmont's vitalistic ideas involving *archei*. Willis reasoned that "many works of nature" came about through fermentation: for example, animal digestion, plant growth, alcoholic fermentation, and putrefaction of bodies. Particles of ferments cause "heat, alterations and motion in materials with which they are in contact." From this foundation Willis developed a theory of fevers. Fever is a general symptom in a variety of diseases where the patient exhibits excessive heat. Fever was present in all the named diseases of the time, smallpox, typhoid fever, plague, and many others unnamed, which suggested that all these diseases arose from a common cause—that is, from fermenting blood. Excessive heat or fever was thus an effect of a cause, "a fermentation or immoderate heat . . . brought into blood and humors, which can spread to the entire body . . . like wine in bottles." Under these circumstances the chemical components of the blood were altered. Willis formulated the process this way:

> "the blood is stirred up . . . is coagulated and corrupted . . . the spirit and sulphur are put into a rage . . . and cause the blood to grow . . . hot . . . the components . . . are changed. This accounts for . . . pestilential fever, the plague, small-pox and measles".

In the case of plague the salt and "sulphur", "being exalted" (stimulated) by the pestilential ferment, "grow together into tumors of various kinds."

Where do these ferments come from? To understand their origin it is important to review Willis' ideas about contagion. He defined contagion as "that force or action by which any distemper residing in one body excites its like in another." He provided three ways contagion could occur: the first by direct contact; the second, "mediately and at a distance," transferred from one house to another house; the third by a garment brought from an infected house. These modes of transmission are characteristic of the way plague is spread. Willis posed three crucial questions: What is it that "streams from an infected body?" What happens to it during its passage from one body to another? How does it cause the same disease in another body?

The answer to question one is that it is a poison; the poison contains a ferment whose particles can attack the blood. When this occurs the blood and the whole body experience fermentation. "All the symptoms of the disease occur." Where does this poison-ferment come? He contended that the seed of poison is in the air:

> "The air is composed of vapors and fumes which are . . . breathed from the earth containing salt and sulphur; altogether they constitute . . . a thick cloud, where their various configurations cause them to combine in different ways. Thus, these active, moving atoms enter animals and impart their motion and "shake the vital flame [of the recipient body]."
>
> If these bodies "swimming in the air" have the right configuration and "power" to be harmful to the "spirits . . . in living creatures," they "pervert the motion in the bodies they enter, and the bodies undergo putrefaction." Willis applied this complex theory to all living beings, so that
>
> the tops of trees, or of corn, being struck with a blast, suddenly grow dry or wither; hence among the cattel (sic), the Murrain often rages, which kills at once whole flocks by reason of this kind of cause".

Willis summarized how the disease agent passed from one body to another, what happened to it while it was being transferred and how it created the disease in another body. From every body,

> "effluvia of atoms constantly fly away . . . of remarkable virtue and energy; these little bodies . . . retain the contagion of the pestilence. These particles retain their identity during passage. . . . With its ferment it imbues the next little bodies, and so acquires new forces . . . ; from whence it lurks a long while in some nest, and after a long time when it assaults a convenient subject . . . imparting the taint of its power to another".

Robert Boyle

Robert Boyle furnished a critique of Van Helmont's theories in "The Sceptical Chymist." Boyle held that there are elements that make up mixed bodies but there are no certain number of elements, neither the four of Aristotle, the three of Paracelsus, or the one (water) of Van Helmont. By rejecting Van Helmont's theory of matter, the basis of his medicine, Boyle undermined his medical theories. Boyle substituted for Helmontian philosophy a mechanistic philosophy. For Boyle all matter was composed of corpuscles, a particle theory. Boyle held there was no transmutation in the Helmontian sense, but rather that elements that were combined in various ways could be recovered in their original form by some other chemical reactions. These particles had shape and size and exhibited motion and in this way could cause disease. From this theoretical base one could distinguish different objects in the world. Objects, including disease-producing agents, are different not because they contained different seminal principles, *archei* in the Helmontian philosophy, but because there are material differences based on the different combination of corpuscles. For Boyle all matter was composed of corpuscles, a particle theory. From this unitary view the entire philosophy of elements, principles, and qualities was discarded and the concept of substantial form was eliminated. Prime matter was also eliminated. Helmontian notion of ideas and *archei* were discarded. The only reality that remained was particulate matter that has size, shape, and motion. Boyle did retain a dualist notion with regard to humans; he allowed soul and body. But for the body itself there was only particulate corpuscles, matter that was identical to the matter in the outside world of nature.

In "The Sceptical Chymist," in a section on the "Causes of the Wholesomeness and Unwholesomeness of the Air," Boyle noted that the incidence of "epidemic diseases" in many places proceed from some excessive heat, moisture, or other palpable quality of the air. The condition of the air depends on the kind of "subterraneal effluvia" it receives. The air that is brought by these subterraneal effluvia may contribute to epidemic diseases.

(ROBERT BOYLE, *The Philosophical Works of the Honorable Robert Boyle;*
Causes of the Wholesomness and Unwholesomness of the Air:
The Air consider'd With regard to Health and Sickness
(abridged), 1738, 2nd edn., vol. 3, Peter Shaw,
pp. 521–542, London)

"Sometimes . . . earthquakes, send into the air . . . . . . venomous exhalations that produce new and mortal diseases . . . . . of a particular species"

(Robert Boyle, vol. 3, 2nd edn., p. 85, 1738)

The propagation, effects, and phenomena of plague come from unseen particles, corpuscles, from the earth that diffuse into the air and transmit disease. These emanations (exhalations) or effluvial particles (corpuscles) from different subterranean places could account for the epidemic spread of plague through the air. Here we have echoes of miasms from decaying matter in the earth. Boyle proceeded to characterize what was contained in these effluvia and concluded that these unseen particles could "propagate." This was not a biological theory but a chemical one, since he compared it to the causative agent of fermentation.

Plague in England

In 1665, in Volume 1 of the Philosophical Transactions of the Royal Society London," the following advertisement appeared:

"The reader is hereby advertised that by reason of the present contagion (Plague) in London, which may unhappily cause an interruption as wel (sic) of Correspondencies, as of Public Meetings, the printing of these Philosophical Transactions may possibly for a while be intermitted".

In 1665 an epidemic of plague broke out in London and in surrounding areas as far north as Durham, about 350 km from London. It was the seventh plague epidemic between the years 1563 and 1666. In Europe in the sixteenth and seventeenth centuries mortality among the young from contagious disease was between 40 and 50 percent by the age of 15. The plague in England in 1665 did not discriminate among children and adults and killed more than 20 percent of the population of London. This plague came at the end of a period of turmoil in England. There had been a civil war and in 1649 Charles I had been executed. After Cromwell's death in 1658, the Stuart monarchy was restored. And then the plague came, and 70,000 people died between December 1664 and December 1665.

This plague and the ones preceding it occurred at the time of philosophical changes in the sixteenth and seventeenth centuries, which, of course, included the development of a new particle matter theory, atoms or corpuscles, and the introduction of a particle theory of disease. Were these new philosophies employed to explain the cause of this disease, and was the treatment of the disease affected in any way by this new philosophy? Two writings, among many, analyzed the cause of plague and demonstrated the influence of Boyle's corpuscular philosophy. William Boghurst (1630/1–1685), an apothecary, described the plague using the terminology of Boyle. Boghurst contended that the air was polluted by vapors, pestiferous effluvia, that contained "invisible bodies" or "atoms" that in some way corrupt the corpuscles or particles that constitute the blood. The result of these interactions was putrefaction, which in humoral theories was associated with disease and death and is now used in the same way with chemical and mechanical philosophies of medicine.

Nathaniel Hodges (1629–1688), a physician, wrote about a "plague poison" made of chemical particles that are contagious and are spread by air.

Needless to say, the treatment of the disease remained the same as it was in Galenic medicine.

An Important Note on Isaac Newton

The impact of Newton on natural philosophy (science), medicine, and disease theory in the last quarter of the seventeenth century was enormous. He created a number of "revolutions" whose contents can be characterized as Newtonianism. It occurred in mathematics with the invention of calculus, which is at the heart of physics and much of basic science. He contributed to the physics of optics and the nature of color. He developed a mechanics that included the three laws of motion. He defined the concept of mass, measured by gravitation and inertial motion. He discovered the principle of universal gravitation and the law of gravity, and within a mathematical system he could account for the motions of celestial bodies in the universe and bodies on earth. This material was contained in his masterpiece, the "Principia," published in 1687. Newton accepted the mechanical philosophy and chose the matter theory of Gassendi over that of Descartes, perhaps influenced by Charleton's "Physiologia," which he read at Cambridge during his student days, and later by his friendship with Robert Boyle. In the "Principia" he addressed the question of forces between bodies and between particles of matter; these are discussed in Query 31. In his work on optics he included discussions designated as Queries. These were speculations, with some experimentation, that revealed his thinking about the mechanical philosophy. Query 31 contains the view that matter is made of particles. They have defined, necessary powers that enable them to "act at a distance," to account for "reflecting, refracting, and inflecting" rays of light, and interacting with each other to produce the works of nature:

"For it is well known, that Bodies act one upon another by the attraction of Gravity, Magnetism, and Electricity; and these Instances shew the Tenor and Course of Nature, and make it not improbable but that there may be more attractive Powers than these . . . How these Attractions may be perform'd I do not consider here. The parts of all homogeneal hard Bodies which fully touch one another, stick together very strongly, and for explaining how this may be, some have invented hooked atoms".

Remarkably, Newton had referred to Epicurus, although he preferred a different model of interaction:

"I had rather infer from their Cohesion that their Particles attract one another by some Force, which in immediate Contact is exceedingly strong, at small distances performs the chymical Operations above mentioned".

This universal force in nature is responsible for phenomena in the inanimate and animate world:

"Seeing therefore the variety of Motion which we find in the World is always decreasing, there is a necessity of conserving and recruiting it by active Principles, such as the cause of Gravity, by which Planets and Comets keep their Motions in their Orbs, and Bodies acquire great Motion in falling; and the cause of Fermentation, by which the Heart and Blood of Animals are kept in perpetual Motion and Heat; the inward parts of the Earth are constantly warm'd, and in some places grow very hot; Bodies burn and shine, Mountains take fire, the Caverns of the Earth are blown up, and the Sun continues violently hot and lucid, and warms all things by his light. For we meet with very little Motion in the World, besides what is owing to these active Principles. And if it were not for these Principles, the Bodies of the Earth, Planets, Comets, Sun, and all things in them, would grow cold and freeze, and become inactive masses; and all Putrefaction, Generation, Vegetation and Life would cease, and the Planets and Comets would not remain in their Orbs".

What created all of this and set things in motion was

"God in the beginning form'd Matter in solid, massy, hard, impenetrable, moveable Particles, of such Sizes and Figures, and with such other Properties, and in such Proportion to Space, as most conduced to the End for which he form'd them".

ISAAC NEWTON, *Opticks: or, a treatise of the reflections, refractions, inflections and colours of light,* 1730, The fourth edition, edited by Sir Isaac Newton, pp. 364, 375, London

Newton's mathematical physics, his dynamic corpuscularity, exerted a powerful influence on a cadre of medical physician-theorists in England near the close of the seventeenth century. Some of them were already adherents of the mechanical philosophy of a distinguished Italian school, which determined how they speculated about the cause of contagious disease. I will examine their contributions to contagious disease theory in Chapter 10. Chapter 9 deals with the discovery of microscopic living *animalcules* in the last half of the seventeenth century.

CHAPTER 9 | The Discovery of Microscopic Life
Antony von Leeuwenhoek and Robert Hooke

"In 1609 . . . Galileo had set a telescope in the garden behind his house and turned it skyward. Never-before-seen stars leaped out of the darkness to enhance familiar constellations; the nebulous Milky Way resolved into a swath of densely packed stars; mountains and valleys pockmarked the storied perfection of the Moon; and a retinue of four attendant bodies traveled regularly around Jupiter like a planetary system in miniature".

DAVA SOBEL, *Galileo's Daughter*, 1999, p. 6, Walker & Co./HarperCollins

IN HIS INFORMATIVE AND fascinating book on Antony van Leeuwenhoek, Brian Ford began his introductory chapter with a statement of profound importance. He wrote that he has in his hand:

"A small piece of metal, brown, mellowed with the passing of time, and utterly prepossessing. It would be difficult to imagine that any man-made object could be more ordinary.

Yet this [it is actually about four inches in length] is one of the most important instruments of scientific discovery. It changed our lives, altered our self-image, revolutionized our understanding of the world in which we live".

BRIAN FORD, *Single Lens: The Story of the Simple Microscope*, 1985, p. 1, Harper and Row

The object Ford was holding was a microscope fabricated by Leeuwenhoek about 1700 CE. This instrument, however "primitive," enabled Leeuwenhoek to visualize bacteria, in addition to a variety of living creatures larger than bacteria but nevertheless invisible to the unaided eye, whose size is in the range of one to three micrometers (one micrometer is 1/1,000,000 of a meter or 0.0000394 inches).

Leeuwenhoek's first published discoveries using microscopes appeared in a letter in the Philosophical Transactions of the Royal Society of London in 1673 entitled "A speciman of Some observations made by a Microscope, contrived by M. Leewenhoeck in Holland, lately communicated by Dr. Regenerus de Graaf." The work originated in Delft, Holland, on April 28, 1673, and was prefaced by remarks by de Graaf, who contended that Leeuwenhoek's microscopes excelled "those that have been hitherto made."There was also an accompanying letter addressed to Henry Oldenburg, Secretary of the Royal Society, urging him to encourage the microscopic studies of Leeuwenhoek. This report contained a description of a "mould" (we would now classify it as a fungus), describing the development of upright filaments, comparing them to "stalks of Vegetables" with rounded knobs on some and "blossom-like leaves" on others at the tip of stalks. The last line of the paper contained an editorial comment by de Graaf predicting that more observations like these would be forthcoming, which would illustrate the power of the "Glasses" (microscopes) created by Leeuwenhoek.

From September 1674 to mid-1676 Leeuwenhoek wrote notes to Oldenburg describing little "animalcules," many of them motile, of various sizes and shapes, obtained from pond water, rain, and well and canal water. On October 9, 1676, a letter was published in the *Transactions* that could serve as the founding document for the field of microbiology.

Leeuwenhoek related that on May 24, 1676, he was examining water samples and

. . . "found in it the oval animals in much greater abundance. And in the evening of the same day, I perceived so great a plenty of the same oval ones, that 'tis not one only thousand which I saw in one drop; and of the very small ones, several thousands in one drop".

There is indeed an asterisk inserted by Oldenburg, who interrupts the narrative to exclaim

"This Phaenomenon and some of the following ones seeming to be very extra-ordinary, the Author hath been desired to acquaint us with his method of observing, that others may confirm such Observations as these".

Oldenburg did this to alert contemporary readers that something radically new had been discovered. He did not do this when Leeuwenhoek, via de Graaf, provided a description of a "mould" in the communication of 1673, since Robert Hooke had already provided pictorial evidence of a fungus in his classic text, Micrographia.

The May 24 letter was translated into English by Oldenburg (Leeuwenhoek knew only Dutch), and during the translation process Oldenburg was living with these findings for some time. Nevertheless, he retained what was probably his initial reaction to the paper, that the animals Leeuwenhoek described had never been seen before.

Following the publication of this paper Oldenburg wrote to Leeuwenhoek that the "Philosophers" (members of the Royal Society) liked his observations but were skeptical that such a large number of "little animals" could be present in a drop of water. Leeuwenhoek responded and described how he estimated this number. Leeuwenhoek's response reveals that he had to invent these counting procedures since there were no precedents for such an appraisal. The letter, reproduced by Clifford Dobell in "Antony von Leeuwenhoek and his "Little Animals," Russell and Russell, 1958 (Dover, 1960) also gives us an insight into his character:

"Yet I can't wonder at it, since 'tis difficult to comprehend such things without getting a sight of 'em.

But I have never affirmed, that the animals in water were present in such-and-such a number: I always say, that I imagine I see so many.

My division of the water, and my counting of the animalcules, are done after this fashion. I suppose that a drop of water doth equal a green pea in bigness; and I take a very small quantity of water, which I cause to take on a round figure, of very near the same size as a millet-seed. This latter quantity of water I figure to myself to be the one-hundredth part of the foresaid drop: for I reckon that if the diameter of a millet-seed be taken as 1, then the diameter of a green pea must be quite 4 ½. This being so, then a quantity of water of the bigness of a millet-seed maketh very nearly the 1/91 part of a drop, according to the received rules of mathematicks (as shown in the margin). This amount of water, as big as a millet-seed, I introduce into a clean little glass tube (whenever I wish to let some curious person or other look at it). This slender little glass tube, containing the water, I divide again into 25 or 30, or more parts; and I then bring it before my microscope, by means of two silver or copper springs, which I have attached thereto for this purpose, so as to be able to place the little glass tube before my microscope in any desired position, and to be able to push it up or down according as I think fit (p. 169).

I showed the foresaid animalcules to a certain Gentleman, among others, in the manner just described; and he judged that he saw, in the 1/30th part of a quantity of water as big as a millet-seed, more than 1000 living creatures. This same Gentleman beheld this sight with great wonder, and all the more because I told him that in this very water there were yet 2 or 3 sorts of even much smaller creatures that were not revealed to his eyes, but which I could see by means of other glasses and a different method (which I keep for myself alone). Now supposing that this Gentleman really saw 1000 animalcules in a particle of water but 1/30th of the bigness of a millet-seed, that would be 30000 living creatures in a quantity of water as big as a millet-seed, and consequently 2730000 living creatures in one drop of water.

My counting is always as uncertain as that of folks who, when they see a big flock of sheep being driven, say, by merely casting their eye upon them, how many sheep there be in the whole flock. In order to do this with the greatest exactness, you have to imagine that the sheep are running alongside one another, so that the flock has a breadth of a certain number: then you multiply this by the number which you likewise imagine to make up the length, and so you estimate the size of the whole flock of sheep (pp. 170–171).

And just as the supposed number may differ from the true number by fully 100, 150, or even 200, in a flock of 600 sheep, so may I be even more out of my reckoning in the case of these very little animalcules: for the smallest sort of animalcules, which come daily to my view, I conceive to be more than 25 times smaller than one of those blood-globules which make the blood red; because I judge that if I take the diameter of one of these small animalcules as 1, then the diameter of a blood-globule is at least 3 (p. 171).

These, Sir, are the trifling observations which I have shown to divers curious persons, to their great satisfaction; but the other things that I have seen, and my particular microscope, I cannot yet resolve to make public: which I beg you, Sir, and your fellow philosophers, not to take amiss" (p. 171).

C. DOBELL, *Antony von Leeuwenhoek and his "Little Animals,"*, 1960, Dover

Almost 50 years later, in the Transactions issue of November–December 1723, Martin Folkes (1690–1754), Vice-President of the Royal Society, wrote, on the occasion of Leeuwenhoek's death and the donation of many of his microscopes to the Society,

"It were endless . . . to give any Account of Mr. Leeuwenhoek's Discoveries; they are so numerous as to make up a considerable part of the *Philosophical*

Transactions . . . And of such Consequence as to have opened entirely new Scenes in some Parts of Natural Philosophy" (vol. 32, p. 449).

Leeuwenhoek

For 50 years, beginning in 1674, Leeuwenhoek published almost 200 letters in the *Transactions,* establishing the presence of microscopic life in a vast variety of environments; he called these microscopic entities *animalcules.* The discovery of these animalcules was a revolutionary event that revealed for the first time life forms whose existence was never imagined, nor derived from philosophical arguments or mathematical reasoning, nor by simple observation with the unaided eye of the natural world surrounding humans. Their discovery depended on the technology leading from telescopes to microscopes.

Leeuwenhoek had already reported previously in his September 7, 1674, letter a great variety of shapes and sizes of animalcules in pond water. They included the phytoflagellate *Euglena*, an organism that has a chloroplast, "green in the middle." This is probably the first time that the organelle responsible for photosynthesis in a photosynthetic organism was seen. In addition, he described the filamentous green alga Spirogyra, many ciliates and flagellates, and rotifers. All of these organisms are easily observed today by microscopic analysis of fresh water of ponds. A lake near Delft was the source of these organisms for Leeuwenhoek. This letter was the first of many that revealed the world of single-celled animals now known as protists and previously called protozoa. These were among the larger animalcules he observed. In letter 23, published in 1678, he reported on a pepper/rain/water infusion that after four days contained "so great a many of such inconceivably small creatures that a mans (sic) mind may not contain them all." He went on to describe "extraordinary little animalcules, little eels." This description indicates that he saw spirilla for the first time, another form of bacteria.

In letter 39 of September 17, 1683, Leeuwenhoek again reported on animalcules obtained from a human, in this case the human mouth. For the first time Leeuwenhoek provided drawings of the bacteria he had visualized: they are recognized as bacilli, micrococci, Leptothrix, and a spirochete. Oldenburg recognized the exceptional nature of these discoveries.

With his microscopes Leeuwenhoek examined a great variety of materials, among them water extracts of peppers, canal and ditch water, rainwater cisterns, fluid in oysters and mussels, bile of cows, rabbits, fowl, and turkey, feces of cows, horses, hens, and pigeons, urine of farm animals, blood of humans and animals, the saliva of humans, and the material from between their teeth. This wonderful curiosity enabled him to discover a world of microorganisms that included protozoans and bacteria.

Leeuwenhoek's observations were received with great interest at the Royal Society, and Robert Hooke was asked to prepare a demonstration experiment following Leeuwenhoek's procedures to allow members of the society to view the "little animals." At the meeting of the Society on November 15, 1677, Hooke exhibited the pepper water experiment in which were observed "exceedingly small animals swimming to and fro." Hooke had earlier verified the presence of microorganisms in pepper water and also demonstrated their presence in infusions of wheat, barley, oats, and peas. Present at this meeting were Christopher Wren (1632–1723), who became President of the Society in 1680, and John Mapletoft (1631–1721), who became a Fellow in 1676. Mapletoft was a colleague of Thomas Sydenham, the premier physician of the seventeenth century in England. Among Leeuwenhoek's unpublished letters to the Society that have been preserved was one written to Robert Boyle in 1676. A number of other letters to Robert Boyle exist. I mention these illustrious persons to point out how well known Leeuwenhoek's work was known in England among the learned establishment. Following these early communications, Leeuwenhoek was elected as a Fellow of the Royal Society in 1680; this confirmed the esteem he was accorded by the Society. Leeuwenhoek's work was also known on the Continent, and he received many visitors to his residence in Delft, Holland, including Gottfried Wilhelm Leibniz (1646–1716), who was a Foreign Fellow of the Royal Society. Leibniz developed differential calculus and was a philosopher with deep interests in everything that included physics and biology. Leeuwenhoek's work was known by Herman Boerhaave (1668–1738), a Dutch physician, who was Professor of Medicine, Chemistry, and Botany in Leiden and was widely known and respected in Europe. In short, Leeuwenhoek's work was acknowledged and honored throughout the learned community in Western Europe.

The fascinating question is this: Who was Antony van Leeuwenhoek? He began his publishing career at the age of 42 with a series of letters, written in his own hand, and only six years later was elected as a Fellow, not simply a Corresponding member, of the Royal Society. Shortly thereafter, he was known throughout Europe. During his lifetime he built some 500 microscopes whose magnifying powers were not equaled for more than 100 years. He never wrote a book or a treatise, and he is acknowledged as founding the science of protozoology and bacteriology.

Leeuwenhoek was born on October 24, 1632, in Delft and died on August 26, 1723 in the same city. He is buried in the old church cemetery where Jan Vermeer (1632–1675) is also buried. Starting in 1654 he made his living as a fabric merchant and later was appointed a civil official. His organized scientific training was nonexistent compared to the academic schooling received by distinguished contemporary microscopists like Jan Swammerdam (1637–1680), who was educated at the University of Leyden; Marcello Malpighi (1628–1694), a Professor of Medicine at Bologna; and Robert Hooke (1635–1703), a founding member of the Royal Society, a member of the Oxford University research group that assisted Robert Boyle, and the author of a landmark text, *Micrographia*. Leeuwenhoek had no university training and did not speak English. He had access to Hooke's book, published in 1665, but he could not read it since it was written in English. Oldenburg suggested in a letter that Leeuwenhoek use French or English in his correspondence, but Leeuwenhoek replied on January 22, 1676, "I must confess . . . I don't know any tongue but the Nether-Dutch." He went on to say that he had friends who could translate French and Latin but did not have a trustworthy person who could handle English and Dutch (at this time de Graaf had been dead four years).

Although the images in *Micrographia* inspired Leeuwenhoek, the production of microscopes was completely the product of his own extraordinary inventiveness. He taught himself to grind superb lenses that offered greater magnification than any of the microscopes of the time. The compound microscopes of Hooke and Swammerdam magnified about 20 to 30 times, while the best of Leeuwenhoek's microscopes magnified about 300 times. He not only prepared magnificent lenses but also fabricated the metallic components that held the magnifying lens (that is, the stage that held the specimen and the mechanisms that adjusted its position with respect to the lens). These microscopes had one lens; they were in fact a magnifying glass. The entire size was about four inches, and to observe the specimen the microscope had to be held by hand close to the eye of the observer.

Leeuwenhoek had no training in natural philosophy (science) and his writing was described as ungrammatical. Since he was observing for the first time microscopic entities, to estimate their size he had to select visible objects as standards for comparison. These standards included two kinds of sand grains, the human red blood corpuscle, and a millet seed.

By the early 1670s, before any of his work was published, an accurate description of Leeuwenhoek would have included his work as a shopkeeper, a municipal official, and an amateur microscopist with no university training. Yet his first letter appeared in the *Transactions*, the publication of the great Society devoted to natural philosophy. How did this happen? At some time after 1665 Leeuwenhoek read *Micrographia* and began making lenses. It is Brian Ford's reasonable suggestion that Leeuwenhoek was inspired by a description of a magnifying glass by Hooke. It was the kind of microscope that he could fabricate rather than the compound microscopes already in use. In 1667 Reinier de Graaf (1641–1673) moved to Delft. He was 26 years old. He was a brilliant physician-scientist who came to practice medicine in Delft. It is instructive to describe briefly de Graaf's training to illustrate the possibilities of medical education in the mid-seventeenth century in Europe and how far removed from such training was Leeuwenhoek. De Graaf began his higher education at the University in Louvain and studied medicine at the University of Utrecht and at Leiden, where he was a student of the illustrious Francis de le Boë Sylvius (1614–1672). In 1665 he was in France doing research on the pancreas, and that same year he was awarded a Doctorate with honors from the University of Angers. In Delft he did not simply practice medicine but carried out a research program so that by 1672 he had described the structure of the graafian follicle that now bears his name. This follicle is contained in the human oviduct, the fallopian tube described by Fallopius in 1562, and is a fluid-filled circular structure about 75 microns (0.0029 inches) across that contains the developing egg, the oocyte. De Graaf was an accomplished anatomist, but he needed some method of magnification to discover the follicle, which is invisible to the unaided eye. It was during these studies that de Graaf established a relationship with Leeuwenhoek, and one may speculate that Leeuwenhoek helped de Graaf in the follicle studies. Whatever help Leeuwenhoek may have contributed to de Graaf's work, de Graaf was certainly knowledgeable about Leeuwenhoek's microscopic studies, for on April 28, 1673, de Graaf wrote a letter, in Latin, to Oldenburg concerning these findings. A portion of the letter was published in the Transactions, and the letter was read at a meeting of the Society on May 7, 1673. This sequence of events makes it obvious that since initiating the construction of microscopes some seven years earlier, Leeuwenhoek had achieved a level of quality of lenses and had accumulated enough observations to have profoundly impressed de Graaf. De Graaf had grown close enough to Leeuwenhoek to have persuaded him that these findings were important enough to be made generally known through the pages of the Transactions. C. Dobell translated a portion of de Graaf's letter to Oldenburg, and we reproduce it here:

"That it may be more evident to you that the humanities and science are not yet banished from among us by the clash of arms [England and Holland were at war at that time] I am writing to tell you that a certain most ingeneous person here, named Leeuwenhoeck, has devised microscopes which far surpass those which we have hitherto seen . . . The enclosed letter from him, wherein he describes certain things which he has observed more accurately than previous Authors, will afford you a sample of his work: and if it please you, and you would test the skill of this most diligent man and give him encouragement, then pray send him a letter containing your suggestions".

C. DOBELL, Antony von Leeuwenhoek and his "Little Animals,"
1960, p. 40, Dover

Oldenburg wrote to Leeuwenhoek and Leeuwenhoek replied on August 15, 1673, in Dutch. Dobell's translation of that letter follows, and the contents reveal some of Leeuwenhoek's thoughts about his work and himself:

"I have oft-times been besought, by divers gentlemen, to set down on paper what I have beheld through my newly invented Microscopia: but I have generally declined; first, because I have no style, or pen, wherewith to express my thoughts properly; secondly, because I have not been brought up to languages or arts, but only to business; and in the third place, because I do not gladly suffer contradiction or censure from others. This resolve of mine, however, I have now set aside, at the intreaty of Dr. Reg. de Graaf; and I gave him a memoir on what I have noticed about mould, the sting and sundry little limbs of the bee, and also the sting of the louse. This memoir he (Mr. De Graaf) conveyed to you . . . I see that my observations did not displease the Royal Society . . . As I can't draw, I have gotten them drawn for me, but the proportions have not come out as well as I hoped to see 'em; . . . I beg you, therefore, and those Gentlemen to whose notice these may come, please bear in mind that my observations and thoughts are the outcome of my own unaided impulse and curiosity alone; for, besides myself, in our town there be no philosophers who practice this art; so pray take not amiss my poor pen, and the liberty I here take in setting down my random notions".

C. DOBELL, 1960, p. 42

It is clear that sometime between 1667, when de Graaf arrived in Delft, and 1673, when he wrote to Oldenburg, de Graaf and Leeuwenhoek had established a relationship and probably some personal ties. De Graaf was already a productive scientist, well known in Europe, and a corresponding member of the Royal Society, and thus his letter of introduction carried great weight with the Society. What appears most decisive was that he was close enough to Leeuwenhoek, perhaps on a personal as well as scientific level, to persuade him to submit his work to the society. Without de Graaf Leeuwenhoek's work may have remained as unknown as it was before de Graaf wrote his letter to the Society in 1673.

Leeuwenhoek's animalcules were extensively documented in the *Transactions*. However, the relationship between microscopic, living entities and contagious diseases, or for that matter between any living agents and contagious diseases, was not part of the discourse in the seventeenth century, which was dominated by the mechanical philosophy. A brief description of Robert Hooke's 1665 *Micrographia* will illustrate this fact.

Micrographia was filled with microscopic observations, with drawings of various living objects and inanimate materials, revealing structural details of the various objects. It was the first illustrated book using the microscopes of the time. Most of the materials he examined were biological specimens. This book was a contribution to the efforts of the Royal Society to support the mechanical philosophy and the experimental philosophy promoted vigorously by Robert Boyle. Hooke wrote that he would "return to the plainness and soundness of Observations on material and obvious things" rather than having the "Science of Nature dependent only a work of the Brain and the Fancy."

Hooke was at one time an assistant to Robert Boyle during the latter part of the 1650s at Oxford and was among the original members of the Royal Society, where he was Curator of Experiments from 1662 to 1677. He was a master in mechanics, an inventor, and a city planner after the London fire of 1666.

The introduction to fungi occurs in a section of the book where he describes a rose bush whose leaves exhibit a number of symptoms:

"I have for several years together, in the months of June, July, August, and September (when any of the green leaves of Roses begin to dry and grow yellow) observ'd many of them especially the leaves of the old shrubs of Damask-Roses, all bespecked with yellow stains, and the underside . . . to have little yellow hillocks of a gummous substance, and several of them to have small black spots in the midst of those yellow ones, which to the naked eye, appear'd no bigger than the point of a Pin, or the smallest black spot . . . of Ink one is able to make with a very sharp pointed Pen".

ROBERT HOOKE, *Micrographia*, 1960, p. 121, Dover

He examined the gummy substance microscopically and saw "multitudes of small black bodies, some attached to stalks." The bodies ranged in size from 1/500 to 1/1000 of an inch. Hooke speculated and decided that they were a form of blight or mildew, a simple kind of vegetation, that emerges from the leaves due to "putrefactive and fermentative" heat in combination with the air. In essence, the blight comes from the complex, decaying, non-living components of the plant by a process of "equivocal generation." Equivocal generation, also referred to as spontaneous generation, indicates that different microscopic life forms can be generated from the same non-life. This is in contrast to univocal generation, where progeny come from a defined identical "parent." Hooke recognized that the rose plant had a disease whose origin was not specified, but he assumed the fungi arose from decaying plant material. He observed microscopically the rounded bodies, spores, which are part of the structure of the fungi, but did not suggest they were vegetative seeds. In fact, he went on at great length on the ability of "moulds" to be produced without seeds:

"First, that Moulds and Mushrooms require no seminal property. But the former may be produc'd at any time from any kind of putrifying Animal or Vegetable substance . . .

Next, that as Mushrooms may be generated without seed . . . for having considered several kinds of them, I could never find . . . to be the seed of it . . . they . . . seem to depend merely upon a convenient constitution of the matter out of which they are made".

ROBERT HOOKE, *Micrographia*, 1960, p. 127, Dover.

Hooke proposed that these "moulds" are spontaneously generated. Why did he come to this kind of causality? It appears that the philosophical base for these speculations is the mechanical philosophy. He began the justification of this belief by introducing the machine model in which all living things are made of "multitudes of contrivances." These structures (contrivances) remain intact, although the living thing may be destroyed, to become the building blocks for a new animate body. To illustrate how this works, Hooke used the analogy of a clock that is made of various contrivances. Although one part of a clock may be removed, other parts continue to operate "exactly as if it were part of a compound automaton." Hooke's foundational principle was that less complex "vegetable bodies" are living entities but generated from the parts of plant tissues.

Once generated they reproduce for "the Omnipotent and All wise Creator" might provide it with all the contrivances necessary for its own existence.

In the early decades of the eighteenth century microscopic organisms were proposed to cause contagious diseases of humans, plants, and cattle, as explained in Chapter 10. This hypothesis was rejected or ignored by contemporary theorists since they were convinced that the cause was rooted in the prevailing mechanical philosophy.

| The Cause of Plague in France
in 1720 CE

RICHARD BRADLEY (1688–1732) AND Richard Mead (1673–1745) wrote treatises on the plague in 1720–21. They agreed that this disease and a number of others can be spread by air and furthermore that there is something in the air that caused infection. The nature of that something, speculated about since antiquity, and discussed in a vast literature in the sixteenth and seventeenth centuries, was the object of contention between Bradley and Mead.

The plague occurred in the years 1720–21 in Marseilles and surrounding areas. The Lords of the Regency in England commissioned Mead to prepare a report on the cause of the disease and ways to prevent its spread to England. Mead was retained by the Lords "as a man best qualified" to perform that task. He was reputed to be "one of the foremost physicians of the day" and was considered an authority on the spread of infectious diseases. He was wealthy, a scholar with a vast personal library, and a Fellow of the Royal Society. Such was his renown that his patients included members of the Royal Family and Isaac Newton. He had already published two theoretical treatises that included the causes of contagious diseases: *A Mechanical Account of Poisons* (1702) and *A Treatise Concerning the Influence of the Sun and the Moon upon Human Bodies, and the Diseases Thereby Produced* (1704). In 1721 he published his report on the plague in France. Mead invoked an inanimate agent working through chemical and mechanical effects to explain the cause.

Bradley was a Fellow of the Royal Society and was about to become the first Professor of Botany at Cambridge in 1724. He wrote extensively on the various aspects of gardening, horticulture, plant breeding, and, most importantly, plant diseases. He argued that blights of trees are the result of infection by living agents both visible and microscopic. Bradley postulated that a living agent was the cause of plague, one of four individuals to ascribe contagious diseases to living agents. The others were Jean-Baptiste Goiffon (1658–1730) in France, who proposed a living agent as the cause of plague; Benjamin Marten in England, who speculated that consumption (tuberculosis) is caused by a living agent; Carlo Francesco Cogrossi (1682–1769) proposed, in 1714, somewhat ambiguously, that a microscopic living agent, an insect (animalcule), was the cause of a contagious disease of oxen. Bradley did more than just propose a living agent for the cause of plague of humans: he advanced the revolutionary hypothesis, developed in his writings between the years 1714 and 1721, that contagious diseases of animals, humans, and plants are caused by living, microscopic agents. He wrote in his plague treatise in 1721 "we may learn, that all Pestilential Distempers, whether in Animals or Plants, are occasion'd by poisonous Insects [Bradley's sometime term for Leeuwenhoek's animalcules] convey'd from Place to Place by the Air."

This theory is the core of Bradley's proposal that differentiated him from Mead and almost all of the medical establishment in Europe. The question to be answered is: Why did Mead and Bradley arrive at such different causes? Mead was a follower of an eminent group of English and Italian physicians who had adopted the mechanical philosophy and its most contemporary theoretical base, the mathematical physics of Newton, to understand the operation of the human body and the cause of contagious disease. In contrast, Bradley had a biological theory of the cause of contagious disease. He had no recourse to the mechanical philosophy when discussing plant processes or diseases of plants. His works were replete with the empirical findings of gardeners, horticulturalists, owners of botanical gardens, and the microscopic discoveries of Antony van Leeuwenhoek and Robert Hooke, and his own experience with plant diseases. His philosophical position was based on three fundamental observations, the first two found in the preface to the first edition of *New Improvements of Planting and Gardening* (1717) and the third contained in the Introduction to A Philosophical Account of the Works of Nature.

1. I shall then advance (what I think) a new System of Vegetation, and endeavor to prove that the Sap of Plants and Trees Circulate much after the same manner as the Fluids do in Animal Bodies, which may be one Argument to shew the beautiful Simplicity of Nature in all her Works.

RICHARD BRADLEY, *New Improvements of Planting and Gardening*, 1717, p. 11, London

2. The Generation of Plants will next be consider'd, and the manner how their Seeds are impregnated, a Discovery which I made some Years since; this will be of great use to all Planters, by directing them in the proper choice of their Seeds.

3. Besides that, there are infinitely more Species of Creatures, which are not to be seen without, nor indeed with the Help of the finest Glasses, than of such as are bulky enough for the naked Eye to take hold of. However, from the Consideration of such Animals as lie within the Compas (sic) of our Knowledge, we might easily form a Conclusion of the rest, that the same Variety of Wisdom and Goodness runs through the whole of Creation, and puts every Creature in a Condition to provide for its Safety and Subsistence in its proper Station.

RICHARD BRADLEY, *A Philosophical Account of the Works of Nature*, 1721, Introduction, London.

The first principle is that there is a structural unity in nature. Analogous structural components convey the major fluid, blood or sap, in animal and plant bodies. The second axiom is that sexual reproduction occurs in animals and plants. The third truth that Bradley accepted is of great importance to a biological theory of contagious disease. There is a microscopic world of living "insects," sometimes called animalcules or creatures, that are not spontaneously generated but reproduce as all living visible organisms do in the proper environment. With these empirically derived maxims it is possible to account for the origins of Bradley's biological cause of disease. But first, we will review the philosophy and theory of Richard Mead.

"If Newton could, with a small number of basic laws, enable us, at least in theory, to determine the position and motion of every physical entity in the universe, and in this way abolish in one blow a vast, shapeless mass of conflicting, obscure, and only half-intelligible rules of thumb which had hitherto passed for natural knowledge, was it not reasonable to expect that by applying similar principles to . . . the analysis of the nature of man, we should be able to obtain similar clarification and establish the human sciences upon equally firm foundations"?

ISAIAH BERLIN, *Concepts and Categories*, 1981, p. 139, Penguin

The origin of Mead's hypothesis about the cause of contagious diseases is stated in "A Treatise concerning the influence of the Sun and Moon upon Human bodies and the Diseases thereby produced", in *The Medical Works of Richard Mead*, 1765, p. 160, Edinburgh.

"As the study of physic [medicine] has in all ages undergone various changes, according to the different opinions of philosophers. That some of the moderns, particularly Galilei, Kepler, Toricelli and Sir Isaac Newton, have made vast improvements in natural philosophy, by joining mathematical reasonings to their inquiries".

This Treatise appeared in 1712. Mead accepted that medicine's roots and trunk are mathematics and physics. He used these tools to understand "the causes of diseases" and "for finding proper remedies for them."

To understand how Mead arrived at this theoretical base, it is obligatory to understand that his teacher, Archibald Pitcairne (1652–1713), and his colleagues had enthusiastically embraced Newton's version of the mechanical philosophy, which included a corpuscular (particle) theory of the composition of matter. In addition to Newton, the influence of the Italian School is apparent in Mead's *Poisons* text. In Italy, there was a distinguished group of individuals led by Giovanni Borelli (1608–1679), a mathematician and student of Galilean mechanics who proposed a mechanical approach to the study of the function of the various organ systems of the human body. A partnership with Marcello Malpighi in 1656 provided the microscopic anatomical expertise to uncover the structure of lung tissue that provides the structural basis for enhancing the circulation of blood. Malpighi furnished an eloquent rationale for the mechanical philosophy employing the argument from design:

"For just as a man born in the forest and ignorant of all human acts, if shown a clock . . . would only wonder at the delicacy and elegance of its structure . . . regularity and spontaneity of its motions and would never guess how the little machine could be made so perfect, so do our powers fail completely when we are confronted by these achievements of Nature, at the elaboration of which we were present neither as spectators nor as participants, and like untutored woodsmen we can only be struck with wonder but cannot divine or conceive by what artifice they have been accomplished; for indeed, each one of them is a little machine within which are enclosed in a way impossible to comprehend almost innumerable [other] little machines, each with its own little motions".

BERTOLINI MELI, *Mechanism, Experiment, Disease*, 2011, p. 44,
Johns Hopkins

Borelli had a similar influence on Laurentius Bellini (1643–1704), who worked on the structure of the kidney and concluded that the secretion of urine was a consequence of the configuration of the vessels of the kidney. Nicholas Steno (1638–1686) came from Denmark to study medicine in Leiden and later moved to Florence, where he conducted research and published in 1667 *Myologiae,* which is a structure/function account of muscle physiology discussed within a mathematical framework. The works of Borelli, Bellini, and Newton were read by the mathematician David Gregory (1627–1720), a professor and a colleague of Pitcairne at the University of Edinburgh. The impact of Newton's work on medicine appeared to depend on the interaction of Gregory and Pitcairne during the 1680s. Gregory was a student of Newton's mathematical physics. He was a colleague of Pitcairne and was instrumental in getting Pitcairne and Newton together. Pitcairne visited Newton at Cambridge early in 1692, at which time Newton gave him his latest work on alchemy, chemistry, and the theory of matter. Pitcairne become Professor of Medicine at Leiden, although he remained only a year and returned to continue as Professor of Medicine at Edinburgh. Pitcairne was a teacher of Mead. It is abundantly clear what are the influences on Mead.

Pitcairne, dissatisfied with qualitative descriptions of the operation of the human body, argued for the application of geometry (mathematics) to medicine to ensure that certain and not speculative knowledge would be achieved. He subscribed to the Newtonian theory of matter and stated:

. . . "Isaac Newton; since we must justly hope, that by the Assistance of the Principles demonstrated by that Great Man, the Powers and Properties of Bodies serviceable to Medicinal Uses and the Comfort of Mankind, may be discovered with greater Ease, and reduced to a greater Certainty".

Archibald Pitcairn, *The whole works of Dr. Archibald Pitcairn*,
1727, p. 56, London

Pitcairne thought that the most vital function (mechanism) in the human body was the circulation of the blood, driven by the beating heart. Consequently the key to understanding human physiology is to understand the dynamics of circulation. As a corollary the fundamental cause of disease comes from obstructing the circulatory system. There are rules that govern this mechanical system, a system composed of fluids and channels that could be described mathematically. The construction of a mathematical medicine was carried on by George Cheyne (1671–1743), one of Pitcairne's students in Leiden and a colleague of Richard Mead. By doing this he would emulate Newton and develop

"A Principia of Medicine . . . would consist of Newtonian principles applied to the more minute . . . Appearances of Nature".

In 1701 Cheyne published a

"New Theory of Continual Fevers Wherein Besides the Appearances of such Fevers, and the Method of their cure; occasionally, the structure of the Glands, and the manner of secretion, the operation of Purgative, Vomitive, and Mercurial Medicines, are Mechanically Explained" (1733 edition, London).

In the preface he wrote,

. . . "the wiser part of Mankind are now persuaded, that this machine we carry about, is nothing but an infinity of Branching and Winding Canals, fill'd with liquors of different Natures . . . Continual Fevers, are only a Complication of Symptoms, which naturally follow upon a general Obstruction of these Canals or the Glands which they constitute".

He continued,

"That where a Machine is disordered, if we should see it righted by adjusting such a Particular Part, we might . . . affirm, that it was some injury done to that part, which disordered the machine. Thus, if we should see a watchmaker, by adjusting only the balance of a watch, make her go right; we might say the distortion of the Axe thereof had occasion'd her going wrong".
G. CHEYNE, New Theory of Continual Fevers, 1722, p. 38, London

The analogy between the human body as a machine takes us back to the earlier writings of Descartes:

"I say . . . these functions . . . follow naturally . . . entirely from the disposition of the organs, no more nor less than do the movements of a clock or other automaton, from the arrangement of its counterweights and wheels".

In this work Cheyne provided diagrams of canals and mathematical calculations of blood flow in the body from which he generated the cause of all fevers:

"The general and most effectual Cause of all Fevers, is the obstruction or Dilation of the complicated Nerve and Arterie, the Excretory duct . . . Other things may occur, but these are the most powerful causes".

G. CHEYNE, p. 46 , 1722, London.

Cheyne accepted the thesis that the processes of the human body are mechanical events that can be expressed in quantitative terms and interference with these mechanisms leads to disease.

Mead "followed Cheyne's lead" in 1702 with the publication of *A Mechanical Account of Poisons* that went through five editions until 1756. From the first to the last edition Mead retained the same general theoretical base for causality. In the preface Mead wrote,

"My design, in thinking of these matters was, to try how far I could carry mechanical considerations in accounting for those surprising changes which poisons make in the animal body.

It is very evident, that all other Methods of improving Medicine have been found ineffectual these two or three thousand years; and that of late Mathematicians have set themselves to the study of it. Men do already begin to talk so Intelligibly and Comprehensibly, even about abstruse Matters, that it may be hop'd in a short time, if those who are design'd for this Profession, are early, while their Minds and Bodies are patient of Labor and Toil, initiated in the knowledge of Numbers and Geometry, that Mathematical Learning will be the distinguishing Mark of a Physician from a Quack; and that he who wants this necessary Qualification, will be as ridiculous as one without Greek and Latin".

RICHARD MEAD, *A Mechanical Account of Poison*, 1702, Preface, London

Shortly after this quote, Mead acknowledged the writings of Bellini, Pitcairne, and Cheyne. He concluded this preface with a statement that he was absolutely certain of his methodology:

"They who have no smattering of Mathematical Knowledge are incompetent Judges of what Service I have done towards the Improvement of the Theory, or Practice of Medicine".

RICHARD MEAD, *Preface*, 1702, "A mechanical account of poisons in several essays". Preface, London

Laying down this new base, Mead proceeded to explain the lethal effects of viper venom in the *Poisons* text. He created a complex series of interactions initiated by *spicula*, his description of the crystal structure of the venom he observed by examining dried venom microscopically. These structures destroy the surface of the globules of the blood, releasing the animal juices (an elastic fluid) and changing the nature of the fluid of the blood, whereupon it will be fermented. Diseases are the result of the improper state of these fluids. The mechanism of lethality appears to result from the disruption of the balance of forces of the blood, which forces obey "the great Principle of Action in the Universe," the Newtonian force of gravity. He offers no evidence for these contentions but claims that his "hypotheses enjoy the protection of mathematical vigor." Mead followed with another speculative treatise, two years later, on the effect of the sun and moon on human health. The work alleges that the sun and moon can cause changes in animal bodies that result in periods of fevers. How do we go from the moon to fevers? This theory is based on the generally accepted idea that the atmosphere is important for the transmission of disease. Any condition that affects the pressure of the atmosphere will affect the health of humans. According to Sir Isaac Newton the sun and the moon affect the oceans and the atmosphere, affecting the gravity and elasticity of the air and culminating in disease.

Before the appearance of Mead's treatise on the plague, in 1717 a series of essays was published by James Keill (1673–1719), revealing the powerful grip that Newtonian philosophy had on the theoretical foundation of medicine. Keill presented, with its historical antecedents, a philosophical and methodological basis for understanding human physiology. In the preface he stated the core foundational statement that the whole of medicine stands upon the principles of the mechanical philosophy. Its methods are the only way to obtain knowledge of the operation of the human body. Once more the analogy of the human body to a machine is invoked:

"The Animal body is now known to be a pure machine, and many of its Actions and Motions are demonstrated to be the necessary Consequences of its Structure".

JAMES KEILL, Essays on several parts of the animal oeconomy,
1717, Preface, p. iii, London

The evidence for this conclusion, according to Keill, are the works of Newton on vision in the *Opticks,* Borelli on mechanical motion, Harvey on the circulation of the blood, Bellini on its motion and velocity, Pitcairne's description of the structure of the lungs, Cheyne's book on fevers, and Mead's work on poisons. Among all these citations the ultimate source of authority is Newton.

The third work of Mead, in response to the request to understand the cause of plague in France, was "A Short Discourse Concerning Pestilential Contagion," which was initially published in 1720 and had appeared in eight editions by 1723. Plague is a contagion and Mead presented the conventional modes of transfer described by Fracastoro in the sixteenth century. Infection is transmitted through the air or contact with a diseased person or goods from infected places. For the air to acquire the ability to infect, there must be preconditions that render the air putrid. Heat, considerable rain, and southerly winds play a role in establishing these conditions. There are additional sources: "stagnating waters in hot weather, putrid exhalations from the earth" or from "corruption of dead carcasses lying unburied." There is nothing new here, nor did he make that claim. Mead did add, however, more contemporary reasons why the air is in an infectious state. He writes of putrefaction as a form of fermentation:

"All bodies in a Ferment . . . emit a volatile active spirit . . . In that way they can change the nature of other fluids which it enters".

RICHARD MEAD, *A Short Discourse Concerning Pestilential Contagion and the methods to be used to prevent it.* 3rd ed., 1720, p. 11, London

This "volatile active spirit" can cause fermentation of the blood that emits "active particles" that act on the body and can be transferred and act on nearby bodies to cause disease.

Thus contagion proceeds in the following manner:

. . "the blood in all Malignent Fevers, especially Pestilential ones, . . . does like Fermenting Liquors throw off a great Quantity of active Particles. . . . . . particularly upon. . . . the Mouth and Skin. . . . . . These, in Pestilential Cases. . . will generally infect those who are very near to the sick person". . . . (p. 12)

These theories are an amalgam of mechanical and chemical ideas, including the Newtonian version of particulate matter (corpuscular matter).

Mead turned to infections transported by goods, and in this section he dismissed living agents as the cause of disease:

"It has been thought so difficult to explain the manner of this, (how goods retain the seeds of contagion) that some authors have imagined Infection to be performed by the Means of Insects, the eggs of which may be conveyed from place to place . . . **As this is a supposition grounded upon no manner of Observation, so I think there is no need to have recourse to it** (my emphasis). If, as we have conjectured that the matter of contagion be an active substance . . . perhaps in the Nature of a Salt . . . generated chiefly from the Corruption of a human Body". [Emphasis added] (pp. 16–17)

This paragraph appeared in the 1720–22 early editions and, except for the last sentence, appeared in all editions, including the ninth, in 1744. In the 1744 edition there was an addition to the footnote alluding to a living agent that reveals that Mead was responding to a suggestion by Athanasius Kircher (1601–1680). This citation appeared 60 years after Kircher's death! In all these years Bradley was not included in the discussion.

In the Mead quote above, I have highlighted the statement that disparages the suggestion of a living agent as the cause of disease because no such "observation" has occurred. This statement is incompatible with Mead's conclusion 18 years previously in his report on an infectious disease called the Itch. During his time in Italy Mead obtained a report by a Dr. Bonomo, who sent a letter to Francisco Redi, ascribing the cause of the Itch to worms. Mead prepared a short, compelling report of this work that appeared in the *Philosophical Transactions* of the Royal Society in 1703 that did indeed ascribe the disease to a "worm." The worm hypothesis was based on the "observation" that every time this disease is present in an individual, irrespective of age or sex or time of the year, a living microscopic *animalcule*, the term consistently used by Leeuwenhoek and now used by Mead, is always present in the pustules that accompany the disease. Mead concluded,

. . . "that a Rational Account of the cause of this contagious disease is, these animalcules . . . and not the Humour of Galen . . . the corrosive acid of Sylvius . . . the Ferment of Van Helmont . . . or the Irritating Salts in the Serum or Lymph of the Moderns". (vol. 23, p. 1297)

Mead decided against the ancient cause, a humoral cause, or a chemical or physical cause. The cause is a living agent that, according to Mead, is transported from one person to another by inanimate objects (i.e., towels or sheets) in contact with persons who have the Itch. Mead used the constant presence of the animalcule when the disease was present to hypothesize that it is the causative agent rather than the result of the disease.

Some 18 years later Mead affirmed the infectious nature of the Itch as part of his discussion of the plague epidemic in France.

Thus, the Itch is a contagious disease, like the plague, and according to Mead's original report in 1702–1703 is caused by a microscopic living animalcule. Nevertheless, Mead abandoned this possibility when writing about the plague, and later, writing about smallpox and measles. Mead ignored his microscopic animalcule and the constant flow of letters from Leeuwenhoek that appeared in the *Philosophical Transactions* about animalcules. In the issue that included Mead's report of the Bonomo paper on the Itch, volume 23, there are three papers by Leeuwenhoek on animalcules in various environments. There is also an earlier paper demonstrating the presence of animalcules in the human mouth. Mead ignored them all; he decided that he had the proper explanation.

Mead compared the effect of snake venom and the plague "poison." He concluded that a poison, such as rattlesnake venom, works its effect in the same way as the plague poison. In the case of plague Mead was initially confounded because he recognizes that plague poison does not kill immediately as rattlesnake venom does. But he rationalized this real difference by claiming that there is at times *rapid* lethality—that is, death from plague occurs within a few hours after the appearance of infection. Although he was at a loss to explain this time disparity, he did offer a probable conjecture based on the writings of "our great philosopher" (referring to Newton's *Optics*, Queries 18–24) who provided the means to explain the phenomenon. Here is Mead's version invoking a particle theory of disease working on the fluid of nerves:

"It is commonly thought that the blood is affected by the morbid effluvia. But I am of the opinion, that there is another fluid in the Body, which is especially in the beginning, equally, if not more, concerned in this Affair: I mean the liquid of the Nerves usually called the Animal Spirits. As this is the immediate instrument of all Motion and Sensation, and has a great Agency in all the glandular Secretions, and in the circulation of the blood itself".

RICHARD MEAD, *The Medical Works of Richard Mead*,
1763, vol. 2, p. 43, Edinburgh

"If therefore we allow Effluvia or Exhalations from a corrupted mass of humours in a body that has the Plague to be volatile and fiery Particles, carrying with them the Qualities, of those fermenting juices from which they proceed; it will not be hard to conceive how these may, when received into the nervous Fluid of a sound person, excite in it such intestine Motions as may make it to partake of their own Properties, and become more unfit for the Purposes of the animal Oeconomy".

Richard Mead, p. 45, 1763

The hypothesized particles of plague, sometimes referred to as Contagious Atoms, and the particles of venom kill in the same way because they affect the fluid of nerves. It is of some interest to contrast Mead's position and that expressed by Robert Boyle, who recognized the problem of differentiating between a particle that is poisonous and one that is pestiferous: Poisons kill the persons they immediately invade, but do not infect others.

Boyle explained why they are different. He did not retreat from mechanical explanations for the effect of a poison or that of a disease-causing agent. However, to account for the difference between the outcomes he suggested that they are different entities. He hypothesized that there are a variety of minerals "and other bodies" present in subterranean regions that may produce various combinations in that fiery environment. In short, many different bodies can be produced from particles having different sizes, shapes, and motion. This is, of course, Boyle's basic corpuscular philosophy. Some combinations of corpuscles (atoms) become "pestiferous" and other combinations become poisons.

Mead, not an original theorist of disease, accepted the philosophical position that infectious disease and the physiology of the human body can only be explained in mathematical/physical terms. His commitment to this *a priori* belief was now so unshakeable that he ignored his own living agent theory of disease, since it had no place in hypotheses dominated by macro- and micro-mechanics.

He was not alone in taking this monolithic philosophical position, as reflected in a number of new treatises published in England during the short period 1720–22. Prominent among these books is Laurentius Bellini's *A Mechanical Account of Fevers*, representing the mechanical philosophy of the Italian School. The title and the context of the text are not surprising, with its unambiguous emphasis on mechanical philosophy and its axiomatic, geometric, way of presenting his argument. In 1720 it was translated into English and received an enthusiastic reception by the translator (not named on the title page), reflected in the preface to this edition:

"The translator states that the world is indebted to the Italians for their Advancement of the most substantial Philosophy, leading into the only Means of arriving to the Knowledge of Nature, by Experiments and Mechanical Reasonings thereupon.

And Malpighi, Redi Steno, Borelli . . . laid a sure Foundation in Anatomy . . . and Borelli in particular, by his application to Mechanicks, and the laws of Motion, taught him to account for the Powers of the Muscles. But it was this Scholar Bellini, the Author now before us, who first taught, upon the same Principles and Conduct, to reason demonstratively about the more minute and more unheeded Agency of the Animal Oeconomy".

LAURENTIUS BELLINI, A Mechanical Account of Fevers,
1720, Preface, London

Bellini developed a theory of fevers based on the mechanics of the circulatory system. It was laid out as a series of propositions or axioms. Proposition 1, "there is no fever without some fault in the blood", led to Proposition 3, "There is no fever without some fault in the Motion, or Quantity, or Quality of the blood, or in some or all of them together." The final summary, numbered Proposition 35, is that every fever "is a Fault of the blood in its Motion, Quantity, or Quality."

What causes the alteration of the blood, what is the cause of pestilence or plague, is not new. Bellini included "venomous exhalations from the earth upon Earthquakes or sinking Pits" containing morbid matter. The morbid matter alters the physical properties of the blood, to increase its heat, and that increases the pressure on the nerves, which translates to an increased pulse and an increase in fever. Mead in his Mechanical Account of Poisons enthusiastically supported Bellini's theory: ". . the great demonstrator has proven that pestilential fevers are owing to a . . . tenuous lentor or slime which obstructs the capillary arteries."

It is clear that the matter is not a living agent.

John Quincy in 1720 added an essay to a text written during the plague epidemic in England in 1665. He acknowledged that to discourse about "agents so extremely minute and subtle" is difficult, as is accounting for how these "Effluvia" can have such lethal consequences for populations. Quincy suggested that for more information about these issues readers should consult Bellini's *Fevers* text, which would aid in understanding the contraction of a pestilence, and Dr. Mead's Mechanical Account of Poisons.

Quincy made it clear what clarified his view on the cause of contagious disease, including plague. There are many natural causes of disease, but all have been

. . . "very obscure and perplexed until the present Age, when Sir Isaac Newton first taught me to think justly, and talk intelligibly about the Motions and

Influences of those remote "Bodies upon our Atmosphere. And upon his Theory Dr. Mead has since further proceeded to determine their Efficacies upon humane Bodies."

He defined specific causes such as "steams and exhalations from putre-fying bodies," from putrefaction of stagnant waters, and from "Mineral Eruptions and subterranean Exhalations." He also referred to *la Grotto de cani* near Naples, Italy, in a volcanic area, and urged readers to consult Dr. Mead, who had presented "a very particular and rational Account of this place, and the Manner of its killing." It is indeed very important to com-pare Mead's "rational Account" of the Grotto and Bradley's account. They have contrasting theories. These contrasting theories derive from different philosophical positions. We will discuss this significant disagreement when we engage the views of Richard Bradley below.

JohnQuincy was certainly aware of the possibility that living agents, microscopic agents, could cause disease. But rather than simply ignoring the possibility he disparaged those who propose a living agent theory of contagious disease:

"Yet either from the wantonesses of a light Imagination, and a false philoso-phy; or from a Vanity to be taken Notice of, the Publick has upon this sad Occasion been amused with the Figments and conceits of Naturalists, who from the Casualties and Distemperatures incident to Plants, and other inani-mate Productions of nature, have drawn conclusions to support very wild conjectures concerning a like Procedure in the Diseases peculiar to mankind".

JOHN QUINCY, *An essay on the different causes of pestilential diseases, and how they became contagious. With remarks upon the infection now in France*, 1721, The third edition, with large additions. London, p. iv, Preface.

This rebuke of "Naturalists" is harsh and puzzling. The one naturalist who had written on plant diseases and proposed "wild conjectures" about the cause of human diseases was Bradley. The other great plant naturalists, John Ray (1627–1705), Nehemiah Grew (1641–1712), Marcello Malpighi (1628–1694), and Stephen Hales (1677–1761), were adherents of the me-chanical philosophy. Grew had been Secretary of the Royal Society, suc-ceeding Henry Oldenburg, and was a Newtonian. None suggested that living agents caused contagious disease of humans. This was an attack on Richard Bradley. Therefore, it is time to consider the writings of Bradley on contagious disease to understand the reasons for his theory.

Richard Bradley: A Living Agent Is the Cause
of Contagious Disease

R. Hooykaas, an eminent historian of science, describes "practical men," which may be an accurate depiction of Richard Bradley:

> "Practical men are, in general, too pragmatic to let themselves be shut up in a system, all the more so if this system shows signs of the limitations of an outsider who, although he delineates a marvelous general program, does not know the whimsical tricks of nature from his own experience".
>
> R. HOOYKAAS, "The Rise of Modern Science: When and Why?"
> *British Journal for the History of Science*, 20 [1987], pp. 453–473,
> Cambridge University Press

In 1720 Bradley, a Fellow of the Royal Society since 1712 and later to become the first Professor of Botany at Cambridge University in 1724, wrote a short treatise titled *The Plague in Marseilles Consider'd* in which he proposed that a living agent was the cause of the disease. He had previously proposed such a theory for animals and plants in his treatise *New Improvements in Planting and Gardening*, which went through seven editions between 1718 and 1739. He arrived at his revolutionary hypothesis based on his extensive practical experience with plant diseases and his microscopic studies of infusions like those examined by Leeuwenhoek and Hooke and his broad comparative view of the biological world that ascribed similar properties to plants, animals, and humans. This biological theory that included a role for microscopic organisms was obviously in contrast to the view expressed by Richard Mead and an array of disease theorists in England and the continent.

Origins of Bradley's Theory

Bradley's theory originates from his work with plants. He claimed that his encounter with the living world through his study of plants provided him with a view of "Nature in all her Works" and convinced him that there is a "beautiful simplicity in nature." This assessment of the living world encouraged him to recognize similarities among plants and animals. For example, there are corresponding structures that contain fluids performing parallel functions:

> "The many curious Observations which have been made concerning the Structure of Animal Bodies, and what Malpigius (sic), Dr. Grew, and my self

have remark'd in the structure of Vegetables, may ascertain to us, that Life, whether it be Vegetable or Animal, must be maintain'd by a due Circulation and Distribution of Juices in the Bodies they are to support".

RICHARD BRADLEY, *New improvements of planting and gardening,*
both philosophical and practical, 1724, (4th edn), p. 2, London

Another important similarity for Bradley is that plants have a mode of generation analogous to that of animals. Julius Sachs (1832–1897), in his *History of Botany 1530–1860*, stated, that Bradley was probably the first to experiment on hermaphrodite flowers establishing the sexuality of plants. Bradley reported that it was Thomas Fairchild who first produced hybrid carnations:

"The Carnation and Sweet William are in some respects alike, the Farina (pollen) of one will impregnate the other, and the seed so enlivn'd will produce a plant differing from either, as may be seen in the Garden of Mr. Thomas Fairchild, a Plant neither Sweet William nor Carnation, but resembling both equally, which was raised from the seed of a Carnation that had been impregnated by the Farina of Sweet William".

RICHARD BRADLEY, *New improvements of planting and gardening,*
both philosophical and practical, 1726, 5th ed., p. 18, London

Bradley did carry out controlled experiments with a variety of plants to prove that pollen is necessary for the production of seeds. These similarities between plants and animals influenced Bradley's discourse about contagious disease of humans. Here is how the logic works, as contained in his treatise of 1717. Plants and humans, and other animals, have similar structures, reproduce in similar ways, and are afflicted with contagious diseases. Contagious diseases of plants are caused by living, visible, and microscopic "insects." Therefore, contagious diseases of humans are caused by living, visible, and microscopic agents.

The essential, experimental evidence that implicates a microscopic, living agent as the cause of contagious disease is contained in a three-page report on the fortuitous appearance of molds on a melon, published in 1714. Bradley's initial goal was to study the structure of "the Vessels which composed the Membrane of each Ovary of a Melon-Fruit." He opened the fruit but did not immediately examine the internal structure. When he returned four days later he discovered "several spots of Moldiness" which were green near the rind and "a paler Color towards the Middle of the Fruit." He did not discard the melon. Over the next five days the whole fruit was covered with these molds. Bradley was "curious" whether there were differences, apart from color, in the various regions of the fruit. He observed the material microscopically and found that there were two morphological types. One fungus developed stalks and both types produced numerous seeds (spores). His observations over time convinced him that these "plants . . . began to vegetate;" the spores had spread over the fruit and in time the melon was entirely covered with growth. After a week the fruit was completely "putrified" and only "stinking water" remained. In a number of days maggots appeared, which developed into flies. After about four weeks there remained only fruit seeds, some vessels that were part of the structure of the membrane, the fruit rind, and the excrement of the maggots. Bradley found that all of this material was one-twentieth the weight of the original fruit. He judged from these observations and others "of the like nature" (see the discussion on "vermin" in *The Gentleman and Gardeners Kalendar* below) that "Vegetable Life is dependent on Fermentation and Animal Life on Putrefaction." By a process of fermentation the material of the fruit has been partially converted to the materials of the molds. Bradley ascribed the destruction of the melon to the activity of the mold, and not that the mold was generated from the destroyed melon tissue.

His interest in microscopic creatures continued in a subsequent publication, *The Gentleman and Gardeners Kalendar*. This work was a practical guide to gardening procedures; for example, when to sow cucumbers with the appropriate fertilizer and when to plant lettuce, carrots, turnips, and onions, but it also contained procedures to "guard" against wasps, ants, other insects, and earthworms. In addition he warned against a class of microscopic agents "classified as vermin. They have the unique ability to propagate to large numbers":

"I the more particularly recommend the Destruction of those devouring Vermin, because of their wonderful increase, especially the smaller kinds of them, such as infect the Collyflower, whose eggs are five hundred times less than the least visible Grain of sand, and as many thousands of them Number lay'd (as I believe) from one single insect; so that from a second generation of them they are so numerous, that if every Egg which one might find on a blighted Collyflower, was a Globe of an inch Diameter, they would fill more Space than the whole Territorial Globe. The wonderful smallness of the Eggs of those Creatures, I confess, may seem strange to those who are unacquainted with the use of microscopes; but this is no more curious than what the late famous Dr. Hook [Robert Hooke] has taught us in his Works, where he mentions the Seeds of Moss, which were so small, that ninety thousand of them being laid together in a strait (sic) line, did not exceed the Length of a Barley Corn, and that 1,382,400,000 of them would weigh only a Grain. But see more of these minute Beings in Mr. Hook's [Robert Hooke] *Micrographia*, Mr Lewenhock's [Antony van Leeuwenhoek] Works in the *Philosophical Transactions*".

RICHARD BRADLEY, New Improvements of Planting and Gardening,
Both Philosophical and Practical, 1720–1, p. 23, Dublin

There was no ambiguity here: he had uniquely taken full advantage of the findings of Hooke and Leeuwenhoek. Bradley stated that there are microscopic living agents and they "infect the Collyflower." Again, it was clear to Bradley that these agents were not the consequence of some disease process but were the initiators of it.

The melon and Collyflower experiments were two of the earliest empirically based arguments for a biological cause of a plant disease. However, this was only the beginning. A more detailed theory of the cause of contagious diseases of plants was presented in Chapter V, Of Blights, in New Improvements in Planting and Gardening in 1718. In the opening paragraph Bradley stated that the intent was to explain the cause of contagious diseases of plants and the benefits that could be obtained from this knowledge:

"As a Blight is the most common and Dangerous Distemper that Plants are subject to, so I shall endeavor to explain by what means Vegetables are affected by it; and if I shall be so happy from the Observations I have made, to discover the Cause of it, the Remedy may be then more easily found out, and the Gardener will with more certainty hope for Success from his Care and Labour".

RICHARD BRADLEY, New Improvements of Planting and Gardening, Both
Philosophical and Practical, 1718, p. 53, London

Bradley offered more than just practical knowledge; he offered something new. (In New Improvements of Planting and Gardening, Both Philosophical and Practical, 1720, p. 34, London)

It becomes obvious what has been "slightly touch'd upon" as we enter the main body of this chapter. A living agent is the cause of disease. Thus, four years after the publication of the melon paper, and less than a year after the *Gardeners Kalendar* work, Bradley was confident and willing to present in a lengthy form, 47 pages of evidence and reasoning, that contagious diseases of plants, animals, and humans were caused by living agents.

The Causes of Disease

There was the popular conception that winds brought on diseases of plants and trees and that there was something borne by the wind. Bradley contended it was a specific biological entity and not a universally destructive inanimate agent because there is a high degree of specificity in the way diseases strike plants. If the air carried some universally contagious entity, every plant that was exposed (and every plant would indeed be exposed) would be infected. Instead, different plants are infected or different parts of the same plant contain disease. The silkworm destroys the leaves of mulberry trees. The bark and wood of many trees are destroyed by several kinds of beetles, many "large enough to be discovered without a microscope." From these observations he concludes "that tis insects which . . . infect Trees . . . that every insect has its own proper Plant or Tribe of Plants."

Why does this happen? Each insect has its own niche,

"which it naturally requires for nourishment, and will feed upon no other. Therefore it is no wonder to see one particular sort of Tree blighted when all others escape."

Richard Bradley, 1718, pp. 55–56, London

Bradley focused initially on visible insects such as caterpillars and then went on to discuss smaller and smaller creatures and eventually arrived at the findings of "Mr. Lewenhoeck" (Leeuwenhoek). He employed Leeuwenhoek and Hooke in two stages in his argument. He reiterated his contention that every herb has its peculiar insect and stated that there are examples of insects that have other organisms that feed upon them. This leads him to conjecture that

. . . "since there are insects which prey upon others, there may be others of lesser Rank to feed upon them like wise, and so to infinity; for that there are Beings subsisting which are not commonly visible, may be easily demonstrated, if there is any truth in a microscope".

Richard Bradley, p. 61, 1718, London

Bradley was speculating but believed he had good reason to do so since he, like Leeuwenhoek and Hooke, had seen these infinitely small animalcules:

"How trifling an object was the Insect I have mentioned in comparison to those discover'd by Mr. Lewenhoeck, in a Quantity of Pepper-Water no bigger than a Grain of Millet, in which he affirms to have seen above 10000 living Creatures, and some . . . witness to . . . 30000, and others above 45000 . . . Now from the greatest of Numbers mention'd, it is inferr'd that in a full Drop of Water there will be 8280000 of these Animalcula, which if their Smallness comes to be compar'd, a Grain of Sand broken into 8,000,000 of equal Parts, one of those Parts would not exceed the Bigness of one of these Insects. These Observations of Mr. Lewenhoeck were not only confim'd by the Famous Mr. Hook, but were likewise improv'd by him: He tell us, that after he had discover'd vast Multitudes of those Animalcula describ'd by Mr. Lewenhoeck, use of other Lights and Glasses, and magnified them to a very considerable Bigness, and that among them he discover'd many other Sorts much smaller than those he first saw, some of which were so very minute, that Millions of Millions of them might be contain'd in one Drop of Water (p. 64) . . . This account may suffice to shew us that there are Beings . . . which are not commonly visible, and therefore it is easy to conceive, that the most gentle Air is capable of blowing them from place to place, and so cause disease". (p. 66)

Richard Bradley, New Improvements in Planting and Gardening, 1718, London

It is clear that Bradley used the word *insect* to describe microscopic agents, many times referred to as *animalcula*. They are certainly not insects like caterpillars or beetles. There were, at this time, no specific names for these microscopic agents. They were hardly differentiated from each other. They were collections of bacteria, protozoa, fungi, and microscopic algae. Bradley's "A Philosophical Account of the Works of Nature" contained an unambiguous presence of microscopic agents such as bacteria, and fungi, creatures referred to as *insects* in previous descriptions:

"In the Lees of Wine, and upon the Outsides of Wine-casks, we find great varieties of living Creatures which are proper subjects for the microscope; and even the Mosses or Mouldiness found in such damp Places as Cellars, and especially where wine is kept, gives us very entertaining Prospects. Vinegar and Pepper-water like wise afford us Abundance of Variety".

Richard Bradley, A Philosophical Account of the Works of Nature, p. 213, 1739, London

Obviously, these living agents were the microscopic organisms first seen by Leeuwenhoek and Hooke. They were the microorganisms present in the pepper-water preparations that Bradley was attempting to grow.

Once again Bradley made the connection between plant diseases and human diseases:

> "I am apt to believe that the most epidemical Distempers of mankind is subject to proceed from poisonous insects which are eaten unregarded, or are suck'd into the stomach with the Breath, as the worthy Gentleman Mr. Balle so curiously observes, in a Letter I have receiv'd from him relating to Infectious Distempers; which I shall annex to this chapter, as it contains many observations, which may help to explain and confirm what I shall offer concerning Blights".
>
> Richard Bradley, New Improvements of Planting and Gardening, Both Philosophical and Practical, 1724, p .245, London

Bradley included at the end of Chapter V a letter from Robert Balle (ca. 1640–ca. 1734), since 1710 a Fellow of the Royal Society. He was one of a wide circle of natural historians, botanists, and garden enthusiasts that included Sir Hans Sloane (1660–1753) and James Petiver (ca. 1665–1718), who were colleagues of Bradley. Balle owned the estate at Campden House where Bradley carried out breeding experiments. He supported Bradley's biological theory of contagious disease applied to humans. Balle wrote,

> "Upon discoursing with you some time since about Blights upon Trees, you seem'd to be of the Opinion that they were the Effect of Insects brought in vast Armies by Easterly Winds, and by lodging upon the Plants proper for their Nourishment they there produc'd that Distemper which is call'd a Blight or Blast . . . You was then desirous of what Observations I had made concerning Pestilential Distempers subject to Mankind, which I believe to proceed from the same cause that produced Blights".
>
> Richard Bradley, New Improvements of Planting and Gardening, 1724, p. 254, London

Bradley and Balle returned to the generally accepted idea that air brings on the plague:

> "It is the common receiv'd Opinion, that the Plague proceeds from an Infection in the Air, and so undoubtedly it does".
>
> Richard Bradley, 1724, p. 255, London

What is in the air? One answer was provided by the corpuscularian version of the mechanical philosophy:

"Some will have it from Steams or Vapours arising from some poisonous Minerals at certain bad Seasons, which being by us taken into our bodies with the Air, ferments and corrupts the Humours of the Blood".

Richard Bradley, p. 255, 1724, London

Bradley and Balle clearly did not believe in such a cause. Consequently their goal was to differentiate between a theory of disease caused by noxious vapors, steams, or damps, which impregnate the air with an inanimate object, and a theory that contends that something living in the air is the cause. To support their contention that air, exhalations, or vapors do not intrinsically cause contagious diseases they described a location where deadly vapors are emitted that do not cause "pestilential distempers," that are not contagious. This location is a grotto near Naples.

The Grotto

Introducing this natural event, the Grotto with its toxic fumes, provides us with a unique opportunity to compare the contagious disease theories of Mead and Bradley because they were based on the same set of objective, empirical observations. Therefore, their contrasting interpretations cannot be accounted for by their use of selected data or different data, but from the way they interpreted the same evidence contained in Mead's account of the Grotto in the "Mechanical Account of Poisons" and Bradley's account of the Grotto included in Chapter V of the New Improvements book.

Mead's philosophical position was stated unequivocally:

"My design, in thinking of these matters, was to try how far I could carry mechanical considerations in accounting for those surprising changes, which poisons make in the animal body".

Mead would search for causes that "are within the Reach of Geometrical Reasoning." That brought him to the Grotto. He asserted that there were

"Venomous Steams and Damps from the Earth the Latins call'd Mephites. I shall therefore, having had the Opportunity of making some Remarks

upon the most famous of all those Parts, give as good account as I can of
that, and its manner of killing. This celebrated Moseta . . . , is about two
Miles distant from Naples, just by the Lago d'Agnano . . . and is commonly
call'd *la Grotta de Cani*".

RICHARD MEAD, Mechanical Account of Poisons, 1708, p. 159, London

Bradley agreed that there are deadly vapors that come from the Grotto
del Cane near Lake d'Averna, adjacent to Naples. Mead reported that a
dog, an ass, and two slaves with their heads in the fume were killed. Bradley attested that nothing burned in this steam, nor did any creature either
of land or water live in it.

Mead continued,

> . . . "from the Ground arises a thin subtle, warm Fume . . . covering the
> whole surface of the bottom of the Cave; and has this remarkable difference
> from common vapours, that it does not, like Smoak, disperse it self into the
> Air, (it rises about ten inches above the surface) but quickly after it rises
> falls back again, and returns to the Earth".

In Bradley's description the steam sits on top of the surface from which
it arises, one or two feet above it. The composition of the steam in this volcanic region is obvious. It consists of two gases, carbon dioxide and hydrogen sulfide, both more dense than air. They do not disperse like smoke
from a fire; they sit at the surface of the water of the grotto, displacing air
(containing oxygen). Any human or animal immersed in this vapor would
be killed almost instantaneously because of the combination of sulfide
with the iron of blood hemoglobin, depriving the body of oxygen. A fire
could not burn for lack of oxygen.

Mead added to the description of the environment of the grotto:

> . . . "the color of the sides of the Grotto being the measure of its ascent [re-
> ferring to the lethal fume]; for so far it is of a darkish green but higher only
> common earth". (p. 160)

In Bradley's description the surface of the water is covered with a green
scum. We know, because of studies in the twentieth century, that the green
coloring is due to the growth of green photosynthetic bacteria that can live
in the absence of oxygen and thrive in an atmosphere of carbon dioxide
and hydrogen sulfide. These bacteria cover the sides of the grotto and constitute the scum on the surface of the water.

These were the identical data about the grotto contained in Mead and Bradley's descriptions. Here is Mead's analysis. There is noxious air whose "elasticity" has been abolished, which replaces normal air and inhibits the normal function of the blood. Elasticity is a physical property of normal air but not further defined. The altered physical properties of the air lead to epidemics:

"And thus I have shewn how Death may enter the nostrils, tho' nothing properly Venomous be inspir'd. It were perhaps no difficult Matter to make it appear, how a lesser Degree of this Mischief may produce effects, tho' seemingly very different from these now mention'd, yet in reality of the same pernicious Nature; I mean, how such an alteration of the common Air as renders it in a manner Mephitical, that is increases its Gravity, and lessens its Elasticity, (which is done by too much Heat, and at the same time too great a Proportion of watery and other grosser Particles mixt with it) may be the Cause of Epidemick Diseases".

RICHARD MEAD, *Mechanical Account of Poisons*, 1708, p. 166, London

Mead used the example of the grotto to declare that the mechanism of killing by lethal fumes is equivalent to the mode of killing by pestilential infections (contagions). Mead stated in the first edition of the *Mechanical Account of Poisons* in 1702 that it is possible to create experimentally lethal air from a mixture of various chemicals without its being contagious. Although Mead appeared to acknowledge that noxious air and contagious air are different, he contended that they can be of the same "Pernicious Nature." Therefore, what causes air to be "Mephitical" (referring to a noxious or pestilential vapor issuing from the earth), which "increases its Gravity, and lessens its Elasticity, . . . may be the Cause of Epidemick Diseases."

Bradley rejected this equivalency. He stated that there are indeed many environments that produce vapors that are lethal and certainly do not cause disease. They include emanations from the grotto, as well as from mines, and from conservatories where corn is stored and putrefying while an intense fermentation is taking place. These vapors are different from those that cause the plague or other "pestilential distempers" for two very good reasons. First, the vapors cause death very quickly, in a matter of minutes, while even the most rapidly acting pestilential infections take a day or two to kill a victim. And secondly, people who die from inhalation of these deadly vapors are not contagious. This is similar to the argument made by Robert Boyle (Chapter 8). Consequently, for Bradley there must be a different element present in contagious vapors that is not present in the noxious air emitted from a place like the grotto. He proposed that the element of contagion is a living agent of extraordinary small size that can be carried by the wind and taken in with the breath.

Bradley and the Plague

At the time that Bradley wrote "The Plague at Marseilles Consider'd" in 1721, he had already committed himself to the theory that all contagious diseases are caused by living, microscopic organisms based on his studies of plant diseases. The plague treatise consisted almost entirely of information pertaining to cattle and plant diseases, largely excerpted from his "New Improvements in Planting and Gardening" book of 1718. The treatise was dedicated to Sir Isaac Newton but did not contain a trace of the mechanical philosophy, although Bradley viewed his theory of disease as conforming to the "rules in nature." He addressed the short treatise to Sir Isaac Newton, President of the Royal Society:

> "Sir, To act under Your Influence, is to do Good, and to Study the laws of Nature, is the Obligation I owe to the Royal Society, who have so wisely placed Sir Isaac Newton at their Head. The following Piece therefore, as I design it for the Publick Good, Naturally claims Your Patronage, and, as it depends chiefly upon Rules in Nature, I am doubly obliged to offer it to the President of that Learned Assembly, whose Institution was for the Improvement of Natural Knowledge. (Following title page)

Bradley made the claim that his concept of disease was an example of a "rule" or "law." Plague, by the measure of this treatise, was a special case of a general class of contagious diseases of humans, plants, and animals. All contagious diseases are caused by living microscopic agents. This is a rule of nature.

In dedicating this work to Sir Isaac Newton it was, perhaps, Bradley's attempt to receive public recognition from Newton and thus gain some renown in the medical establishment of England. If this was his goal, it was a great failure. Nevertheless, Bradley was not unknown. His book of the same year, "A Philosophical Account of the Works of Nature," contained a list of subscribers. The book was in the hands of three booksellers in London. Newton obtained six copies. There were seven medical doctors, including Hans Sloane, Secretary of the Royal Society, and a prodigious collector whose materials became the basis for the British Museum; a Professor of Experimental Philosophy and Astronomy at Cambridge; a Professor of Botany at Pisa, Italy; and Robert Balle, also an Fellow of the Royal Society, a long-time friend of Bradley. In total, there were more than 300 subscribers. Bradley was, of course, widely known as a distinguished contributor to the gardening and horticulture community in Britain. He knew everyone in this group, which is amply documented by Blanche Henry in "British Botanical and Horticultural Literature Before 1800" and by U. R. Roberts, "Bradley, Pioneer Garden Journalist," published in the Journal of the Royal Horticultural Society of London. Nevertheless, Bradley and his theory disappeared from the medical literature.

The Plague Treatise

The preface of the plague treatise offered the hope that the disease would not cross the Channel and at the same time stated unequivocally that the disease is conveyed by a living agent:

> "There is room enough to hope, the approaching cold, which we naturally expect at this Season, may prevent its spreading amongst us for some Months, 'til the Air begins to warm, but the seeds of that Venom maybe brought over in Merchandizes even in the coldest months, and according to the Nature of Insects will not hatch, or appear to our Prejudice, 'till the hotter Seasons. For to suppose this Malignant Distemper is occasion'd by Vapours only arising from the Earth, is to lay aside our Reason, as I think I have already shewn in my New Improvements of Planting, etc. to which my reader may refer".
>
> (Preface) Richard Bradley, The Plague at Marseilles, 1721, London

Apart from this reference there is very little in this work on the cause of plague except near the end of the text. There, Bradley related comments from reports of the 1665 epidemic about the "signs of distemper and Methods of cure." As a final comment he related that the claim was made that plague is caused by "corrupted and unwholesome air;" however, Bradley objected, for if this were the cause "all mens bodies would be so infected."

In the body of the text Bradley discussed diseases of cattle from the years 1682 to 1714 in Switzerland, Poland, Padua, and London and diseases of plants in New England, Russia, and Ukraine, using references to his chapter on *Blights*, to drive home his basic philosophical principle that all contagious diseases arise from a common cause. For example, he had learned of an infectious episode among cows in the year 1714 around London. Mortality was high that summer, and later in that same summer fresh cattle were brought in to graze in the same field where death occurred. These new cattle died. The next summer new cows feeding in that same field did not die. However, cows placed in sheds where sick cows had been housed the year before were "seiz'd with distemper and died." This is how Bradley explained these events. The disease is caused by "insects"—that is, a living agent. These agents remained in the open fields during the summer, and when a fresh group of animals were introduced they succumbed to the disease. The next summer no cattle died in the open field since over the winter these disease-causing agents were destroyed by the cold. On the contrary, the cowsheds continued to contain the disease agents since the alleged living agents survived the winter in the "warmth ....of the sheds."

It is instructive to compare Bradley's speculations about the cause of this disease of cattle and an analysis of the cause of the same disease (most probably) reported in 1714 by Thomas Bates, surgeon to His Majesty's Household, who was retained by the Lord Justices to verify the contagious nature of the disease. Bates, in consultations with veterinarians, described the disease as a murrain and speculated that it was contagious. From his many dissections of deceased animals Bates concluded that the "sole" cause of the disease is a "want of Natural Purgation; that obstruction terminates in putrefaction. . . . Putrefaction apparently leads to "Infectious Effluvia." These effluvia could pass from animal to animal of the same species but seldom affected different animals. They might also adhere to fabrics and in that way could be spread to other bodies far removed from the source. The disease resulted in a mechanical obstruction, but no indication was given what constituted the "Infectious Effluvia," and certainly nothing was mentioned to suggest that a living agent was present in these effluvia.

There is one ambiguous discussion about the cause of a disease of cattle published in 1714 in Italy by Cogrossi. The equivocation becomes clear as we read the author's conclusions. Initially, the disease of oxen is ascribed to a microscopic, living agent. These "slender insects" multiply by gaining nourishment from the host. The author noted that a "famous Dutchman" (obviously Leeuwenhoek) had demonstrated the presence of such life. However, in his conclusion he was reluctant to discard the possibility of a corpuscular (physical particle) theory of contagious disease according to Robert Boyle:

"Notwithstanding all this I am quieted in this matter by the spirit of Robert Boyle. He . . . did much for experimental philosophy, and from his most noble treatise "On the Atmospheres of Solid Bodies, the Strange Subtilty and Great Efficacy of Effluviums" I can understand fully that from these can proceed readily all the phenomena of a contagious disease".

c. f cogrossi, Nuova idea del male contagioso de'buoi (New theory of the contagious disease among oxen). Milan. Facsimile, with English translation by Dorothy M. Schullian 1953, p. 33, Rome (6th Internat. Congr. Microbiol).

Bradley finished his comparative discussions of contagious disease in the plague tract with the recommendation that if the reader wishes a "larger Account of Animals and Plants, how they have been particularly Infected," they should refer to his *New Improvements of Planting and Gardening.* In closing Bradley presented two unambiguous summary statements of his contagious disease theory that contain nothing specific about plague but presume that it is one example of a general phenomenon, namely that all contagious diseases are caused by living agents:

"By the foregoing Accounts we may observe, that Mankind, Quadrupeds and Plants seem to be infected in the same manner, by unwholesome Insects; only allowing this difference, that the same Insect which is poisonous to Man, is not so to other Animals and Plants.

All Pestilential distempers, whether in Animals or Plants, are occasion'd by poisonous Insects convey'd from Place to Place by the Air".

richard bradley, The Plague in Marseilles Consider'd, 1721, p. 57, London

The Promise and Disappointment of Microscopic Studies

Bradley's microscopic observation of animalcules and his reliance on the visual evidence of Hooke and Leeuwenhoek were indispensable elements of his living agent theory of contagious disease. The influence of microscopic studies was great in the latter half of the seventeenth century, characterized as the *Classical period*, which included the pioneer microscopists Hooke, Grew, Malpighi, Leeuwenhoek, and Swammerdam. There was an enthusiastic reception of microscopic investigations by the philosophical community, conveyed by Hooke in the preface to *Micrographia* in 1665:

"Microscopic studies will reveal the subtil . . . the composition of Bodies, the structure of their parts, the various texture of their matter, the instruments and manner of their inward motions".

There is implicit in this grand vision that the microscope will reveal the particle basis—in Boyle's term, the corpuscular basis—of living matter. The reviewer of *Micrographia* in the *Philosophical Transactions* of volume 1 in 1665 is amazed by

> . . . "a new visible world . . . so that in every little particle of matter, we may now behold almost as great a variety of creatures, as we were able before to reckon up in the whole Universe it self".

This optimism was echoed by Joseph Glanville (1636–1680), a strong supporter of the Oxford research group of Robert Boyle, who wrote that the microscope would reveal "the minutes and subtilties of things." Henry Power (1623–1668), a physician who wrote on microscopy and the particle (corpuscularian) theory of matter, declared that the microscope would enable us to see "what the illustrious wits of the atomical and Corpuscularian Philosophers durst but imagine." Nicolas Malebranche (1638–1715), a Cartesian philosopher who discoursed on metaphysics, theology, and physics, was amazed at what was revealed with magnifying glasses: "The imagination boggles at the sight of such extreme smallness." Bernard Fontanelle (1657–1757), also a Cartesian philosopher, appeared convinced by the studies of Leeuwenhoek to conclude that the magnitude of numbers of microscopic animals must be as great as the visible animals. However, in the eighteenth century the use of the microscope declined as a tool of investigation among the mechanical philosophers because it failed to reveal the working of atoms. It survived as a tool of investigation among a disparate number of individuals in England and on the Continent, Bradley and others, who discovered and were fascinated by a world of infinitely small living organisms whose origins were in doubt and whose role in the world of living things was completely unknown. Nevertheless, among a small number of these individuals a microscopic agent became the cause of contagious diseases. To arrive at this theory Bradley had to adopt a number of principles about microscopic life. They are living organisms that can multiply under the proper conditions. These *insects* or *animalcules* are present in many environments that appear to be their natural habitat. These environments provide nutrients that are essential for their generation. When they occupy an animal, plant or human, they may cause disease. They are not spontaneously generated.

Bradley appeared immune to the mechanical philosophy, although he was well aware of the mechanical account of disease and the hostility to his theory, explicitly conceded in his short publication of 1722: Precautions against infection; containing many observations necessary to be consider'd, at this time, on account of the dreadful plague in France, . . . In this work he referred to

> . . . "a Treatise Concerning Contagion, Publish'd by a Learned Member of the College of Physicians, wherein the Author espouses the Other Opinion, Viz. That infection is communicated from one Person to another, by means of Vitiated Air; and that there are no Insects in the case". (p. 2)

The "Learned Member" was probably Mead. Bradley found his views, and others, not surprising since

> . . . "their Discourse is grounded only upon such subtil Bodies as the minute Parts of Air; or, which are call'd in other Terms, Aereal Attoms (sic)". (pp. 2–3)

Support for a Living Agent Theory of Disease

There were other individuals struck by the presence of microscopic organisms who speculated that they caused disease. Among them was Benjamin Marten a physician in London who wrote a treatise on the cause of tuberculosis, and Jean-Baptiste Goiffon in France, who ascribed the cause of plague to a living, microscopic organism.

In 1720 Marten published "A New Theory of Consumption: More Especially of Phthisis, or Consumption of the Lungs". What is clear from the text is that he was knowledgeable about Hippocratic and Galenic medicine and current contagious disease theory. He referred to Paracelsus, Van Helmont, Sylvius, and Willis from the sixteenth and seventeenth century and contemporaries like Leeuwenhoek, among others working in the early eighteenth century. At the heart of Marten's theory is that an animalcule, a living microscopic agent, is the cause of consumption. He discounted explanations of disease based on humors and ferments and instead suggested certain species of animalcula that may come from the air and enter the lungs to cause disease:

> "This opinion of Animalcula or exceeding minute animals, that are inimicable to our Nature, being the Cause of a Consumption of the Lungs, will doubtless seem strange to abundance of Persons, and more especially to those, who have no idea of any living creatures besides what are conspicuous to the bare Eye; but the Curious who have not only employ'd themselves in the

Philosophical Studys of the vast Machinerie of the Universe, but have also turn'd their Thoughts Upon the Admirable Works of Omnipotence in the Minima Natura, and have consequently considered the new World of Wonders, that Microscopical Observations have opened to our view, will easily conceive the Possibility of very minute Animals being not only the Original and Essential Cause of this, but of many other diseases hitherto inexplicable; and that they are perhaps, the very Malignity so much complain'd of in many Distempers, but so little understood".

BENJAMIN MARTEN, *A new theory of consumptions: more especially of a phthisis, or consumption of the lungs*, 1720, p. 52, London

He carried out his own microscopic studies and found himself, like Leeuwenhoek, amazed by the "infinity of animals, in all parts of the natural world." There were, he concluded, many kinds of living creatures of various sizes and shapes, and he speculated that there was "no such thing as Equivocal Generation" (that is, spontaneous generation) since "the learned world now all agree; every . . . minute . . . creature must be produced from an ovum or egg." It is possible that these animalcula have affinity for different parts of the body, and thus the agent that causes consumption is specific for the lungs. Because they are living creatures, it is possible that they can develop into vast numbers. Since they are of microscopic size, they could pass easily from person to person in the air or across larger distances on various objects. In that way they could cause epidemics.

The restrained character of Marten's writing and his references to contemporary medical practitioners can be viewed as an appeal to the medical establishment to consider seriously his theory of infectious disease. It is expressed in this way:

"But I would not be understood to advance that all Distempers are caused by them [animalcules]: I would urge only the possibility and likelihood of their being the Essential cause of the Plague, Pestilential and other malign Fevers, Small Pox, and some other diseases . . . and particularly of that dreadful one, a Phthisis, which is the peculiar subject of these Papers". (p. 76)

Jean-Baptiste Goiffon

Jean-Baptiste Goiffon was a physician who served in the French and Italian armies as a doctor. He returned to his native Lyons, where he practiced medicine. During his years in Lyons there was an outbreak of plague in

Marseille, some 240 km south. This is the disease outbreak that alerted the authorities in England to engage Richard Mead to report on the disease. There was considerable concern that the plague would come to Lyons, and therefore quarantine was established. Goiffon had studied how the plague was spread as well as its effects on victims. He made a number of interesting arguments and observations that have the characteristics of an epidemiological survey. He concluded that the agent that caused the plague "is something which has life." He wrote:

> . . . "it is impossible to explain the effects of the plague and especially its mode of spread and its recurrence by assuming the cause to be inanimate. One must of necessity suppose some small invisible insect, which by successive multiplication and reproduction continues to produce an agent as powerful as that from which it arose. Such a property belongs only to animate things. If an inanimate substance, as a grain of salt, for example, were divided a thousand times, no one would believe that each could be as powerful as the original grain. Ordinarily the initial cause of the plague is small and confined, in order to spread throughout a town, province or a realm, through the air or by other common medium, it must multiply. If a subdivision of its substance occurred as one must postulate for the ferments (LeVains) and all inanimate causes, instead of its effects increasing they would diminish in proportion to the subdivision, but far from diminishing the number of persons affected by the plague increases. This seems to me certain proof that the cause of the plague cannot consist of saline atoms, nor of ferments nor any other substance of like nature, Only something which has life, which can multiply throughout successive generations and constantly be renewed can be considered as a cause".
>
> R. WILLIAMSON, *Annals of Science*, 1955, 2:44–57; citation, pp. 55–56

Goiffon's writing, like all work in this field, is a theoretical argument based on the characteristics of the disease process, which he experienced as a physician, the cause of which, he alleged, could be explained only by postulating that the disease agent is a living thing.

Bradley's Theoretical Position: Summary

There are rules that apply to biological systems, a uniform law for the natural world of living things. Plant and animal systems are intrinsically the same structurally and functionally. They reproduce in similar ways.

Reproduction is the property of a living being and not that of a machine. All plant diseases are caused by living agents, visible and microscopic. It follows, by analogy, that all animal diseases have the same cause. Bradley and Mead had different, incompatible causes of contagious disease. Proof for either theory was impossible to obtain at that period in the eighteenth century. Although neither cause could be justified, Bradley's theory and the person of Bradley disappeared from the discourse about contagious diseases. Marten and Goiffon are equally missing. Mead's writings were reproduced into the last half of the eighteenth century. Obviously, in competition with a physical theory of the cause of contagious disease, a living agent theory had no chance.

CHAPTER 11 | Plant Diseases Are Caused by Living Microscopic Cells (Fungi) That Are Not Spontaneously Generated

[On] August 16, 1845, The Gardeners' Chronicle and Horticultural Gazette described "a blight of unusual character" in the Isle of Wight . . . without undue alarm. But a week later on August 23 [the editor] Dr. John Lindley was telling his readers in consternation. A fearful malady has broken out among the potato crop. In Belgium the fields are said to be completely desolated . . . As for cure of this distemper, there is none . . . We are visited by a great calamity which we must bear.

About one month later Lindley delayed the publication of the *Chronicle* to announce on September 13, 1845:

"We stop the Press with very great regret to announce that the potato Murrain has unequivocally declared itself in Ireland. The crops about Dublin are perishing . . . where will Ireland be in the event of a universal potato rot"?

CECIL WOODHAM-SMITH, The Great Hunger: Ireland 1845–1849,
1962, pp. 39–40, Hamish Hamilton

Between January 7, 1854, and October 3, 1857, Miles Joseph Berkeley, the recognized British authority on all matters connected to mycology [the study of fungi] . . . nothing concerning fungi seemed to escape his net, and once President of the Biological Section of the British Association,

wrote a series of 173 articles in the *Gardeners' Chronicle and Agricultural Gazette* on vegetable pathology. At mid-nineteenth century, when there was a vigorous debate about the cause of human contagious disease and plant diseases, and an equally contentious argument about spontaneous generation, Berkeley wrote an unambiguous statement that fungi, which are living microorganisms, cause plant diseases and are not spontaneously generated from diseased tissue since they thrive on healthy plants.

Berkeley wrote that there are cases

"In which the parasite makes its first growth entirely within the substance of the plant, and only makes its appearance externally after it has become deeply incorporated with its tissues. The more or less complete destruction at least of the part on which such parasites are developed is always consequent on their presence, and some of the most important diseases of vegetables depend on their growth. These belong almost entirely to fungi, which are capable of growing in plants which were in perfect health previous to being attacked. This fact has frequently been denied, but there is not the slightest foundation for the notion that fungi cannot grow on healthy tissues. The presence of the spores may indeed at once produce disease, in consequence of which the fungi when developed may flourish the more vigorously: but if so, the disease is still produced in the first instance by the spores".

Berkeley could not have been more definitive. Nevertheless, the issue remained controversial and was not resolved for 20 years.

In March 1839, in Berlin, Theodore Schwann wrote, in the preface to his classic text, *Microscopical Researches into the Accordance in the Structure and Growth of Animals and Plants,*

"It is one of the essential advantages of the present age, that the bond of union connecting the different branches of natural science is daily becoming more intimate, and it is to the contributions which they reciprocally afford each other that we are indebted for a great portion of the progress which the physical sciences have lately made. This circumstance therefore renders it so much the more remarkable, that, notwithstanding the many efforts of distinguished men, the anatomy and physiology of animals and plants should remain almost isolated, though advancing side by side, and that the conclusions deducible from the one department should admit only of a remote and extremely cautious application to the other. Of late, the two sciences have for the first time begun to be more and more intimately allied. The object of the present treatise is to prove the most intimate connexion of the two kingdoms of organic nature, from the similarity in the laws of development of the elementary parts of animals and plants"

London, SYDENHAM SOCIETY, 1847, p. ix

Had Schwann read Bradley's works? There is no evidence that he had; Bradley had disappeared from the literature.

Contagious Diseases of Plants

Discoveries of the causes of contagious diseases of plants by microscopic living cells predated the definitive work on the causes of human diseases and experimentally demonstrated that spontaneous generation did not exist. Neither conclusion significantly influenced human contagious disease theory, nor did it make an impact on the spontaneous generation controversy surrounding the origin of the microorganisms hypothesized to be responsible for human diseases and the process of fermentation. Simply stated, here are the controversial issues of this period.

There are entities variously named smuts, rusts, blights, molds, blasts, and mildews, present on plants whenever a disease is present. Are they the cause of disease? How can we account for their presence? There are logically three possible responses. The named entity, for example smut, is the cause of the disease; it is generated from the diseased tissue; it is fortuitously present, growing at the expense of the decayed tissue of the plant. If the agent is a consequence of the disease, generated in some way from the components of the plant, that phenomenon is defined as spontaneous generation. Therefore the entity is not the cause of the disease. The resolution of this issue is of supreme importance to the validation of a living microscopic agent theory of contagious disease—what is later popularly called a germ theory of disease.

The phenomenon of fermentation is intimately tied to the discourse on the cause of contagious disease. Microscopic agents are present when fermentation occurs just as they are present when there are plant diseases. The same questions are raised. Is the microscopic entity the cause of fermentation or the product of the fermentation process? This issue will also be discussed.

The aim of this chapter is to trace the development of disease theory of plants focusing on the period from the opening of the eighteenth century (1700 CE) to the third quarter of the nineteenth century. We will deal with the cause of human contagious disease. An integral part of this discussion will involve the issue of spontaneous generation.

It has been understood since antiquity that plants contract contagious diseases. In the writings of the period covered in Chapter 1, the Homeric–Hesiod–Hebrew Bible eras, two of the major afflictions of populations were famine and disease that can appear independently. Famine can result from a variety of causes, such as poor weather in the planting, growing, or harvesting seasons; no rain or too much rain; and diseases of cereal grains and grasses that destroy food crops for human and domesticated animals. In the Greek and Roman sources there are recognized diseases of wheat designated as rusts and smuts. Pliny the Elder wrote of a festival time, Robigalia, when the population entreated the rust god, Robigus, to refrain from sending disease. Diseases of the major food grains, wheat for example, cultivated in countries around the Mediterranean Sea, were recognized by Theophrastus in Greece (fourth–third centuries BCE). In the fourth century CE the Bishop of Caesarea described a disease of wheat. He observed that although "clean" seeds had been planted there were produced black kernels in the wheat head. He stated this was not a change in the nature of the plant "but instead a sort of disease." The wheat "merely blackens like being burned." The cause is excessive heat. The name *smut* was used to describe the disease of wheat that resulted in the catastrophic destruction of the kernel, resulting in the conversion of its contents to a fine black powder. Some 2,000 years later the microscope revealed that the powder was composed of the vegetative spores of a fungus. When these diseases were present, fungi, which are living, microscopic agents, are always *associated with* the plant. At the time of the great Irish Hunger of 1845 there was no definitive proof that fungi caused the disease, although Berkeley was convinced they did. For 100 years preceding the famine, starting in the mid-eighteenth century, a number of important experiments were carried out that established a connection, not proof, between fungi and plant diseases. Fungi are living agents of incredible variety. Their growth is generally initiated from microscopic spores, and when fully developed they are visible. The most obvious examples to the casual observer are the many fruiting bodies, mushrooms, visible on tree trunks and decaying wood, on the ground after damp and rainy weather, and on decaying fruit.

The Origin of Fungi

The characterization of fungi and their mode of reproduction was a matter of speculation since antiquity. Theophrastus, Plutarch, and Pliny believed truffles were produced from the action of thunderstorms and lightening. The physician-poet Nicander (*ca.* 185 BCE) said the fungus was generated as "an evil ferment of the earth." It was observed that fungi did not produce flowers or seeds, obviously a mystery since they were considered plants.

Some 1,600 years later, in 1552 CE, a treatise appeared in Germany that repeated earlier observations that they suddenly materialized from the earth:

> "Fungi and truffles are neither herbs, nor roots, nor flowers, nor seeds, but merely the superfluous moisture of earth, of trees, of rotten wood, and of other rotting things. This is plain from the fact that all fungi and truffles . . . grow most commonly in thundery and wet weather".

What was confirmed was a belief in spontaneous generation as a way of explaining the origin of these entities, often designated as plants. Andrea Cesalpino (1519–1603), the great botanist of the sixteenth century, affirmed that "some plants have no seeds, . . . and spring from decaying substances."

The Microscope, Fungi, and Robert Hooke

A new era in the study of fungi was initiated when microscopes became available to study their morphology. In 1665 Robert Hooke (1635–1703) published *Micrographia,* filled with microscopic observations, with drawings of various living objects and inanimate materials. This book was a contribution to the efforts of the Royal Society to support the mechanical philosophy and the experimental philosophy promoted vigorously by Robert Boyle. Hooke wrote that he would "return to the plainness and soundness of Observations on material and obvious things" rather than having the "Science of Nature dependent only a work of the Brain and the Fancy."

The introduction to fungi occurs in a section of the book where he describes a rose bush whose leaves exhibit a number of symptoms:

> . . . July, August, and September (green leaves of Roses . . . dry) . . . have little yellow hillocks of a gummous substance, and several of them to have small black spots in the midst of those yellow ones, . . . to the naked eye, appear'd no bigger than the point of a Pin, or the smallest black spot (p. 121) (see Chapter 9) Micrographia, 1960, Dover

He examined the gummy substance microscopically and saw "multitudes of small black bodies, some attached to stalks." Hooke recognized that the rose plant had a disease whose origin was not specified, but he assumed the fungi arose from decaying plant material. He observed microscopically the rounded bodies, spores, which are part of the structure of the fungi, but did not suggest they were vegetative seeds. In fact, he went on at great length on the ability of "moulds" to be produced without seeds:

> "First, that Moulds and Mushrooms require no seminal property. But the former may be produc'd at any time from any kind of putrifying Animal or Vegetable substance . . . Next, that as Mushrooms may be generated without seed . . . for having considered several kinds of them, I could never find . . . to be the seed of it . . . they . . . seem to depend merely upon a convenient constitution of the matter out of which they are made".
>
> (p. 127) Robert Hooke, Micrographia

Hooke proposed that these "moulds" are spontaneously generated (again, see Chapter 9).

Marcello Malpighi (1628–1694) presented a different view of fungal reproduction. Malpighi had convinced himself that spontaneous generation did not occur through his work on chick and insect development and observations on the origin of insects in oak galls where he showed that the insects came from eggs deposited in the leaf. Malpighi wrote a book on plant anatomy in the period 1675–1679, describing fungal spores and spore stalks, regarding stalks as the flowering part of the plant and the spores as florets from which fungi could arise. In addition, he stressed that fungi are reproduced from fragments of the network of filaments that are part of the structure of the organism.

The two incompatible theories of generation became the battleground for the origin of fungi and their role in diseases of plants. The problem simply stated is this: Are fungi the cause of disease, and as a corollary are they independent living organisms that are reproduced from preexisting fungi, or is there an antecedent disease process that results in the generation of the fungus on the decaying material of the plant? The spontaneous generation controversy persisted for some 200 years and turned out to be a major impediment to the acceptance of a living agent theory of the cause of contagious diseases in plants and humans. In this chapter we shall discuss how this problem was resolved in the case of plant diseases; in the following one we shall discuss how it was solved with regard to human diseases.

Pier Antonio Micheli and Richard Bradley

In the opening decades of the eighteenth century a number of individuals took sides on this issue. Micheli (1679–1732), born in Florence, was an outstanding student of plants. In 1706 he was appointed by Cosmo the Third, of the Medici family, to be Director of the Botanical Garden of Florence. In 1729 he published a treatise that contained a section on the origin and growth of fungi. At this time fungi were classified as plants, and Micheli included 900 fungi! Prior to this work he had already carried out growth experiments, using fungi, similar to an experiment carried out fortuitously by Richard Bradley in 1714. Bradley, intending to study the structure of melon membranes, cut the melon in half and left it exposed to the air for some time before he returned to the study. Fungi grew on the exposed melon surface. Micheli had taken "seeds," actually spores, from a previous source of a specific mold and placed them on the surface of fruits, including freshly cut melon, to ensure as much as possible a surface free of other contaminants, and then he placed the melon in an enclosed area. Why would he place the melon in a protected area? One could reasonably surmise he acted on the belief that microscopic agents are present in air, and this procedure would prevent the introduction of extraneous seeds on the melon. Furthermore, he believed that a fungus could only grow from a seed from a preexisting fungus. Here is his description of the procedure to cultivate an individual fungus:

"On the fifth of November of the same year [1718] I took a piece of Melon . . . about four inches long and two inches wide and thick. Next, with a very soft brush, I collected from some other place the seeds from the dark sparkling heads of *Mucor*.

I then smeared them on to the surface of the piece of melon on one side only, and put it in a place in no wise exposed to the wind or sun. On the tenth day of the same month, the infected part appeared everywhere white and strewn with a very thin down-like white cotton, which on the twelfth attained almost an inch in height and assumed a greyish color; and some of the filaments of the down began to appear with white heads. On the fourteenth, the other filaments bore heads of the same kind. Finally on the fifteenth, all the heads became black, and after that the seeds came to maturity.

With the seeds which had been produced in the heads of the Mucor in the previous experiment, on the sixteenth day, I smeared another portion of the same melon on one side, and on the other side I placed the seeds of the (capitate) Aspergillus with glaucous heads and rounded seeds. On both sides, within the same interval of time as I have mentioned above, they sprang up, grew in the same manner as before, and produced seeds of their kind. When I had done this several times, always using those seeds which each new crop of plants produced in its turn, I still observed no difference in the plants which sprang up".

Cited by G.C. AINSWORTH in *Introduction to the History of Mycology*, 1976, p. 86, Cambridge University Press

It is clear that he had dismissed the idea that fungi could be generated from the components of the melon. In about two weeks the mold that grew was the same one that gave rise to the spores he had used to initiate the culture. They produced a mycelium (thin strands), stalks, and spores, which he gathered and then initiated another round of culturing on the fruit surfaces. He obtained the same results and thus demonstrated that spores were the vegetative seeds of the fungus and bred true. These experiments, their number and variety, attest to the commitment of Micheli to three important principles he had accepted as fact:

1. The separate unalterable identities of fungi
2. They grow only from seed. As a corollary, there is no spontaneous generation.
3. Fungal seeds can be transmitted through the air.

One more comment about Micheli's methodology. He had the reasonable expectation that the solid fruit surface was a sterile medium and would support growth. Because it was a sterile surface it was only the spores that he placed on the surface that would be responsible for the visible growth. It is possible that the sequential culturing of the fungus may have achieved a pure culture of that fungus.

Richard Bradley, Again

We have already summarized the seminal experiment by Bradley that connected fungi to the disease process. However, a summary cannot adequately convey Bradley's response to a serendipitous event, revealing his thinking and methodology—in short, his philosophy. Here is the communication as it appeared in 1714 in the Philosophical Transactions of the Royal Society (four years before Micheli's experiment) under the title "Some Microscopical Observations, and Curious Remarks on the Vegetation, and exceeding quick Propagation of Moldiness, on the Substance of a Melon.", communicated by the same Richard Bradley:

"I Had lately a large Melon-Fruit, which I split length-ways thro' the Middle, in order to observe the Vessels which composed the Membrane or Tunick of each Ovary; but my affairs at that time not permitting me to Continue the Work I had began (sic), I lay'd by the one half of the Melon, to be examin'd when I might have more Leisure.

At the end of four Days, I found several Spots of Moldiness began to appear on the fleshy Part of the Fruit, somewhat Green towards the Rind; and of a pale Colour towards the Middle of the Fruit. These Spots grew larger every Hour, for the space of five Days; at which time the whole Fruit was quite cover'd.

This surprising Vegetation made me Curious to examine, if there was any difference between those Parts which were Green and the others, besides their Colour. The first being seen with the Microscope, appear'd to be a Fungus, Whose Cap was fill'd with little Seeds, to the number of about Five Hundred; which shed themselves in two Minutes after they had been in the Glasses [this indicates microscopic examination].

The other Sort had many Grass-like Leaves, among which appear'd some Stalks with Fruit on their Top. Each Plant might well enough be compared to a sort of Bull-Rush. They had their Seed in great Quantities, which I believe were not longer than three Hours more, before the Plants were wholly perfected: for, about seven of the Clock one Morning, I found three Plants at some Distance from any others; and about four the same Day, I could discern above Five Hundred more growing in a Cluster with them, which I supposed were Seedling-Plants of that day. The Seed of all these were then Ripe and Falling.

When the whole Fruit had been thus cover'd with Mold for six Days, this Vegetable Quality began to abate, and was entirely gone in two Days more.

Then was the Fruit putrified, and its fleshy Parts now yielded no more than a stinking Water, which began to have a gentle motion on its Surface, that continued for two Days without any other Appearance. I found then several small Maggots to move in it, which grew for the Space of six Days; after which they laid themselves up in their Bags [pupal stage]. Thus they remain'd for two Days more without Motion, and then came forth in the Shape of Flies. The Water at that time was all gone, and there remain'd no more of the Fruit than the Seeds, the Vessels which composed the Tunicks [membranes] of the Ovary, the outward Rind, and the Excrement of the Maggots; all which together weigh'd about an Ounce. So that there was lost of the first weight of the Fruit when it was cut, above twenty Ounces.

We may Judge from this, and other Cases of the like nature, how much Vegetable Life is dependent on Fermentation, and Animal Life on Putrifaction".

RICHARD BRADLEY, Philosophical Transactions of the Royal Society,
1714, vol. 29, p. 490

Hooke and Leeuwenhoek established in the seventeenth century the reality of microscopic life, and in the early decades of the eighteenth century Bradley and Micheli believed that these microscopic creatures, some of which were recognized as fungi, reproduced from seeds. There was another view, more generally held, that they arose by "equivocal generation," as Hooke phrased it. Stephen Hales (1677–1761), a pioneer figure in plant physiology, held a position similar to Hooke. Hales graduated from Cambridge in 1696. The great influence on his work was Isaac Newton's *Opticks* and the mechanical philosophy of the Royal Society. Hales' method for obtaining knowledge of the works of nature required experimentation, and he was a brilliant experimentalist. In his work "Vegetable Statics," which can be characterized as plant physiology or plant physics, he performed experiments to determine the amount of water taken into the roots of plants and "exhaled" (transpired) by the leaves, among other topics. In the work on movement of water in hops, he stated his theory on the origin of disease. When these processes, uptake of water and transpiration, are in balance, the hops remain in a healthy state. However,

. . . "in a rainy moist state of air without a due mixture of dry weather, too much moisture hovers about the hops, so as to hinder in a good measure the kindly perspiration of the leaves, whereby the stagnating sap corrupts and breeds moldy fen . . . This was the case in the year 1723, when ten or fourteen days almost continual rains fell . . . upon which the most flourishing and promising hops were all infected with mold or fen".

Subsequently,

"The planters observe, that when mold or fen has once seized any part of ground, it soon runs over the whole . . . Probably because the small seeds of this quickly growing mold . . . are blown over the whole ground".

STEPHEN HALES, Vegetable Staticks, 1727, pp. 33–34, London

Hales believed moldy fen are the result of corruption, spontaneously generated, supported by a later comment that excessive heat may lead to "blasts." Once generated molds appear to propagate; however, Hales did not claim they are living entities.

The question remains: What is the cause of disease if it is not a living agent? Hales contended that a property of air is elasticity. Any event that destroys the elasticity of the air is lethal to life. Mead had the same theory. These events include burning of sulfur, explosions in mines that fill the air with fuliginous vapors, burning of rags dipped in melted brimstone, and pestilential and other noxious epidemic infections conveyed by the breath to the blood. Hales constructed a theory that equates the products of burning or explosions to material that causes contagious disease. According to this philosophy, the causes of infection are equivalent to chemical or physical entities.

Spontaneous Generation

"It remains that rightly has the earth won the name of mother since out of earth all things are produced. And even now many animals spring forth from the earth, formed by the rains and the warm heat of the sun".

LUCRETIUS, *De rerum natura*, 1947, p. 473, Oxford

"Why we may see worms come forth alive from noisome dung, when the soaked earth has gotten muddiness from immoderate rains".

LUCRETIUS, *De rerum natura*, 1947, 283, Oxford

"Menocchio [a miller who lived in the late sixteenth century CE in Italy] explained his cosmogony to the inquisitors . . . Amidst such a great variety of theological terms one point remained constant . . . the stubborn recurrence of the apparently most bizarre element: the cheese and the worms born in the cheese. . . . the repeated mention of the cheese and the worms was intended to serve simply as an explanatory analogy. He used the familiar experience of maggots appearing in decomposed cheese to elucidate the birth of living things . . . from chaos came the first living beings . . . by spontaneous generation . . . produced by nature. The doctrine of spontaneous generation of life from inanimate matter, was fully accepted by all the intellectuals of the day".

C. GINZBURG, *The Cheese and the Worms*:
The Cosmos of a Sixteenth-Century Miller,
1992, pp. 56–57, Johns Hopkins,University Press

At the start of the seventeenth century the occurrence of spontaneous generation appeared to be self-evident. The sudden appearance of fungi and mushrooms following wet weather, the presence of flies and worms on decaying meat, and the generation of plant and animal life in ponds attested to this belief. The mechanical philosophy of Robert Hooke, for example, provided a theoretical base for spontaneous generation. The recycling of non-living matter endowed with motion could lead to new life. Opposition to spontaneous generation however, came from a number of sources, philosophical, experimental, and religious.

In antiquity, theories of generation and development were live issues with a strong metaphysical base. A most visible example of generation and development was the fertilized chicken egg. Aristotle recognized that the process involves two sexes, both of whom contribute to the process, however in different ways. The male supplies the original motion and reproduction, Aristotle's efficient cause, while the female furnishes the matter, the material cause. The reproductive event begins with an undifferentiated small area on the yolk of the egg and in time there emerges a chick. The philosophical question, phrased in Aristotle's terms, is this: Is "the form of the adult already present in the egg at the beginning, at the moment of fertilization?" Recall, the form is one of Aristotle's four causes, which makes a thing what it is.

Or, does the form develop over time? If the form, the ultimate adult, is present at the origin of development, the process is designated *preformation*. If the adult emerges over time the process is labeled *epigenesis*. In Aristotle's philosophy there is an additional component in the epigenetic process; the development is directed, caused, by the "soul," a postulated life force.

In 1651 William Harvey, a devoted Aristotelian, published *On Animal Generation.* From his own studies and his reviews of other investigators. he marveled at the production of large things from small beginnings:

> . . . I cannot but express my admiration that such strength . . . should be reposed . . . . in such insignificant elements, . . . it has pleased the omnipotent Creator out of the smallest beginnings to exhibit some of his greatest works. In the seeds of all plants there is a gemmule . . . in so small a particle does all the plastic power of the future tree seem lodged!
>
> WILLIAM HARVEY, *The Works of William Harvey*, 1847, . . . trans. from the Latin by Robert Willis, pp. 320–321, Sydenham Society

Such a view led to the axiom of *ovism*, the principle that all living beings are derived from eggs, a universal law of nature. There is no spontaneous generation of animals and plants. It is important to note that, for Harvey, all of these events were examples of epigenesis. This certainly applies to the generation of the chick.

Near the end of the century additional experimental evidence was added to the critique of spontaneous generation by Jan Swammerdam, Malpighi, and Francesco Redi.

Jan Swammerdam

A strong opponent of spontaneous generation was Jan Swammerdam (1637–1680). He received his medical training in Holland and was a contemporary of Regnier de Graaf, who introduced Leeuwenhoek to the Secretary of the Royal Society. He was an outstanding experimentalist, adopting the "Baconian principle that favored empirical observation and experimentation," and he used the microscope in his study of insect anatomy in order to understand insect development. He adopted Harvey's rule, *Omne vivum ex ovo* ("everything comes from a seed or egg") and went even further to conclude that only one kind of animal would emerge from one kind of seed. He argued for the regularity of the conversion of insects, from eggs, to larvae, to pupae, to adults. Swammerdam described the major changes that take place in the nervous, digestive, and muscle systems during pupation.

He included in his study of insects the butterfly. Their organ systems are different from other insects for they have structures that appear adult-like before the pupal stage. They have wings, legs, and antennae that "gradually appear within the caterpillar prior to pupation," although they are not fully formed structures. From these findings he concluded that the various structural stages of the butterfly represent different forms of the same organism. This conclusion would also apply to the egg at the very start of development. There is no preformed adult organism in the egg, but the same *species* is present throughout all developmental stages. In short, the end product of development is determined and is not a spontaneous event. Swammerdam was also opposed to spontaneous generation on religious grounds. The delicate, stunning insect structures revealed by his microscopic studies demonstrated the power of the Creator. They could not have been the product of chance; they were intelligently designed. For Swammerdam the designs of insects and humans were of one piece, and in arguing against spontaneous generation of insects he was opposing such generation of humans and all forms of life. Belief in spontaneous generation was "the short path to atheism." He was a strong proponent of the experimental philosophy and Christian theology.

Swammerdam, like Malpighi, worked on plant galls, which are abnormal outgrowths on trees. When the gall is opened, insect larvae are revealed. Were they spontaneously generated? Swammerdam maintained they were not, providing evidence similar to that gathered by Malpighi:

> ... "it is not possible to prove by experience that insects are engendered out of plants. But on the contrary we are certain that these small animals are

enclosed there only to take nourishment and there is even the probability that these same plants are created only for this purpose. It is even true that by a constant and immutable order of nature we see regularly several sorts of insects attached to specific plants and fruits . . . they come from the seed of the animals of their own species which were laid there earlier. These insects insert their seed or eggs so far into the plant that they consequently unite with them and the aperture closes. The eggs nourish themselves within".

JOHN FARLEY, *The Spontaneous Generation Controversy from Descartes to Oparin*, 1977, pp. 15–16, Johns Hopkins University Press

According to Swammerdam, nothing happens by chance.

Francesco Redi

"These creatures living in the early days of the world were, according to Empedocles and Epicurus, born all at once, hastily and in disorder from the womb of the earth, still unused to motherhood. Such haphazard generation resulted in great confusion . . . But at last the great mother, perceiving that such monstrosities were neither good nor likely to endure . . . succeeded in producing men and animals according to their species . . . still she retained enough vigor to bring forth (besides plants that are presumed to be generated spontaneously) certain small creatures such as flies, wasps, spiders, ants, scorpions, and all other insects . . .

I shall express my belief that the Earth, after having brought forth the first plants and animals at the beginning by order of the Supreme and Omnipotent Creator, has never since produced any kinds of plants or animal, either perfect or imperfect; and everything which we know in past or present times that she has produced come solely from true seeds of the plants and animals themselves . . . And, although it be a matter of daily observation that infinite number of worms are produced in dead bodies, . . . I feel, I say, inclined to believe that these worms are all generated by insemination".

FRANCESCO REDI, *Experiments of the Generation of Insects*, 1909, translated from 1688 edition by Mab Bigelow, pp. 22–23, 27, Open Court Publishing Co.

Redi was born in 1626 in Tuscany, educated at a Jesuit school in Florence, and studied medicine and philosophy at the University of Pisa. He returned to Florence to practice medicine. He was characterized as having "never failed in deference to the Jesuits." His family was of provincial nobility, and soon thereafter he became court physician and head of the Medicean laboratory. The Grand Duke "was a bigot, whose mind was chiefly occupied in prosecuting religious offenders. Redi's training admirably fitted him for this position."

Redi was in an environment where Galilean mechanics had replaced Aristotelian motion theory; however, Aristotle's theory of spontaneous generation of lower animals persisted in different forms.

Redi was the first person to experimentally test the hypothesis that maggots and flies could arise from decaying flesh by spontaneous generation. In 1668 he published Experiments on the Generation of Insects, in which he acknowledged the general view that spontaneous generation occurs: "living creatures are . . . generated by chance, that is . . . without paternal seed." In contrast to this view, Redi wrote that he was

> . . . "inclined to believe that these worms are all generated by insemination and that the putrified matter in which they are found has no other office than that of serving as a place, or suitable nest". (p. 27)

In short, dead matter does not generate worms but provides the substrate on which they thrive. How did he demonstrate this view? The basic starting point was to incubate rotting flesh that breeds flies. Within a short time maggots were present, and after some days they turned into pupae and later there emerged various kinds of flies. All of these stages, he noted, were well known to hunters and butchers, who "protect their meats in Summer from filth by covering them with white cloths." This procedure was the inspiration for Redi's experiments. He placed meat and fish in a large vase covered with a "fine Naples veil," which allowed air to enter the vase. For further protection against flies, "I placed the vessel in a frame covered with the same net." He stated that he "never saw any worms in the meat, though many were to be seen moving about on the net-covered frame." Redi turned to another popular material from which worms were supposed to be generated, namely cheese—"this opinion being held not only by common people, but even by men of science." Redi, of course, believed that the same process that "creates" worms in meat and fish will occur with cheeses, but "if [cheeses] are kept in a place secure from flies this will not happen." What these experiments proved is that flies, and generalizing to all insects, are derived from eggs laid by parental flies; there is no spontaneous generation of these animals.

Additional opposition to spontaneous generation came from two works, one by the botanist John Ray in 1691, "Wisdom of God manifested in the works of creation," and in 1713, a discourse by William Derham, "Physico-Theology or a demonstration of the Being and Attributes of God from his works of Creation." The message, simply put, was that all living creatures were created by God, at the beginning, which established the fixity of species from then on. There was no spontaneous generation.

The physico-theological argument had been initiated earlier in the seventeenth century (see Chapter 8) and received a passionate defense by Robert Boyle and support from Isaac Newton. Boyle's position was supported by two arguments; the important one was the "argument from design," which was the motif in the Ray and Derham works:

> If we consider "the Vastness, Beauty and regular Motions of heavenly bodies, the excellent structure of animals and plants, besides a multitude of other phenomena of nature, and the subserviency of most of these to man, may justly induce him, as a rational creature to conclude, that this vast, beautiful, orderly and admirable system of things, which we call the world, was framed by an author supremely powerful, wise and good".
>
> ROBERT BOYLE, *The philosophical works of the honorable Robert Boyle*, 1738, p. 239, 2nd ed., vol. 2, London

Ray expressed the equivalent meaning in different words:

> "Animals can be form'd by Matter divided and mov'd by what Laws you will or can imagine, without the immediate Presidency, Direction and Regulation of some intelligent Being".
>
> JOHN RAY, *The Wisdom of God Manifested in the Works of the Creation*, 1714, p. 52, London

Newton permitted his account of the solar system and gravity to be included in a general design thesis. The force of gravitation, without physical contact, "cannot be innate and essential to matter." Therefore, "gravity . . . is not itself mechanical, but the immediate fiat and Finger of God, and the execution of the divine law."

The Process of Development: Preformation or Epigenesis?

If all living things come from seeds, it becomes a matter of great interest and speculation as to what is in the seed. There are two theories, one characterized as *preformation* and the other as *epigenesis*. Nicolas Malebranche stated the most radical version of the preformation theory in 1673: All future generations exist in the egg.

"We may with some sort of certainty affirm that all the trees lie in miniature in . . . their seed. Nor does it seem less reasonable to think that there are infinite trees concealed in a single [seed] since it not only contains the future trees whereof it is the seed, but also abundance of other seeds, which may all include in them new trees still, and new seeds of trees . . . and thus in infinitum . . . .

We ought to think that all the bodies of men and of beasts, which shall be born or produced till the end of the world, were possibly created from the beginning of it".

JOHN FARLEY, *The Spontaneous Generation Controversy from Descartes* to Oparin, 1977, p. 17, The Johns Hopkins University Press

In the model of epigenesis germ cells, seeds, have a substance that does not have within it the final structure of the fully developed entity. The microscopic studies of Swammerdam and Malpighi on the development of insects from eggs, where there is successive differentiation into unique structures, provide examples of such a phenomenon.

A number of experimental findings argued against a theory of preexistence, leaving the field open to an epigenesis theory and possible support for a theory of spontaneous generation. Regeneration in Hydra, a freshwater polyp, indicated that the development of an animal could result without seeds or eggs. Plant hybrids, new kinds of plants, were produced by crossing pollen of one species with the female parts of another species. In general, plant classification showed variations in species that undermined the notion of preexistence. There were also a series of experiments with microorganisms that indicated that spontaneous generation occurred, which could be explained by a theory of epigenesis. These experiments were carried out by John Turbeville Needham (1713–1781) a Catholic priest in England. He was elected to the Royal Society in 1747 and subsequently went to Paris to carry out experimental work in collaboration with George, Comte de Buffon (1707–1788). A sense of Buffon's philosophy of nature and epistemology can be gained from the following:

"There exists in nature an infinity of little organized beings, and that these little organized beings are composed of living organic parts which are common to animals and vegetables . . . that organized beings are formed by the grouping of these parts, and that reproduction and generation are therefore nothing but a change of form which comes about simply by the addition of these similar parts".

JOHN FARLEY, The Spontaneous Generation Controversy
from Descartes to Oparin, 1977, p. 23

"Buffon's alternative to pre-existent germs was his theory of organic molecules and organic molds, ... universal components of living bodies. The distinctive combination of organic molecules acts as an internal mold. By internal mold Buffon means a constellation of active . . . forces, analogous to gravitational force.

Within lower grades of life the activity of the organic molecules depend . . . upon heat. Gravity, which is attractive, and heat, which is repulsive in its action, are the ultimate agents in nature for Buffon".

M. J. S. HODGE, "An Alternative to the Dominant Historiographic
Tradition," in *Companion to the History of Modern Science*,
1990, pp. 378–379, Routledge

Buffon envisions a recycling process that evokes the argument for equivocal generation by Robert Hooke where "contrivances," physical materials derived from degraded living material, are used to create new living matter. The theory that matter displayed dynamic properties, that it could be moved by the Newtonian universal forces acting on particles of matter (for example, the forces of gravity and heat), provided a philosophical base for a theory of epigenesis. It is therefore no surprise that Buffon was an adherent of spontaneous generation. In addition, his method of inquiry was empirical:

"A series of similar facts, or, if you prefer, a frequent repetition and uninterrupted succession of the same events constitutes the essence of physical truth . . .

One goes from definition to definition in the abstract sciences, but one proceeds from observation to observation in the sciences of the real. In the first case one arrives at evidence, in the latter at certainty".

PHILLIP R. SLOAN, "Natural History: 1670–1802" in *Companion
to the History of Modern Science*, 1990, p. 304, Routledge

This takes us to his collaboration with Needham. Needham's primary focus was to explain the origin of animalcules, among them the organisms originally revealed by the microscopic studies of Hooke and Leeuwenhoek.

Needham's experiments, published in 1748 in the *Philosophical Transactions*, were designed to differentiate between the two theories of generation—whether the microorganisms were preexistent in his broths, or whether they developed from undifferentiated complex materials contained in the broths, epigenesis. Preformation is incompatible with the notion of spontaneous generation. The latter kind of generation, epigenesis, is compatible with a materialist matter theory where "contrivances," or any sort of basic physical particles, can be rearranged to create a new living entity.

Needham was a proponent of spontaneous generation and epigenesis, persuaded by the philosophy of Buffon. For his experimental material Needham studied the generation of microscopic entities contained in organic broths. Surprisingly, however peripherally, out of this discussion of the origin of microscopic agents in these media, Needham presented a theory of "contagious epidemical distempers," a living agent theory of disease, that included references to fungi that cause disease of plants and humans.

In his first experiment he placed almond seeds in water, corked the mixture, and waited more than a week, after which time he saw microscopic, moving bodies. He called these bodies *atoms*, whose movements he reasoned were deliberate. If these "atoms" were living beings, from what material were they derived? Were they preexistent (preformed) or did they develop from undifferentiated material of the almond seed preparation, an example of epigenesis? He resolved to enlarge these studies, and in collaboration with Comte de Buffon he repeated these experiments with 15 different kinds of infusions and found they "swarmed with clouds of moving atoms so small and so prodigiously active." Their presence in the broths could be explained in two ways: they were already present at the start of the experiment, or they arose during the experiment. To counter the argument that they preexisted in the infusions, he treated his organic soups with heat to eliminate any living forms. After removing the soup "hot from fire," it was placed in a vial and essentially hermetically sealed. Needham commented,

"I thus effectually excluded the exterior Air, that it might not be said that moving Bodies drew their origin from Insects, or eggs floating in the Atmosphere".

JOHN NEEDHAM, *A Summary of some late Observations upon the Generation, Composition, and Decomposition of Animal and Vegetable Substances*, Philosophical Transactions of the Royal Society, vol. 45, pp. 615–666, 1748

Needham relied on premises that heat destroyed all living matter in the broths and all living organisms possibly present in the air would be excluded by the sealing procedure. After some time the vial's contents "swarmed with life, the largest and smallest agents he had seen, multitudes perfectly formed and animated." He concluded that they were a class of living microscopic agents generated from a non-living substance. Hooke had labeled it "equivocal generation." Needham and Buffon agreed that there was a "real productive force in nature" that worked, according to Buffon, by changing and recombining organic parts. As Needham stated, this "force" resided in every "microscopical point" so that living beings were brought into existence. Needham and Buffon asserted that these experimental results supported a theory of epigenesis: the microscopic entities were derived from non-living matter. There were no preformed living microscopic entities in the numerous infusions that could give rise to the innumerable numbers of agents, "perfectly formed and animated."

Needham carried out additional experiments studying infusions of wheat and rye components that are relevant to our discussion of plant diseases. Wheat grains were crushed in a mortar, water was added, and in time they turned into a gelatinous mass. Microscopic examination revealed the presence of filaments that developed into moving globules designated as "perfect Zoophytes teeming with Life, and Self-moving." All of these structures were fungi, apparently generated from the wheat infusions. These observations reminded Needham that he had observed years ago "those very eels in blighted wheat" in the field. He made the perceptive conclusion that diseases of grains and other vegetables, called blights, are "no longer mysterious . . . . in an atmosphere charged with humidity," there was produced a "new kind of vegetation," producing "filaments which cause blighted wheat." He obtained similar results with rye and labeled the diseases associated with these filaments "contagious epidemical distempers." These maladies, he proposed, were caused by these microscopic entities that arose spontaneously from the material of the grain. Needham recognized that the allegedly spontaneously generated agents in the wheat and rye infusions resembled in structure the agents found in diseased wheat and rye in the field. Could these agents cause illness in humans? He answered yes, suggesting that these agents were the cause of "contagious . . . distempers." He posed this question because of information given to him by members of the Royal Academy of Science (France), who told him that "blighted rye is . . . full of filaments." When this blighted rye exists and is made into bread it produces "very strange effects on poor country people . . . which causes their limbs to drop off." The symptoms described are accurate. Farm animals exhibit the same devastating effects, including gangrene and loss of limbs, when the infected grain is ingested. The disease of rye principally affected poor people because they ingested contaminated rye bread, which was less expensive than bread made from uninfected rye. It is not a contagious disease among humans but is contagious for various cereal grains, particularly rye and triticale, a wheat/rye hybrid. Although it seemed that the agent of the disease of grains appeared spontaneously, Needham considered them contagious; once produced, they could infect other plants.

Needham also identified another disease in which the grain crumbles and is turned into a black powder. We will encounter a similar disease in the experiments of Mathieu Tillet (below) some six years later in France.

Mathieu Tillet (1714–1791)

"Even strictly measuring science could hardly have got on without that forecasting ardour which feels the agitations of discovery beforehand, and has a faith in its preconception that surmounts many failures of experiment".

GEORGE ELIOT, *Daniel Deronda*, 1979, p. 467, New American Library

The eighteenth century in Europe and Colonial America has been characterized as the period of Enlightenment. There is a vast literature on the meaning of the Enlightenment, and one of its important consequences was the spread of scientific learning. A significant contribution toward this new learning had been made in the previous century, the seventeenth, with the growth of the mechanical philosophy culminating in the Newtonian revolution. Nature was now to be understood by a combination of experimentation, mathematics, and physics, and the phenomena in the world would be explained by mechanical and chemical causes. In Germany new universities were founded, like Goettingen, where natural philosophy and its practical applications, such as agriculture, engineering, and mining, were taught. In France and Italy various other institutions were created in many localities, established by educated groups, which played an important role in intellectual life, providing a place for public lectures and establishing libraries that allowed access to the larger public of the large number of books that were being published. These institutions served the Enlightenment ideal that the general public would be able to understand societal issues, not the least of which had to do with science. One such local resource was the Royal Academy of Literature, Sciences, and Arts of Bordeaux, France.

In 1750 in the region of Bordeaux there was an epidemic disease of wheat that caused the wheat kernels to become swollen and blackened. In place of kernels there were "bunt balls," which when broken open yielded a fine black powder. That same year the Royal Academy of Literature, Sciences, and Arts of Bordeaux offered a prize for the best dissertation on the cause and cure of the blackening of wheat. Mathieu Tillet responded and in 1755 published a treatise, "Dissertation on the cause of the corruption and smutting of the kernals of wheat in the head and on the means of preventing these untoward circumstances." Tillet was awarded the prize by the Royal Academy.

He was the first person to carry out a controlled experimental program to determine the cause of the disease of wheat or any plant. He invented a controlled field experiment that ensured that the most important possible causes of the disease of wheat were simultaneously being tested. This would be a remarkable achievement for any agriculturalist or botanist, but particularly for Tillet, since his real job was Director of the Mint of Troye. He had no training in botany or agriculture, although he did have a strong interest in natural philosophy (science). He had no collaborators; he assimilated the information on diseases of wheat from antiquity to contemporary times through his research of the extant literature. His philosophical posture guided his research, a relentless experimental approach to the problem of contagious disease that was unique in the eighteenth century. Tillet's preliminary remarks are important since they reflect his philosophical position, the importance of finding the cause of the disease:

"At first glance, to comprehend the nature of the particular disease of wheat does not seem to require arduous labor; it seems to be self-evident. But when we consider the malady from the standpoint of its **cause** it becomes more obscure. It has not been possible to set out from a known point of origin in order to follow the progress of the destruction of the kernal. The derangement of the organic parts of the plant has taken place insidiously; and even though one recognizes it finally by certain external symptoms, it is then too late to disclose its cause".

MATHIEU DU TILLET, *Dissertation on the Cause of the Corruption and Smutting of the Kernals of Wheat in the Head*, 1755, trans. by Harry Baker Humphrey, "Phytopathological Classics" No. 5, p. 16, 1937, American Phytopathological Society, Ithaca, NY

Tillet is clear that the "point of origin" and the "cause" are the same. Furthermore, it is possible to access a "point of origin" by "observation and experimentation" rather than relying on philosophical principles:

"The experiments a man himself conceives and performs have something to their advantage . . . he follows them closely. These experiments with which he becomes very familiar, strike him from the different aspects in which he may view them and afford him to multiply his observations by increasingly stimulating his curiosity."

MATHIEU DU TILLET, p. 70, 1937.

Tillet understood intuitively, or at least from experience, the importance of the interplay between experimental results and reason for they lead to future experimentation. He believed that experiments were the "exact method of compelling nature to explain herself." This was his theoretical base, which had its roots in the empiricism of the Baconian–Boyle–Lockean philosophy of the seventeenth century. He carried out two sets of experiments, the first in 1751 and the second in 1752–53.

To determine a cause, if not the cause, Tillet needed to take seriously the various theories of disease that had been inherited from antiquity, from Theophrastus and Pliny, to the most recent ones by his contemporaries, and include these causes as variables in the complex, controlled experiment he created. These previously proposed causes included too much soil moisture, a theory promoted by Jethro Tull (1674–1741) in 1733, dangerous mists, sheep manure spread on soil with fresh seeds, faulty conformation of the grain, burning rays of the sun on heads already wet with rain, and an ethereal volatile oil on the plant activated by the sun. Another theory involved moist air containing nitrous, sulfurous components that permeate the grain and cause fermentation. The latter two causes contain language used in chemical theories of human diseases of the seventeenth century.

Tillet also cited M. Duhamel, who "has had some slight suspicions . . . of bunt being a hereditary disease." To this long list Tillet added one more possible cause that clearly and radically set his theorizing apart from the conventional contemporary theories of diseases of plants: the black powder that is found in kernels is included as a cause. This possibility had come to him, he stated, from his observations of bunt balls. Using an excellent microscope, he noted that the black mass present in infected wheat comprised perfectly round objects of equal size and resembled the structures of the black powder found enclosed in puffballs of Lycoperdon.

Lycoperdon is a fungus. Its puffballs, when mature, are filled with a mass of powdery, dark spores. No previous theories of the bunt suggested the black powder as a cause, and it is reasonable to conclude that Tillet had already decided that the black material was one possible cause of the disease, if not the only cause. Why do I say this? He wrote that he "has commenced to have some inkling of it." We should accept Tillet's "inkling" as an example of the *a priori* development of a hypothesis, or the invention of a hypothesis, as Hempel phrased it, to explain a phenomenon. The experiment is explicitly designed and dominated by the inclusion of the black spores as a variable because he seems to have decided that the black powder caused the disease. As a corollary Tillet rejected the spontaneous generation of the fungi.

The Great Outdoor Field Experiment

The field, 540 ft. long and 24 ft. wide, was organized in the following way:

1. It was divided longitudinally into four lanes (1, 2, 3, 4) with a walking space of one foot between lanes. Winter wheat seeds, treated in the following ways, were planted in each lane. All of lane 1 received seeds blackened with the dust of diseased bunted kernels. All of lane 2 received the dusted seeds used in lane 1 that, however, had been treated with a saturated solution of salt and niter intended to destroy the black powder. All of lane 3 received untreated seeds. All of lane 4 received the dusted seeds used in lane 1 but treated with a saturated solution of lime or lime and niter, another concoction designed to kill the black dust.
2. The lanes were divided horizontally into five blocks (each block was 108 feet), A, B, C, D, and E. Four received various manures to test the effect of various nutrients (A, pigeon; B, sheep; C, night soil; D, horse and mule); block E received no manure.
3. Each block of 108 feet was divided horizontally into six strips (a, b, c, d, e, and f); the various treated seeds were sown in each strip across the entire block on different days of the month to test the effect of different weather conditions at time of sowing: strip a, October 16; strip b, October 22; strip c, October 27; strip d, November 3; strip e, November 10; strip f, November 22, all in 1752.

Tillet also carried out experiments where infected seeds were grown in pots, where he could control more carefully the amount of water and sunlight a plant received.

The Results

In the field experiment, only in Lane 1 of all sections, A, B, C, D, and E, was there a significant number of diseased plants. No other factor had as much influence on the level of disease as exposure to the black powder—not the quality of fertilizer, the time of planting, nor the amount of sunlight or watering, results obtained from the pot experiments. Procedures designed to destroy the black powder on dusted seeds, such as treatment with lime, produced plants with a low incidence of disease. The low level of diseased plants in Lane 3, where nondusted seeds were planted, was inevitable since it would have been impossible to prevent entirely wind and insects from transferring bunt powder to these plants.

Tillet's conclusions, in his own words, convey accurately the only conclusion that could be derived from these experiments:

> . . . "on the testimony of many experiments, (1) that the common cause . . . resides in the dust of the bunt balls of diseased wheat; that the clean healthy seed, inoculated with this dust, receives through rapid contagion . . . the poison peculiar to it; that it transmits the poison to the kernels, once infested become converted to black dust and become for others a cause of disease.
>
> MATHIEU DU TILLET, p. 127, 1937

Tillet introduced another factor in the disease process, suggesting that the spores carry a poison. Nevertheless, the nature of the disease process is unchanged. The plant must be invaded by the black dust to contract the disease.

It is important to stress that Tillet's conclusions were not based on metaphysics or some form of the mechanistic, corpuscular philosophy, but on experimentation, a methodology that had not been applied previously to the study of plant diseases, and certainly was impossible to use in the study of human diseases. Despite his conclusions Tillet was sensitive to the anomalies in biological experiments. Let us return to his major conclusion that the cause of the disease is due to the presence of the black dust. If this is the case, he asked,

> "Why . . . in a field . . . when only clean, selected . . . seed had been sown are there nearly always **some** (my emphasis) bunted [diseased] heads to be seen?"

His answer is that:

.. "no matter what attention one may give to the selection of seed it is all but impossible not to find some kernals either wholly bunted or partly so. In spite of the great precautions,
    some bunt balls slip in".

MATHIEU DU TILLET, p. 128, 1937

The causal connection between the bunt balls and the disease is reinforced by the results obtained by two farmers:

... "one of the most skillful ... of the neighborhood ... who devotes scrupulous attention to the choice of seed [and] never has any diseased wheat

and

... the farmer whose bunt infected field ... has had his wheat constantly damaged in several successive years through the practice of sowing only the same seed, seed that ... was damaged by smut.

MATHIEU DU TILLET, p. 128, 1937

Tillet concluded this case history with a query for the reader:

"Does not the condition of the wheat of these two farmers, so different for the sole reason that the one sowed only selected seed and the other for a long time had used only contaminated seed, lead you to suspect that smut is not of spontaneous origin, . . . and does not occur except through contagion"?

MATHIEU DU TILLET, p. 128, 1937

Tillet suggests that plant diseases exhibit characteristics of human diseases, and the severity of the disease depends also on the response of the host. Here are these comparative remarks by the Director of the Mint, a person certainly not trained in medicine, in his own words. They follow directly after the comment "except through contagion"

"In that respect it is similar to a formidable and unfortunately too common disease, which from the time it appeared in Europe, passed for an epidemic disease caused by bad atmospheric conditions; whereas today it is regarded as very certain that it does not develop at all unless the virulent principle is provided by a host already infected with the same disease and that, once the principle is eliminated, the disease is never produced by any other circumstance".

MATHIEU DU TILLET, p. 129, 1937

Tillet referred to the writings of a French physician, M. Austre, who described the onset and development of syphilis, which arrived in a virulent form in Western Europe at the end of the fifteenth century and was for more than 40 years not recognized as a contagious disease (Chapter 7):

"In reading the work of the scholarly physician whom I cite here, one will note a sort of analogy between the disease of which he speaks and that of bunted wheat. For example, with reference to the former, has it not been observed that an ordinary disease principle is not invariably peculiar to such and such host; or that, if this principle, without exception, produces its effect, it is more widespread and marked on certain hosts than in others? Is it a question of the latter disease? One observed that not all the wheat grains that are uniformly blackened with smut are affected by the virus [*sic*] and that among the number of those that are affected, some produce completely bunted plants and others produce plants in which the disease extends only to a part of the heads or to some kernals within a single head. Whether one finds the virus innocuous or only partly effective or, again, wholly so, undoubtedly always depends relatively on the peculiar nature of the inoculated kernals and on the special disposition of the germ. **Here we have a condition not unlike that peculiar to human beings exposed equally to attack by pests and without protection against them. Whether this or that terrible plague produce the effect of which it is capable or only a mild attack or none of any noticeable degree, will depend relatively on the condition of the body at the time or on its particular temperament** (my emphasis)".

MATHIEU DU TILLET, p. 129, 1937

Tillet commented on the fact that only one part of a plant may be affected by a disease, which was previously pointed out by Richard Bradley, and once again made the analogy to human disease where one part of the body may be involved.

It is difficult not to admire Tillet's biology from our contemporary perspective. He developed a comprehensive theory of contagious disease. Although Tillet did not demonstrate, and actually could not have demonstrated unequivocally, that the bunt disease is caused by a fungus, the correlation between the presence of the black powder and the incidence of the disease convinced him and might have suggested to all subsequent investigators, at the very least, that a cause could be ascribed to the black powder. In Italy there was an ambiguous response.

The year 1766 was, unfortunately, a good year for diseases of grains in Italy. Felice Fontana (1731–1805), a Professor of Philosophy at the University of Pisa, observed the wheat rust during the epidemic in Tuscany. He declared that the spores were semiparasitic plantlets, since their insertion into the plant to cause disease depended on prior damage to plant structures that released humors. The humor emerges on the leaf, where the sun hardens it and transpiration is inhibited, leading to a damaged leaf and the development of mold.

Giovanni Targoni Tozzetti (1712–1783) wrote a treatise, "True Nature, Causes and Sad Effects of the Rust, the Bunt, the Smut, and Other Maladies of Wheat, and of Oats in the Field", with minimal references to Tillet. Tozzetti, originally trained as a physician, was a student of Micheli and inherited his role when Micheli died, becoming the Director of the Botanic Gardens. He described the summer of 1765 as atypical, with excessive rains, stormy winds, and lots of clouds and mists. This led to poor conditions of the soil, which could not be properly prepared for autumnal seeding. The winter was cold and a two-week period in the spring was also cold. Tozzetti asked the readers to forgive his tedious descriptions of the weather, but he provided them because during the first half of this summer (1766) the rust occurred at the time of "damaging" fogs and dews that "ravaged the fields." One might reasonably anticipate that Tozzetti led with the weather to prepare the reader for an environmental cause of disease. Not so: he concluded with a living agent theory. The rust is caused by a microscopic parasitic plantlet. Their seeds can be transported by air and remain dormant for extended periods to germinate at a time conducive to their development. Although Tozzetti's position appeared clear on the cause of the disease, the extensive inclusion of the writings of numerous investigators who did not believe in a living agent theory of disease might have suggested to contemporary readers a degree of ambiguity about the cause of disease. An entire page was devoted to one N. E. Tscharner who believed the disease is caused by fog. F. Ginanni was cited extensively by Tozzetti (by my count 20 times); he believed that dew could acquire a caustic and corrosive nature and produce the rust. Both positions would suggest the formation of fungi by spontaneous generation. Tozzetti's admiration for Ginanni was, perhaps, a matter of national pride. However, he used Ginanni's work to refute Tillet's disease theory. Tozzetti accurately stated Tillet's disease theory but wrote that, Ginanni contradicts "his supposition with the support of conclusive observations."

According to Tozzetti, Tillet had a supposition while Ginanni had conclusive observations. Labeling Tillet's opinion as a supposition, an assumption, suggests it was put forward without any support. Conclusive, on the other hand, indicates a decisive answer that settles the question. In fact, it was Tillet's work that was consistently experimental. The studies of Tozzetti, Fontana, and Ginanni were based on observation. At that time there was no way to arrive at conclusive answers.

The Resumption of Experimental Studies on Plant Diseases

Isaac Benedict Prevost (1755–1819)

Prevost enlarged on the experimental study of a disease of wheat in 1807, some 50 years after Tillet. He was self-educated, in mathematics, physics, and natural history, and more, since he published papers in chemistry and philosophy. He was appointed chair in philosophy in 1810 in a new academy in the city of Montauban, where he had resided since the age of 22.

His work, "Memoir on the Immediate Cause of Bunt or Smut of Wheat, and of several Other Diseases of Plants, and on Prevention of Bunt," is summarized in this way:

> "By all that precedes, I have incontestably established that the immediate cause of bunt is a plant of the genus uredos or of a nearly related genus; that the growth of this plant, as well as the majority of the uredos, begins in the open air and is completed within the plant that it attacks".

> BENEDICT PREVOST, *Memoir on the Immediate Cause of Bunt or Smut of Wheat, and of Several other Diseases of Plants, and of the preventatives of Bunt, 1807.* Phytopathological Classics No. 6, 1939, p. 60, trans. by George Wannamaker Keitt, American Phytopathological Society, Menasha, WI

Uredo is a fungus that Prevost claims to be the cause of the bunt. The term "uredo" has its origin in the Latin word "to burn." This characteristic is associated with diseases labeled "blights" or "blast." Prevost concluded that the fungus could develop independently of the host. It was therefore a contagious agent and not spontaneously generated from diseased plant tissue.

Prevost was the scientific successor to Tillet. Let us recall briefly what Tillet had established. Bunted heads contain a black powder, composed of microscopic oval bodies, designated as spores. This powder can confer the disease on seeds of wheat. Tillet concluded that the disease-causing agents are the spores. Prevost extended Tillet's work, performing for the first time laboratory experiments as well as field experiments. He studied systematically the putative cause of the bunt disease, from its microscopic origin to the development of the full-blown visible entity. He created methods for growing the fungus in water culture, while controlling for the source of water and its purity and sterility. He developed methods for ensuring that the possible cause of the disease, the spores, are in a form, as pure as possible, so that one could make the essential argument that their unique presence was necessary to cause the disease. Prevost recognized that there are many kinds of plant diseases and many kinds of fungi that may cause disease.

Prevost's Experiments

Prevost began by selecting one infected kernel suspended in distilled water. The kernel was washed a number of times with "very pure and recently distilled water." Why was this done? Prevost recognized that the water containing the diseased kernels should

> . . . "not have been long exposed to the open air, from which are always deposited dust, spores of moulds, or other microscopic plants, which grow there and interfere with the experiment".
>
> BENEDICT PREVOST, p. 32,1939

This quote presents explicitly and precisely Prevost's philosophical stance. An external, specific living agent is the cause of the disease. That relationship may be compromised by any number of other living agents that are present in the air. Consequently, the disease does not arise spontaneously from the host. There is no equivocal generation. Now, let us return to the experiment with the washed kernels.

Prevost placed the kernels in water, and when they were disrupted a "black dust" drifted out. It was already recognized that in the field infected kernels contain an "almost black powder" that, when examined microscopically, is composed of spherical bodies, nearly black, with a diameter of about 0.00025 to 0.00034 inches. These are the structures released in water. In time, these spores, at the appropriate temperature, sprouted "stalks" that became elongated and branched. He placed these germinating spores on a bed of wet rags or paper and enclosed them in a bell jar to maintain humidity. The development continued and the stalk produced globules. These data convinced Prevost that the fungus, the cause of bunt, could begin its development independent of the wheat plant. This is an

important conclusion, for it counters the argument that the fungus originates from the components of the decayed plant. Subsequently, he demonstrated that the spore-stalk agent can infect the wheat plant in the field. In
one experiment in which spores were scattered upon the soil with wheat
seeds immediately or very soon after the wheat was sowed, there were
many bunted heads. He concluded,

> . . . "it is only at the time of germination, or a very little while after that the
> introduction of the bunt plant into that of the wheat occurs.
>
> BENEDICT PREVOST, p. 38, 1939

Prevost's careful experiments did not make a significant impact on plant
disease theory for a number of reasons. Although the publication had been
sent to the Institute in Paris for analysis, it somehow became inaccessible
until the mid-century. Even Anton de Bary (1831–1888), the great plant
pathologist of the nineteenth century, did not have access to the original
publication in 1853! In the early decades of the nineteenth century other
attempts to infect plants using spores, according to the methods of Tillet
and Prevost, were not successful, due to differences in the pathway of infection in various hosts, and host sensitivity to the acquisition of infection
at different periods in the development of the plant. These conditions were
not recognized at the time. In addition there was a continuing, formidable
belief in the generation of fungi from plant material that decayed because
of climate, such as temperature and humid conditions. In this theory, fungi
were the consequence of the disease, not the cause of the infection: fungi
are spontaneously generated.

The cause of plant diseases by fungi was a contentious issue in the first
decade of the nineteenth century. A contemporary of Prevost, Joseph
Banks (1743–1820), once President of the Royal Society and Scientific
Adviser to the Royal Garden at Kew, in 1805 published a short paper
where he acknowledged that

> "Botanists have long known that the blight of corn is occasioned by the
> growth of a minute parasitic fungus . . . Agriculturalists do not appear to
> have paid . . . sufficient attention to discoveries of their fellow-labourers in
> the field of nature; for though scarce any English writer of note on the sub-
> ject of rural economy has failed to state his opinion of the origin of the evil,
> no one of them is has yet attributed it to the real cause".
>
> JOSEPH BANKS, *A short account of the cause of the disease in corn, called
> by farmers the Blight, The Mildew, and The Rust*, 1805, p. 3, London

Banks believed the disease was due to the spread of the seeds of the fungus.

The Biology of Fungi

"The **cells** (my emphasis) of plants appear either singly so that each one forms a single individual, **as in the case of some algae and fungi** (my emphasis), or are united together in greater or smaller masses to constitute a more highly organized plant".

E. B. WILSON, *The Cell in Development and Heredity*, 3rd ed., 1925, p. 3, The Macmillan Co., quoting an 1830 statement by F. J. F. Meyen

Contagious disease is a general biological phenomenon. Numerous diseases are caused by living, microscopic cells, a fact definitively established in the nineteenth century. To prove this principle three characteristics of putative contagious agents needed to be demonstrated. They exist, shown by microscope observation. Their structures and physiology are equivalent to cells that constitute the bodies of all plants, animals, and humans. They are capable of independent growth in the appropriate nutrient media. Fungi exhibit all these characteristics and cause contagious diseases of plants. We shall develop the theoretical and experimental evidence for these conclusions.

In March 1839, in Berlin, Theodore Schwann (1810–1881), in the preface to his classic text "Microscopical Researches into the Accordance in the Structure and Growth of Animals and Plants", wrote,

"The object of the present treatise is to prove the most intimate connexion of the two kingdoms of organic nature, from the similarity in the laws of development of the elementary parts of animals and plants".

Microscopical Researches into the Accordance in the Structure and Growth of Animals and Plants, trans. by H. Smith, 1847, p. ix, Sydenham Society

This landmark treatise by Schwann, with a contribution by Matthias J. Schleiden (1804–1881), led to an inclusive cell theory stating that all plant and animal organisms were composed of cells. In addition, it was determined microscopically that the interior of all cells appeared similarly constructed. In 1828 Robert Brown (1773–1858) discovered a phenomenon that came to be termed Brownian movement, the tremulous movement of the materials in the interior of cells. In 1831 he uncovered the nucleus in two kinds of plant cells, confirmed in other plant cells by Schleiden in

1838. Another generalization emerged between 1840 and 1870, through microscopic studies on plant material and free-living animal organisms. The interior of cells exhibit *streaming*—that is, there is back-and-forth movement of the semifluid interior of the cell. The interior of cells contains a living substance! The presence of a dynamic interior of cells led to theories of cell division based on *molecules* or *granules*, presumed to be part of the substructure of cells, from which new cells were derived. Schwann postulated such a mode of cell reproduction. New cells are produced by the aggregation of "molecules . . . from a fluid blastema." Such a process recalls the suggestion of Hooke that cells are derived from "contrivances" within cells. Investigators who adopted such a model of reproduction from molecules or granules that are not living were led to a belief in heterogenesis, spontaneous generation.

An alternate view, that cells are only derived from cells, was supported by experiments of Robert Remak (1815–1865), carried out between 1842 and 1854. He demonstrated that the nuclei of embryonic chick red blood cells duplicate with the coordinate division of the surrounding material. During this period botanists also showed the process of cell division in plants.

Ferdinand Cohn (1828–1898), a 22-year-old botanist relying on previous researches on rhizopods (protozoa) by Felix Dujardin (1802–1860) and the work of Hugo von Mohl (1805–1872) studying the interior of the plant cell, stated that plants and animals were analogous not only because they were constructed of cells, but that these cells were composed of a common substance, identified chemically as containing a nitrogenous substance. The substance was later labeled "protoplasm." By the 1830s it was recognized that fungi have many of these characteristics. Fungi are composed of cells.

Fermentation Is Caused by a Fungal Cell: A Contested Issue

Theodore Schwann in the 1830s decided to use yeast, a fungus associated with the phenomenon of fermentation, to study whether it is the cause of fermentation. Yeast were observed in 1680 by Leeuwenhoek in beer; he noted their morphology and their appearance in strings of two, three, and four particles. Two other individuals studied the physiological properties of yeast, in France Charles Cagniard de la Tour (1777–1859) and in Germany F. T. Kutzing (1807–1892), who described the nucleus of the yeast in 1837. It was at this time that a colleague of Schwann, F. J. F. Meyen

(1804–1840), agreed that yeast was a living organism, and Meyen proposed the genus-species name for it, *Saccharomyces* (Latin for sugar fungus) *cerevisiae* (for Ceres, the Roman goddess of grain). The taxonomy of yeast, however, was a matter of dispute for some time, while there were similar problems in naming fungi because there appeared to be instances of changes of morphology that could be interpreted to indicate that there was no fixity of species. M. J. Berkeley (1803–1889), the leading expert on fungi in England, stated that growing fungi from single spores required a level of experience that few experimenters achieved.

In short, much experience was required in handling fungal material to ensure the uniformity of a culture. Schwann, Cagniard de la Tour, and Kutzing accepted as a foundational principle that yeast is a living cell and is a specific cell.

One of the major characteristic of cells enunciated by Schwann in the 1839 opus is that cells must have "metabolic power" in order for the cell to undergo division. He offered the yeast cell as an example that best illustrates this metabolic power since he accepted as fact that the yeast cell carried out "vinous fermentation." This form of fermentation was known since antiquity inasmuch as one normal habitat of yeast cells is on the surface of grapes and other fruit. When these fruits are crushed and their sugar-containing contents come into contact with yeast, a fermentation occurs. The products are carbon dioxide, which gives the characteristic bubbling to the process, while alcohol is the other product. Schwann proclaimed a living agent cause of fermentation based on the biological characteristics of fungi, which are living cells.

"We have every conceivable proof that *the fermentation-granules are fungi*. Their form is that of fungi; in structure, they, like them, consist of cells . . . They grow like fungi by the shooting forth of new cells at their extremities; they propagate like them . . . Now, that these fungi are the cause of fermentation, follows, first from the constancy of their occurrence during the process; secondly, from the cessation of under any influences by which they are known to be destroyed, especially boiling heat, arseniate of potassium, &c; and thirdly, because the principle which excites the process of fermentation must be a substance which is again generated and increased by the process itself, a phenomenon which is met with only in living organisms".

T. SCHWANN, *Microscopical Researches*, 1847, p. 197, Sydenham Society

Cagniard de la Tour had been observing yeast microscopically for many years, but in 1838 he obtained a better microscope that allowed him to gain an enlarged view of the yeast. With this new instrument he measured them and described "budding," a stage in yeast reproduction. Yeast division begins with a "bud," an earlike protrusion that becomes the new daughter cell. In this way two cells or more may be observed linked together. With such a system he followed the production of gas (carbon dioxide) in a sugar solution and yeast growth in a rich medium used to make beer. In the latter medium, which contained more than all the necessary nutrients required by the yeast cell, the number of yeast cells increased. He carried out a parallel experiment in which he placed yeast cells into a sugar solution with no other ingredients and demonstrated that carbon dioxide was produced (that is, sugar fermentation occurred), but there was no growth of cells. However, as he pointed out, in a medium containing the rich nutrient mixture, also used to make beer, growth and carbon dioxide production, fermentation, occurred. Cagniard de la Tour demonstrated that yeast, under the right conditions, reproduces. A reasonable analysis is that the conversion of sugar to carbon dioxide and alcohol provided the power for the yeast to grow while using the components of the nutrient medium to make new cell material. Cagniard de la Tour also found that similar-sized structures of reproducing yeast cells were present in wines and beer.

Schwann, convinced that yeast reproduced, also carried out microscopic studies. He prepared fresh grape juice and kept it at a temperature of 25 degrees C (77 degrees F). In a few hours a fermentation began. Microscopic observations showed the increase in numbers of globular structures similar to those found in conventional beer fermentations. Schwann characterized the yeast as fungi based on the characteristics of their reproduction. Yeast is a fungus, and Schwann used it as a model cell to reveal the metabolic properties of all cells, animal or plant. It is the cause of fermentation. This was a controversial path to pursue since there was significant opposition to the idea that yeast are the cause of fermentation by Justus Liebig (1803–1873), the premier chemist of the era. To further complicate matters, Liebig also contended that microscopic living agents do not cause contagious diseases of humans, in contrast to an important contemporary version of a living agent theory of the cause of human contagious diseases by Jacob Henle.

G. Jacob Henle (1809–1885)

Henle was an anatomist and a pathologist and a pioneer in the microscopic study of tissues (histology). In 1852 he became Chair in Anatomy and Director of the Institute in Goettingen. Henle was unquestionably influenced by the cell theory of Schwann and Schleiden and the experiments of Schwann and Cagniard de la Tour on the causal connection between yeast and alcoholic fermentation, and mindful of the work by Augostino Bassi (1773–1856) and Jean Victor Audouin (1797–1841) on the cause of the disease of silkworms. In 1835–36 Bassi reported that dead silkworms were covered with a white efflorescence that, on microscopic examination, was revealed to be a fungus. He did not make much progress in understanding the cause of the disease until he explored the relationship between the white fungus and the dead silkworms. He discovered that when the white substance was introduced on the head of a needle into healthy silkworms, at any stage in their morphogenesis, the animals were killed, with the production of white matter. Under the microscope this material resembled a fungus. Bassi concluded that the fungus was a living agent and was the agent of disease because he could use a small quantity of this material to infect healthy worms.

There was a contesting view that the fungus was spontaneously generated, and therefore Bassi requested the formation of a university committee to judge his experimental work. The committee was formed in 1834, its members chosen from the faculties of Medicine and Philosophy of Padua. The report concluded that the white matter transmitted the disease; that it could be inactivated by various chemical means; and that its small size and rapid spread ensured its ability to cause widespread infection.

Nevertheless, there remained the unanswered question: Did the worms die because of the activity of the fungus, or did the fungus proliferate after the worms died? Jean Victor Audouin addressed this crucial issue. He infected worms and followed the course of the disease microscopically, observing the proliferation of the fungus and the destruction of the body of the animal over time until it died. Another commission was appointed in 1838 to review the entire literature, including Audouin's findings. It agreed that the fungus caused the disease. These studies represent the earliest experiments to demonstrate that a fungus, a microscopic, living agent, constructed of cells, can cause a contagious disease of animals.

Meanwhile, Audouin introduced a complication. He concluded, from other studies, that the fungus was initially spontaneously generated, but once generated could continue to infect animals. Henle did not contradict Audouin about his view that the fungus is originally spontaneously generated but was firm in his contention that the disease is only spread by contagion via the live fungi through their spores or parts of their filaments. This is because Henle decided to accept as fact that cells such as the yeast, studied by Schwann and Cagniard de la Tour, and the fungi found in silkworm disease were living agents able to reproduce in their appropriate environment and were the cause of contagious disease. Henle commented,

> "This conjecture [a living agent as the cause of disease] received powerful support through the observations which Bassi and Audouin made recently on a contagious or miasmatic-contagious disease of silkworms, the muscardine. I wish to communicate them coherently here as completely as the importance of the subject demands".
>
> JACOB HENLE, On Miasmata and Contagia, 1840, pp. 36–37,
trans. by George Rosen, 1938, Johns Hopkins University Press

Henle characterized most infectious diseases as "miasmatic-contagious." He provided a different meaning to miasma: it is not something nebulous, labeled bad air, but contains a living agent.

> "The miasma, i.e., that which contaminates, arose as a concept and up to our time, it has remained little more than a concept, since it has neither been perceptibly demonstrable through any of the aids to our senses, nor do we know to which of the kingdoms of nature it belongs or whether it belongs at all to either of them".
>
> JACOB HENLE, 1938, p. 6

The agent that causes the disease can be transmitted through the air or by contact. There are diseases like syphilis and scabies that are contagious but can only be acquired by contact. Henle went on to clarify the concept of a contagious disease: it is one caused by a germ or seed that reproduces itself. He wished that everyone who read his treatise was clear that it was the cause of the disease that reproduced itself; the disease does not reproduce itself. He offered reasons for asserting that causes of contagious diseases were living things, reasons based on analogies.

In fermentations yeast grow to large numbers in a proper medium, initiated by a minute amount of yeast. Various diseases are initiated in identical ways. A small amount of pox on the head of a needle applied to an original site can spread throughout the body. Freshly prepared grape juice does not have visible yeast. If the juice is allowed to remain at a certain temperature for about 36 hours there appear many yeasts and a vigorous fermentation. Henle concluded that all these are examples of a phenomenon carried out by a living agent. Only living agents can reproduce themselves.

Once he had defined the cause of contagious disease, Henle engaged the difficult question of how to experimentally and definitively demonstrate that one unique living agent could cause a contagious disease:

"If under the present conditions of our technical means the question concerning the nature of the contagious agent could be solved with certainty, then a demonstration on theoretical grounds, as I have attempted it, would be superfluous and a very unnecessary, roundabout way. Unfortunately, however, it may be predicted that a strict proof by positive observations is as yet impossible, even if they should speak more in favor of our hypothesis than the former ones.

Even if living, mobile animals or distinct plants are found in contagious materials, then they may have arisen accidentally . . . And even if they would be found constantly in contagious matter and inside the Body, then the objection would still be possible and at first hardly to be refuted, that they were only parasitic . . . that they are elements which may develop in the fluid and may even be significant for the diagnosis without therefore being the active stuff. . . . .

It could be empirically proven that they are actually the active part (the cause of the disease) if one could isolate the contagious organism and the contagious fluid and observe the powers of each separately—an experiment which one must probably abandon (my emphasis).

JACOB HENLE, 1938, p. 42

What Henle did was to formulate a requirement, universally applicable and necessary, to establish a living agent cause of disease. The agent must be obtained in a pure state, so that no other entity is present, to detract from the specificity of the infection, and in that pure state, introduced into a susceptible host to cause the disease. He admitted that the methodology was unavailable to carry out to completion such an experiment for a human disease. Henle rejected the notion that a disease entity contains microscopic life because they are spontaneously generated. In addition he restated his basic principle that every disease is a miasmatic-contagious disease—that is, it is transferred through the air or by contact.

Justus Liebig

Liebig (1803–1873) contended that yeast cells were not the cause of fermentation, nor did microscopic living agents cause contagious diseases. The flavor of his attacks is revealed in the following paragraphs:

"There is no opinion so destitute of a scientific foundation as that which admits that miasms and contagions are living beings, parasites, fungi, or infusoria, which are developed in the healthy body, are there propagated and multiplied, and thus increase the diseased action, and ultimately cause death. A theory of the cause of fermentation and putrefaction, which is utterly fallacious in its fundamental principles, has hitherto furnished the chief support of the parasitic theory of contagion. The opinions concerning the cause of putrefaction, which the adherents of the parasitic theory have formed, are founded chiefly on observations which have been made on the role of yeast in the fermentation of wine and beer".

J. LIEBIG, *Animal chemistry in its application to physiology and pathology*, 1852, pp. 127, 143,145, 3rd edn, ed. by W. Gregory, John Wiley, New York

Liebig did have an alternate theory of both fermentation and contagious disease, describing both in purely chemical terms. Within the same period that Henle published his book on contagious disease, Liebig published two papers in 1839 and 1840, in which he described a new theory of fermentation and a theory of infectious disease. Liebig defined fermentation very specifically as a change in a vegetable material at ordinary temperatures in the absence of oxygen that did not give an unpleasant odor. The yeast, he wrote, is an "insoluble substance" (a nitrogenous substance) that does not cause fermentation. Why did he come to this conclusion? He experimented with yeasts and suspended them in boiled, cooled water. This wash water, presumably free of yeast, carried out fermentation. It seemed clear to him that one did not need the intact yeast cell to obtain fermentation. He believed he had a catalyst that carried out this process. To illustrate this catalytic cascade he also took a very small amount of a fermenting liquor and added it to a very large volume of grape juice, which in a short time was converted to a large fermentation. This procedure was the one used by beer and wine makers. However, instead of a biological explanation provided by Cagniard de la Tour and Schwann, Liebig provided a chemical explanation: he explained fermentation in language used to explain organic chemical changes. A molecule, somehow set in motion, comes in contact with another molecule and imparts its motion to it.

How did he answer the question: Why are yeast always present when alcoholic fermentation occurs? The answer was that yeast is a nitrogenous substance whose molecules impart "motion" to the sugar molecules, causing them to be converted to alcohol and carbon dioxide.

For diseases of various kinds, Liebig appropriated the term *zymotic*. It is derived from the Greek word that means pertaining to or caused by fermentation. A zymotic disease can be a contagious disease produced by some morbid principle or organism acting on the system like a ferment, a catalyst. A catalyst causes fermentation, and a catalyst causes infectious disease. Henle, in contrast, proposed a biological explanation for the analogous phenomena: yeasts cause fermentation while microorganisms cause contagious diseases.

Liebig had considerable support in England for his views. Not only was he an important and influential chemist but also an influential teacher. Many of his English students became imbued with his chemical-physical theories of fermentation and disease and when they returned to England served as a source of support for his chemical theories for both fermentation and infectious disease..

The Mid-Nineteenth Century

Let us briefly review the status of contagious disease theory in the 1840s. There was evidence that plants, animals, and humans are similarly constructed of cells, and the interior of their cells presumably contained the identical components. There was considerable evidence that fungi, microscopic living entities, were present when many plants have diseases. In this climate there might be a favorable response to the possibility that diseases of humans were caused by microscopic, living agents, but it didn't quite work that way. This chapter will continue to follow the controversy about the cause of diseases of plants, and the following chapter will follow the complex controversies concerning the contagious diseases of humans. In each case the fields developed independently of one another. We will consider why this happened.

The Great Hunger

In the August 23, 1845, edition of The Gardeners' Chronicle and Agricultural Gazette, an editorial announced the outbreak of a disease of potatoes in Belgium. The disease was also present in other countries in Europe and was now found in England. "It spread faster than cholera amongst men." The disease was described in this way:

"The disease consists in a gradual decay of the leaves and stem, which become a putrid mass, and the tubers are affected by degrees in a similar way. The first obvious sign is the appearance on the edge of the leaf of a black spot, which gradually spreads; then gangrene attacks the haulm, and in a few days the latter is decayed, emitting a peculiar and offensive odour".

What was the cause? The editor, John Lindley (1799–1865), who with two other individuals had been appointed by the government to inquire into the potato disease, wrote in the same issue of The Gardeners' Chronicle that the disease is generated by environmental factors. We reproduce his argument to understand the substance of his theory:

"The cause of this calamity is, we think, clearly traceable to the season. During all the first weeks of August the temperature has been cold—from 2 to 3 degrees below average;—we have had incessant rain and no sunshine. It is hardly possible to conceive that such a continuation of circumstance should have produced any other result, all things considered.

The Potato absorbs a very large quantity of water. Its whole construction is framed with a view to its doing so; and its broad succulent leaves are provided in order to enable it to part with this water. But a low temperature is unfavorable to the motion of the fluids, or to the action of the cells of the plant; and, moreover, sunlight is required in order to enable the water sent into the leaves to be perspired. In feeble light the amount of perspiration from a plant, is comparatively small; in bright sunshine it is copious; in fact, the amount of perspiration is in exact proportion to the quantity of light that falls upon a leaf. At night, or in the darkness, there is no appreciable action of this kind. During the present season all this important class of functions has been deranged. The Potatoes have been compelled to absorb an unusual quantity of water; the lowness of temperature has prevented their digesting it, and the absence of sunlight has rendered it impossible for them to get rid of it by perspiration. Under these circumstances it necessarily stagnated in their interior; and the inevitable result of that was rot, for a reason to be presently explained. If the first days of July had not been suddenly hot it would not have happened; if we had had sunlight with the rain it would not have happened; and, perhaps it would not have occurred had the temperature been high, instead of low, even although the sun did not shine, and rain fell incessantly. It is the combination of untoward circumstances that has produced the mischief.

The mischief, although very general, is not universal.

Here, then, we find what we look upon as good evidence of the justice of the opinion which we have ventured to express as to the cause of the pestilence in question. The whole crop is not attacked, because, although the land is ill-drained, yet the recent trenching, and the decaying turf just below the Potatoes, have been sufficient to carry off the water. Where it was otherwise, that is where road sand was thrown down to some depth, and consequently the turf drainage (damaged), there, and there only, the disease appears".

Writers in Belgium hypothesized conditions different from Lindley's that, however, led to the same end-result, a moisture-laden potato that undergoes putrefaction.

The Rev. M. J. Berkeley, an acknowledged expert on fungi who knew about the extensive work on the diseases of wheat called rust and bunt that implicated fungi as causative agents of these diseases, and observing the progression of the potato blight in his own parish, proposed "the revolutionary theory that the mould might be the cause and not the consequence of the Potato Murrain." (E. C. LARGE, *The Advance of the Fungi*, 1940, p. 15, H. Holt and Co., New York)

It is somewhat puzzling that the author of this 1940 classic text on fungi, E. C. Large, characterized this suggestion as revolutionary when fungi had been implicated in plant diseases for more than 100 years. Perhaps, in the face of opposition to this idea by Lindley and the Commission, Large considered Berkeley's view revolutionary.

A Dr. Montagne, in France, sent diseased plant material to Berkeley, who found that the infected leaves from France resembled the diseased potato leaves in fields in his home area. Montagne also sent Berkeley detailed drawings of the growth of the fungus on leaves, which Berkeley published in 1846. Berkeley was convinced that the fungi caused the disease. Lindley, a Professor of Botany as well as the editor of the Gardeners' Chronicle, held that the fungi were the consequence of the disease, in short spontaneously generated. Lindley's theory would also accommodate the argument that the fungus is a saprophyte (it grows on dead tissue of the plant), and consequently the presence of the fungus was coincidental and not causative.

Both scenarios, by Lindley and the Belgians, discounted a living agent theory of the disease of potatoes. Berkeley countered with experimental evidence that appeared to invalidate Lindley's arguments and those of the Belgian group. The disease was not "peculiar" (that is, unique) to the year 1845, but was also destructive in 1844, for example, when there was no similarity to the weather of 1845. In fact, in many districts the disease broke out during very dry weather. Moreover, measurements of the amount of moisture in potatoes "taken from the ground in the month of October," when the potatoes were "in an extremely diseased state," did not show any excess of moisture.

Berkeley concluded, "I think then it is at least plain that no supposed peculiarities of season are sufficient without some more specific cause to account for the general prevalence of the disease." Berkeley appeared to have effectively countered Lindley's arguments for an environmental cause of the disease; however, this logic does not prove that a living agent causes the disease. Lindley was aware that there was experimental support for a living agent theory of disease of plants, beginning with the work of Tillet and Prevost and reports of fungal involvement in the potato disease from United States and other European countries. Furthermore, Lindley knew Berkeley's views, since he had consulted Berkeley as a member of the government committee charged with investigating the disease. Why, then, did he take the strong position that the disease was due to a serious environmental disturbance? If one accepts the belief that Lindley was relying on evidence for his position, it is possible to suggest some observations that indicate fungi are not the cause of the disease.

It is true that in certain weather conditions the disease spreads rapidly. In fact, it is generally recognized that the growth of fungi, in contrast to the growth of other "vegetables," depends on atmospheric conditions. It is obvious that the growth of mushrooms, which are fruiting bodies of fungi, is produced after periods of rain and humidity. It also happens that a diseased potato may contain more than one fungus, which suggests that a number of fungi can emerge from the diseased tuber. But perhaps the most influential factors for Lindley are simply the belief that the disease is not caused by a living agent and the conviction that spontaneous generation is a reality, thus leading to the presence of fungi.

In 1845 all would agree that when there is a disease of plants, including the potato blight, one fungus (or more) is always present. Tillet and Prevost had demonstrated this fact. The issue that remained to be demonstrated is whether a specific fungus was a necessary cause and not the consequence of a prior event.

Anton de Bary (1831–1888)

In 1853 de Bary published a treatise, "Investigations of the Brand Fungi and the Diseases of Plants Caused by Them With Reference to Grain and Other Useful Plants." De Bary, convinced that fungi caused numerous diseases of plants, gathered together in this work a substantial body of evidence in support of this position and later, in 1861–63, established experimentally a unique infection.

De Bary was born in Frankfurt, Germany, a son of a physician who was versed in natural sciences and a mother whose family contained scientists. He took numerous field trips with family and among his friends was a botanical gardener. He devoted himself to the study of botany. On an island in the Main River his father created a garden, where they cultivated many plants. During the years of 1848–53 he also studied medicine and received his medical degree in 1853. However, after a short time he abandoned medicine for plant study, becoming a lecturer in that subject at the University of Tubingen. The next year he was appointed a special professor in botany at the University at Freiburg and in 1859 was made full professor. The 1853 review was written while he was a 22-year-old medical student.

De Bary's goal was to establish the individuality of different fungal species as part of his attack on the theory of spontaneous generation. Different fungi display different growth patterns, and they are all composed of cells:

"The forms we have considered . . . identify themselves as fungi by their mycelium, out of which the fertilizing parts are formed later and, with respect to its form and the manner in which the spores develop from it, agrees completely with that of many other growths recognized to be fungi. It has been proved . . . that the growths which bear the collective name [fungi] . . . all have one and the same cell-structure. (p. 54)

These . . . . . . cells, which in the simplest cases such as the yeast fungus are globose and which continually multiply by the "tying off" of a daughter-cell, arising as an outgrowth, a cell bud as it were, from the mother cell". (p. 55)

ANTON DE BARY, *Investigations of the Brand Fungi and the Diseases of Plants Caused by Them with Reference to Grain and Other Useful Plants*, 1969, Phytopathological Classics No. 11, University of Wisconsin

In this work he summarized structures and development of Brand fungi and presented a classification scheme. (Brand is a German word used as a general term applied to blights of plants; its English equivalent is burn.) He also presented evidence that they cause disease in the form of four sets of field experiments. When these diseases, brands and rusts, are present, fungi from various specific groups are involved:

"The smut . . . is distinguished by the fact that it destroys the tissue which it attacks and causes them finally to disintegrate into blackish-brown dust. This consists of the spores of Ustilago species. The . . . forms of diseases which are called rust, or rash, . . . are accompanied by the appearance of fungal forms from the genera [de Bary cited three different genera]. The rust of the various kinds of grain shows either reddish-yellow spots . . . or blackish-brown ones" . . . (p. 64)

De Bary presented these observations to support his contention that specific fungi cause specific diseases, part of his continuing critique of spontaneous generation. However, he had not completed the dismantling of this theory. He described the various manifestations of this hypothesis and proceeded with a blistering criticism:

"The advocates of the views just stated . . . either deny entirely or leave undetermined the Contagiousness of the Brand diseases, relying upon the negative results of their often quite scanty experiments So far as the Brand fungi are concerned, it has been shown in the foregoing that they show the most complete agreement in the essential features of their structure, in development and growth with many organisms proved to be independent, which arise from germs; that they are indeed true fungi. (p. 68)

Moreover, an origin by *generatio aequivoca* [equivocal generation] has by no means been demonstrated for the fungi; . . . and while previously all fungi were believed to arise through spontaneous generation, this has, since the discoveries by Ehrenberg [referring to a paper in 1820] increasingly lost support as investigation proceeded . . . It must now be decided whether the Brand fungi are parasites or products of diseased conditions, in other words whether they, together with the disturbances in the plant's life accompanying their appearance, are to be regarded as cause or as effect. In the latter case they would be incontrovertible proof for a *generatio spontanea*; however, the facts on which the proponents of this view support their argument turn out to be in part the products of delusions and in part to have been the products occasionally of a truly deplorable mania for theories and analogies". (pp. 69–70)

It is clear that de Bary, at the age of 22, had completely lost his patience with regard to spontaneous generation theorists, but he was not finished:

"The mycelium filaments are always the first thing that is found in blighted plant parts; from them the spores arise in various ways; therefore, concerning them it must be decided how they arise, whether and how they get into the plants.

By gathering together all certain, precise observations and experiments of the authors, the entry of these into the plants by growth from outside, even though not absolutely proved, has been put almost beyond doubt". (p. 70)

This last statement is very important since it establishes the independent growth of the fungus, outside the plant. The fungus subsequently infected the plant rather than emerging from the decayed plant material.

One experiment that was evocative of Mathieu Tillet's dealt with treating seeds with agents that are detrimental to fungi. Here are the results, adapted from de Bary:

What was sowed:	What was harvested:	
	Good ears	(a) Smutted ears
1. Wheat alone	806	2
2. Wheat plus smut	210	463
3. Wheat plus smut, lime water, and arsenic	600	44
4. One-half level of wheat plus smut dust	67	375
5. One-half level of wheat, smut dust, lime, and arsenic	443	44

De Bary commented, "these experiments . . . clearly speak for an infection."

De Bary's Experiments With a Specific Fungus

From the late 1840s to the early 1850s a significant change occurred in the identification of the fungus associated with potato disease. Its structural characteristics were recognized. Its growth as a white fringe on the healthy green part of potato leaves clearly provided the argument that its origin was independent of the plant, and certainly independent of the decaying plant tissue.

De Bary devised a method to infect a host with a pure population of one kind of fungus. He gathered spores of the fungi, separating them from all parts of the complex makeup of the organism, suspended them in water, and observed them microscopically. Two things happened to the spores in water. Some germinated and put out a thin tube, while others underwent internal division and released motile bodies. It is obvious that they could undergo division outside the plant. When these spores were deposited on potato leaves in droplets of water they began to divide; they sent out filaments that entered the interior of the leaf. In time there were produced in the leaf many filaments that contained many spores. These results conclusively demonstrated the ability of the potato fungus to enter and multiply in healthy plants. He followed with experiments to show the development of the disease of potatoes. He grew uninfected potato plants in many pots. Half the potted plants were infected with spores, while the others were not and were covered so that there was no possibility of unintended infection. Within a few days the inoculated plants began to show spots of infection on the leaves, identical to the symptoms shown by plants infected in the field. The disease progressed until the plants were completely destroyed. Meanwhile, the control plants remained intact and grew free of disease. The destruction occurred rapidly, reproducing the speed of the infection in the field, where millions of spores were produced.

He subsequently addressed the question: Did spores reach potatoes that grow beneath the ground? He accomplished this infection by scattering spores on the surface of the soil.

De Bary's experiments on potato blight disease, carried out in the years 1861–63, definitively established, for the first time, that a microscopic living organism, a unique species of fungus, caused a specific plant disease. These experiments also proved the contagious nature of this disease. "Contagious" connotes that a disease agent developed in one host can produce the same disease when transferred to a new host. Such a model of contagion explains the reason for epidemics in plants. These results preceded by about two decades the definitive demonstration that a human contagious disease is caused by a living, microscopic agent. The cause of human contagion is discussed in the following chapter.

| **The Nineteenth Century**

The Cause of Human Contagious Diseases
in the Nineteenth Century: Continuation
and Resolution of the Spontaneous Generation
Controversy; Reality of the Analogy That
Living Microorganisms Cause Fermentation,
Living Microorganisms Cause Contagious
Diseases; Bacteria Cause Contagious Diseases
of Humans

"There is no opinion so destitute of a scientific foundation as that which admits, that miasms and contagions are living beings; parasites, fungi, or infusoria, which are developed in the healthy body, are there propagated and multiplied, and thus increase the diseased action, and ultimately cause death

J. LIEBIG, Animal chemistry in its application to physiology and pathology,
1852, p. 127, ed. by W. Gregory, 3rd edn, John Wiley, New York

"This conjecture [a living agent as the cause of disease] received powerful support through the observations which Bassi and Audouin made recently on a contagious or miasmatic-contagious disease of silkworms, the muscardine. I wish to communicate them coherently here as completely as the importance of the subject demands".

JACOB HENLE, On Miasmata and Contagia, 1840, pp. 36–37, trans. by
George Rosen, 1938, Johns Hopkins University Press

These quotes from Liebig and Henle convey unambiguously the contentious nature of human contagious disease theory (all contagious disease theory) in the first half of the nineteenth century. Controversy about the cause of disease was obviously not a new phenomenon; previous eras had contained discordant views. However, in the first half of the nineteenth century, there were new, important experimental ideas in biology and chemistry but there was not yet any substantial evidence about the causes of the major human diseases, such as tuberculosis, smallpox, the newly arrived disease cholera, childbed fever, and many others. In addition there was no general agreement that each contagious disease of humans might have a unique, specific cause. There were unique diseases but no distinctive causes: there were the same numerous different causes for different diseases. This was the faulty reasoning that Jacob Henle criticized in an 1844 essay:

"Only in medicine are there causes that have hundreds of consequences . . . Only in medicine can the same effect flow from the most varied possible sources . . . This is just as scientific as if a physicist were to teach that bodies fall because boards or beams are removed, because ropes or cables break . . . and so forth".

JACOB HENLE, "Medicinische Wissenschaft und Empirie," *Zeitschrift fur rationelle Medizin*, 1844, 1:1–35, p. 25

Henle commented, of course, that things fall when cables break. Henle, however, was reminding medical theorists that there is a general principle or law that governs falling bodies; in the same way there must be a general principle governing the cause of contagious disease that would include a specific cause for a specific effect. In the first half of the century diseases such as cholera and childbed fever were considered to have multiple causes, reflecting the state of theory at this time. The transition between multiple causes of disease and the establishment of an entity that is a specific living agent for one disease took place over a period of three decades during the last half of the nineteenth century. Parallel to these controversies, and integral to them, were the fierce continuing disputes about spontaneous generation and the cause of fermentations. The occurrence of a specific disease dependent on the presence of a particular bacterium, a living, microscopic, cellular agent, represented a revolution in biology and medicine and consigned 2,000 years of theorizing about the cause of contagious disease to the province of historians and philosophers. The change that took place was at once a theoretical and practical development that relied solely on experiments to justify hypotheses proposed for the causes of individual diseases. Such a powerful combination of theory and practice had not happened before.

At the end of the century laboratory methodology dominated the field, a methodology so powerful that within a 20-year period (1880–1900) the causes of many major human diseases were shown to require the presence of a unique bacterium. We shall follow the development of these remarkable discoveries that occurred more than 200 years after Leeuwenhoek discovered *animalcules* and William Harvey expressed perplexity about the cause of infectious disease. It took that amount of time to arrive at the cause.

At the end of the nineteenth century there appeared another mystery, the discovery that there are diseases that are contagious even though no bacteria nor fungi were implicated in the disease. This led to the discovery of viruses whose identity and causal relationship to diseases of plants, farm animals, humans, and bacteria would be worked out over the following 60 years (that story will be narrated in Chapter 13).

Initially, I shall present the discourse about the causes of two human diseases, cholera and childbed fever. Analyses of the debate about these diseases, with their multiple causes, reflect the contemporary state of disease theory in the first half of the nineteenth century.

Cholera

On Sunday evening, October 16, 1831, a woman in Sunderland, England, in seemingly good health, started vomiting at about midnight and was seized with extreme diarrhea. She was dead in about 16 hours. This was the first reported case of cholera in England in the epidemic of 1831–32. The disease spread north to Edinburgh and south to London by January 1832. Cholera had never appeared earlier in Great Britain but was endemic in India and Bengal. The disease moved to Europe and reached Hamburg, Germany, a major port for shipping goods to England. The sudden onset and severity of cholera focused attention on the disease by governmental authorities, medical societies, scientific journals, and the press. The most important response to the disease was by sanitary authorities, the Metropolitan Sanitary Commission and the Board of Health in London, whose members were part of the reform movement in Britain influenced by the philosophy of Jeremy Bentham (1748–1832). One response was to appoint a Royal Commission to reform the current Poor Law, which was some 200 years old and inadequate to cope with the degraded environmental and social conditions created by the increase in population in urban areas stimulated by the Industrial Revolution. The elderly, the poor, and orphans suffered most during the epidemic. Among the members of the Commission was Edwin Chadwick (1800–1890), a disciple of Bentham, who wrote on sanitation in the *Westminster Review*, a publication founded by Bentham. Chadwick investigated the horrendous conditions under which the poor lived in London and was responsible for a major portion of the new regulations in the Poor Law of 1834. For the new commissioners, which included Chadwick, sanitary reform was essential to provide good health for the poor so they could furnish a proper livelihood for themselves. Chadwick believed there was a causal connection between deplorable sanitary conditions and disease, as well as poverty and crime. To understand the grounds of Chadwick's philosophy concerning the cause of disease, a few excerpts from his 1842 publication on the sanitary conditions of the laboring population is useful.

Local reports on the sanitary condition of the labouring population in England: in consequence of an inquiry directed to be made by the Poor law commissioners, 1842, London

First, there is wretched housing:

"broken window panes . . . filth and vermin in every nook . . [walls] black with smoke of foul chimneys, without water, with floors unwashed from year to year". (p. 361)

There is also the outside environment:

. . . "there are streets elevated a foot, sometimes two, above the level of the causeway, by the accumulation of the years, and stagnant puddles, here and there, with their foetid exhalations, causeways broken and dangerous, ash-places choked up with filth, and excrementitous deposit on all sides as a consequence, undrained, unpaved, unventilated, uncared-for by any authority but the landlord, who weekly collects his miserable rents from his miserable tenants . . . **Can we wonder that such places are the hot-beds of disease** (my emphasis)"?

In France there were similar theories by social reformers of the connection between poverty and its attendant environmental decay and disease. In 1834 a report by a cholera commission sponsored by the Academy of Moral and Political Science pronounced the theory that contagion (disease) was a social problem. Social reformers like Louis-René Villermé (1782–1863) made the association between poverty and disease. Poverty itself was a disease that led to immorality, which sustained the syndrome. There were a variety of suggestions for other causes of the disease. Individuals in England whose theoretical grounding of disease was based on sanitary principles wrote of "epidemic atmospheres," a theory of miasms, where the air would be contaminated, poisoned, by putrid exhalations from the filth in the environment. This view was supported by the various reports of the Metropolitan Sanitary Commission and the General Board of Health. There was ambiguity about the possibility that the disease was contagious—that is, it passed from person to person. It was suggested that the evidence for such a view was not sufficient. Needless to say, the evidence for transfer through bad air was equally insufficient.

There were some statistical data on mortality that suggested that the disease was present in certain environments, which argued for the presence of instigators of disease in particular places. However, there appeared to be little effort to engage the issue of the sudden appearance of the disease in various parts of England when the conditions for the outbreak of the disease were already present without the presence of cholera. After 1832 there were sporadic cases of cholera, but no epidemics. Between the first epidemic of 1831–32 and the second one in 1848 there emerged a greater emphasis on the cause of the disease that did not detract from pressure to improve sanitary conditions.

The Return of a Living Agent Cause of Contagious Disease

In January 1848 Charles Cowdell published an unusual and unexpected book with the lengthy and informative title, "A Disquisition on Pestilential Cholera; Being an attempt to explain its phenomena, nature, cause, prevention, and treatment by reference to an extrinsic fungous origin." (Samuel Highley, 1848) Cowdell was a provincial physician living in, as he described it in the preface, an isolated town, Oundle, Northamptonshire, about 45 km northwest of Cambridge. There was no medical institution or library, and he had no opportunity to travel to metropolitan libraries; his only resources were his own books, and he did not have Henle's writings on contagion.

Cowdell believed that his book was different from the other works available to him. These writings contained the contemporary views of the cause of the disease: the environmental causes of the sanitary commissions and the opinions of various physicians who had experience with cholera epidemics in India, for example, like Edmund Parkes (1819–1876) and George Budd (1808–1882) in England. He also had a work by Henry Holland (1788–1873), a theoretical treatise involving animalcules as the cause; however, they were not further identified. Budd was designated as an anti-contagionist, while Cowdell argued for contagion as the cause of disease. Parkes' position was that there was a morbid agent, a poison. He wondered what sort of poison it is and its origin which he acknowledged was unknown.

Parkes' view would be consistent with a chemical theory of disease. Cowdell's theory was obviously different. For Cowdell to arrive at his position he adopted one important basic principle: there is a specific agent that causes the disease, not some complex of environmental conditions. What remained to be identified was the nature of the agent. To do this he had to make choices, just as he took sides among the apparently conflicting evidence about the way disease is spread. Here is the important decision he made:

> . . . "a close analogy exists between the generation and propagation of the pestilential virus [at this point he resorted to this term] and the above-named disease [cholera], and the phenomena observed in the development of the lowest orders of the vegetable creation, as to warrant the conclusion, that between the virus and the germs of some species of protophyta—fungi— there exists a positive identity".
>
> CHARLES COWDELL, *A Disquisition on Pestilential Cholera; Being an attempt to explain its phenomena, nature, cause, prevention, and treatment by reference to an extrinsic fungous origin*, 1848, p. 100, Samuel Highley

Cholera is caused by a fungus. The literature on fungi and their role in nature was compelling and decisive for Cowdell. There are fungi parasitic on wood and fungi that cause blight of corn and cause other diseases known as smut, rust, ergot, and many kinds of mildew. Cowdell also cited the contemporary devastating potato disease that, according to Miles J. Berkeley, was caused by a fungus, although this was disputed by John Lindley, a member of the government commission to investigate the potato disease, who believed the fungi were the result of spontaneous generation and thus were the consequence of the disease and not the cause. Cowdell continued with other analogies. Fungi cause a disease of silkworms. Cowdell asserted that the mode of reproduction of fungi by the large numbers of "sporules" could account for epidemics. Like Henle, he referred to the analogy between fermentation carried out by yeast cells and diseases conducted by fungi. He engaged Liebig's chemical theory of fermentation but concluded that the phenomenon of fermentation "can only be explained on the supposition on their [yeast] being an organized living body."

Cowdell chose two controversial positions: fungi cause fermentation and contagious disease. The dispute was not about their presence in each of these circumstances: they are present during alcoholic fermentation and they are present in a plant disease and in a disease of silkworms. What was in dispute was their role in these processes. There was no conclusive evidence that they played the role that Cowdell assigned to them, particularly for fermentation and plant diseases. Thus, he ultimately made a choice by selecting certain putative facts and rejecting others. He and Henle made similar choices.

Cowdell's fungal theory for the cause of cholera, a biological theory, reflected Henle's theory of contagious disease caused by a living biological agent published some eight years before. This fact was pointed out in a review of Cowdell's book in the *Monthly Journal of Medical Science*. First referring to Cowdell, the reviewer stated, the proposal was not new since a "masterly exposition" was published by Professor Jacob Henle where he noted that microscopical observations had implicated fungi in the diseases of plants and the muscardine of silkworm.

The critique, simply stated, was that Cowdell's theory was explained more expertly by Henle, and the evidence for a fungal cause of disease of plants and silkworms was not definitive. The reviewer stated that the diseases *may* be due to a fungus, indicated by microscopic studies, but appears to ignore the experimental evidence that the disease can be transmitted to uninfected silkworms using fungi. Henle would have agreed that microscopic evidence was not enough to establish causality.

A second cholera epidemic occurred in 1848, again originating in the east and spreading to Europe and the Western Hemisphere. In England, a Board of Health was appointed by the Privy Council. It included Edwin Chadwick and Anthony A. Cooper (1801–1885), the Earl of Shaftesbury. Neither was a physician, but both were relentless social reformers, primarily concerned with sanitary conditions as the primary cause of disease. Consistent with this view Chadwick believed in a miasmic, airborne, conveyer of disease. Poverty and filth were at the heart of the cause of the disease. It was indeed the case that the conditions of the lives of the poor continued to be abominable in the cities and were not much better in the outlying districts. One example of such an atrocity perpetrated on the poor was an asylum for children near London, called Drouets, that suffered an attack of cholera:

"Drouets was one of the infamous child farms where people sent their destitute young . . Here children who were orphans or whose parents couldn't afford to keep them were shut away and largely forgotten . . . the less spent on the children's food, clothing, and care, the fatter the profits to be had.

Fourteen hundred gaunt-faced children, some as young as three, were starved half to death, their bellies swollen . . . their joints deformed, their skins covered with boils and ulcers".

SANDRA HEMPEL, *The Strange Case of the Broad Street Pump, John Snow and the Mystery of Cholera*, 2007, p. 110, University of California Press

News of the children's plight reached local authorities and then London some days later. Chadwick sent Dr. Richard Grainger (1801–1865), a Fellow of the Royal Society and a member of the Council of the Royal College of Surgeons, to investigate. He confirmed the appalling conditions. The children were removed to other institutions, but by the end of January 1849, 180 children had died of cholera. On January 20, 1849, a story appeared on the front page of the weekly newspaper *The Examiner*, which was unsigned but written by Charles Dickens, describing accurately the conditions at the institution. He also believed that the living conditions were responsible for the epidemic:

"The diet of the children is . . . unwholesome and insufficient . . . their clothing shamefully defective. Their rooms are cold, damp, and dirty and rotten . . . of all conceivable places in which pestilence might—or rather must—be expected to break out and to make direful ravages, Mr. Drouets' model farm stands foremost".

SANDRA HEMPEL, *The Strange Case of the Broad Street Pump, John Snow and the Mystery of Cholera*, 2007, p. 125, Univ. of California Press

Grainger's report contended that the disease was noncontagious, that the noxious atmosphere generated from putrid matter and the miserable living conditions of the children were responsible for the disease. The Board's report was subject to some criticism; the most important was that the Board seemed incapable of considering that the disease could be contagious. In the July 13, 1849, issue of the *Medical Gazette*, the Board was given credit for the cause of sanitary improvement but was faulted for having only one physician and being "wedded" to sanitary conditions as the cause. Before later reports were produced, a number of new contributions to the discourse about the disease, in addition to the work of Cowdell, were published.

In England during the 1840s there were established "Microscopical Societies," one in London and the other in Bristol. Early microscopists did not interest themselves in medical subjects, but things changed when John Simon (1816–1904), who became the first lecturer in pathology at

St. Thomas's Hospital and the First Medical Officer of Health for the City of London in 1848, introduced the microscope into medical laboratory work. Although microscopic studies were declared to be inadequate to establish causality, the use of microscopes in the medical laboratory setting created potentially a new body of evidence on the cause of cholera. This story has its origin in another Bristol institution, the Bristol Medico-Chirurgical Society.

Bristol had experienced the cholera epidemic of 1832, and when another epidemic struck in June 1849, the Society appointed a subcommittee to deal with the disease that included William Budd (1811–1880), Joseph Swayne (1819–1902), Frederick Brittan (1823–1891), and several others. All were physicians. Budd had expertise in infectious disease and in 1845 had lectured at the Bristol Literary and Philosophical Institution on the potato disease using microscopic studies. On July 9, 1849, the subcommittee met in Budd's house. He had obtained from his cholera patients specimens of the fluid diarrhea characterized as "rice water evacuations." Brittan and Swayne separately examined this material microscopically, a novel approach since its use in the diagnostic medical laboratory had only recently been introduced. They observed similar structures—"certain bodies in considerable abundance"—and concluded "that they were characteristic of the evacuations of cholera, if not the very agents causing the disease." The entire committee was more circumspect but acknowledged that something new was observed.

On September 27, 1849, Budd published a short treatise, "Malignant Cholera: its causes, mode of propagation, and presentation." He opened the essay with a reference to a report that had appeared in the London Medical Gazette "of last Friday," where Brittan's microscopic observations were revealed. Budd commented that the presence of these peculiar microscopic objects in rice water discharges from persons infected with cholera was a "very important discovery." He presented a theory of disease formulated on the presence of these bodies: that the cause of malignant cholera is a living organism of a distinct species and is ingested into the intestine, where it multiplies just as any living organism can. When this happens, he theorized, the disease occurs. He was "inclined" to identify the bodies as fungi.

This was quite a comprehensive disease theory. Where did all this come from? Like Henle and Cowdell, he provided reasons based on analogies and accepted as fact that yeast cause fermentation and fungi cause diseases of plants. The latter conclusion, he stated, "was a well established field of inquiry" (and indeed it was, but still in some dispute). He did not believe that the disease was transmitted through the air, but rather through the water supply. He arrived at this conclusion independently, but he acknowledged in a postscript the "ingeneous pamphlet on cholera by Dr. Snow [John Snow, 1813–1858] who deserves the whole merit for this finding."

In fact, Snow deserved this priority beginning with his landmark epidemiological study published in August 1849, one month prior to Budd's publication. We will shortly examine Snow's works. Budd, to drive home his message of a fungal cause of disease, also provided in the postscript reference to Cowdell's book:

"The preceding pages were prepared before my attention was called to a work published in the early part of 1848, by Dr. Charles Cowdell, entitled "A Disquisition on Pestilential Cholera." In this work which displays great ability and learning, Dr. Cowdell concludes, from an elaborate review of the symptoms, character, and course of the cholera, that Fungi constitute the morbific agent in the propagation and diffusion of this pestilence. I need not add that this work brings strong and to me unexpected confirmation of the truth of much that has been advanced in the preceding pages".

WILLIAM BUDD, *Malignant Cholera: Its Mode of Propagation and Its Prevention*, 1849, p. 28, London, Churchill

Budd's hypothesis was based on the same writings used by Cowdell to support his theory. There was introduced a new piece of evidence from microscopic studies that alleges the presence of unique bodies labeled fungi in the intestinal contents of cholera victims. Given the complex contents of intestinal discharges in rice water material, the difficulties of characterizing one peculiar body were insurmountable. Even if such an effort were possible, the criticism directed at Cowdell in the Monthly Journal of Medical Science would hold that the presence of an agent, however unique, would not constitute evidence for causality as Henle pointed out in 1840.

Nevertheless, it is significant that a living agent theory for a human disease gained some adherents in the 1840s to the 1850s and that non-human diseases were used as analogies to account for human disease.

Cholera Is Communicated Through Water

"One individual, however, stands out during these years for his conviction and singular belief that contaminated water was the main means of the spread of cholera".

STEPHANIE J. SNOW, Commentary: Sutherland, Snow and water: the transmission of cholera in the nineteenth century *Intl. Jour. of Epidemiol*, 2002, 31: 908–911

John Snow published his first work on cholera in August 1849 ("On the Mode of Communication of Cholera", in the London Medical Gazette) and summarized his views on October 13, 1849, before the Westminster Medical Society. Snow had an unusual training career in medicine: he first passed an exam to be an apothecary, then a surgeon, and obtained his medical degree from the University of London. His only previous contact with cholera was during the 1831–32 epidemic, when he experienced the conditions of miners in a colliery during the epidemic. He noted that the miners ate and defecated where they worked, which created the conditions for the transfer of cholera among them. On the second page of his first communication, Snow stated a principle about the general spread of cholera that defined for him that the disease was communicable, that it was a contagious disease. After briefly surveying its origin in India, he wrote,

"There are certain circumstances, however, connected with the progress of cholera, which may be stated in a general way. It travels along the great tracks of human intercourse, never going faster than people travel, and generally much more slowly. In extending to a fresh island or continent, it always appears first at a seaport. It never attacks the crews of ships going from a country free from cholera, to one where the disease is prevailing, till they have entered a port, or had intercourse with the shore. Its exact progress from town to town cannot always be traced; but it has never appeared except where there has been ample opportunity for it to be conveyed by human intercourse".

JOHN SNOW, *On the Mode of Communication of Cholera*,
1855, p. 2, 2nd edn, John Churchill, London

Snow then offered examples of the spread of the disease that formed the basis of his "mode of communication." For one incident he returned to the epidemic of 1832. John Barnes, an agricultural worker, had been suffering for two days from extreme abdominal cramps and diarrhea. He was attended by a "respectable" physician who attempted to find the source of the infection without success. Barnes died the following day. His wife and visitors contracted a mild form of the disease. As Snow framed the scenario, the "mystery" was "unraveled" when Barnes' son arrived from Leeds, where he was apprenticed to his uncle, whose wife (John Barnes' sister) had died of cholera two weeks before. Her unwashed clothes were sent to John Barnes, who "opened the box in the evening" and on the next day fell sick with cholera. According to Snow, he contracted the disease via an intermediary. Snow offered another example of the communicability of the disease. He had observed that miners suffered from cholera more than "any other occupation." Why? Snow's answer was that miners spend about eight to nine hours per day in the pit and take with them food and drink. "The pit is one huge privy, and of course the men always take their victuals with unwashed hands." The disease could thus pass easily among the men. This view was contested in a report to the College of Physicians: women and children, it stated, who do not work in the mine contracted the disease as often as the men. Why would anyone raise this objection? If one believes that the disease is noncontagious, there could be no direct transfer of the disease from the men to their families. Under those circumstances, the disease must be transferred by a miasm. Snow simply contended that the men brought the disease home at a time when they did not display symptoms. He concluded that "if a special inquiry were made on this point this would probably be found to be the case."

Transmission of cholera via the water supply was reinforced by Snow's study of the occurrence of the disease in a street, Albion Terrace, in the Battersea district of London. This was a suburban neighborhood of middle-class inhabitants, absent the awful environment of the poor districts. During the summer of 1849 more than half the residents of the 17 houses on the street contracted cholera, and 24 people died. Snow considered the extent of mortality unprecedented in this country "at the time." Surrounding streets were free of the disease. All residents received water from the same source. A pipe led to each house into the kitchen. There was a cesspool behind each house under a privy situated four feet from the water source. Snow reported that a Mr. Grant, Assistant Surveyor to the Commissioner of Sewers, found breaks in the water pipes and observed the exchange of fluid between the water and the cesspool. Snow examined water removed by Mr. Grant from tanks behind houses and found it to have an odor of privy soil. Snow stated, "I found in it various substances which had passed through the alimentary canal having escaped digestion."

The official report of the General Board of Health made a connection between the occurrence of the disease and the unhealthy conditions of the water supply. Recall that the Board of Health had been accused of being preoccupied with sanitary conditions such as filth and decay that fouled the air. In its report of 1848–49 the Board added "poisonous" water to those "predisposing factors." Snow's conclusion was more definitive: "It remains evident then that the only special and peculiar cause connected with the great calamity . . . was the state of the water." This conclusion was supported by the General Board of Health.

During another cholera epidemic, in 1854, Snow carried out two classic epidemiological studies, one in the Broad Street region of London and the other in South London. In London's streets there were a large number of public sources of water delivered by local pumps. In his book Snow provided a map of the Broad Street neighborhood (between pp. 44–45) locating seven pumps in the area, three about 230 to 240 yards from the Broad Street pump and the farthest about 400+ yards away. Snow obtained mortality lists from the Registrar-General's office from the area around the Broad Street pump and found that a high percentage of the deaths from cholera were clustered around and near this pump.

Snow cited the circumstances leading to the death of a person who did not live near the pump but had taken water from the pump, "which are perhaps the most conclusive of all in proving the connexion between the Broad Street pump and the outbreak of cholera." On September 2, a 59-year-old widow who lived in the West End died of cholera. There was no cholera present in the West End. This woman had not been in the Broad Street neighborhood for many months, according to her son; however, each day a cart carrying a large bottle of Broad Street pump water was brought to her. On August 31, a Friday, water was brought to her; 24 hours later "she was seized with cholera" and on Sunday she died. A niece had visited the widow and drank the water; she returned to Islington, where she died of cholera. There was no cholera in Islington.

The other epidemiological study carried out by Snow involved the epidemic in South London. In the previous outbreak of 1849 this area of London had been supplied by two water companies, Lambeth and Southwark-Vauxhall. Both obtained water from the same region of the sewage-polluted Thames River. The cholera levels among people using either water supply were the same. During the years late 1849 to August 1853 London did not have any cholera epidemics, and during this time there was an important change in water supply to the southern districts of London. In 1852 the Lambeth company moved its waterworks from the Battersea area to Thames Ditton, near Hampton Court, about 15 miles upstream as the river winds west. The Thames is a tidal river and Thames Ditton is beyond the salt line, so that any pollution near Battersea would not reach this part of the river. The Southwark-Vauxhall company continued to take its water from the same point of origin in the Thames.

Snow once again obtained mortality data from the Registrar-General's report, which provided him with the evidence for a natural experiment. In the areas receiving Southwark-Vauxhall water only, there were 114 deaths per 100,000 population; corresponding figures were 60 for the areas receiving both Lambeth and Southwark-Vauxhall water and 0 for those receiving only Lambeth water. Snow concluded that cholera was a communicable disease contracted by drinking polluted water or eating contaminated food. His diagnosis of the spread of the disease was obviously based on his epidemiological studies but also upon the pathology of the disease. The first symptoms were intestinal distress, which indicated that whatever caused the disease was ingested with contaminated water. For Snow this was analogous to the acquisition of smallpox contracted via a "morbid poison." The poison analogy is misleading since he did not locate the focus of action of the disease agent in the blood but in the intestinal tract. In addition, he believed, ultimately, that the disease agent was a particle and not a soluble chemical.

Snow followed with a book that dealt with the logical question: What is in the water, or in the air, or passed directly from person to person that caused disease? His answer was contained in a theoretical treatise, "On Continuous Molecular Change," whose title appears to support a chemical theory. Snow, however, occupied a middle ground, combining chemical and biological elements. He had already decided to include among the molecular changes processes associated with living phenomena, including fermentation. In this way he was aligning himself with those who viewed fermentation as clearly a chemical process, albeit carried out by a living organism. Other molecular changes were processes occurring in animal and plant bodies, including the development of adult bodies from seeds and ova. Rather than making an absolute distinction between vital and non-vital processes, as in the dispute between Liebig and Schwann, for example, where Liebig stated it was a chemical process while Schwann, and others contended it was a biological process, Snow asserted that fermentation was indeed a chemical change, "yet it has great claims to be entitled a vital process—it is **always** (my emphasis) accompanied by the formation of cells or sporules of the yeast fungus." Snow accepted the principle that "blending . . . vital and chemical phenomena need not surprise us."

This brought Snow to the issue of contagious diseases that involve an increase in the *materius morbi:*

"The material cause of every communicable disease resembles a species of living being in this, that both one and the other depend on, and in fact consist, a series of continuous molecular changes, occurring in suitable materials".

JOHN SNOW, *On Continuous Molecular Changes*, 1853, p. 14, John Churchill, London

In the final pages of "On Continuous Molecular Change" Snow justified his theories and rejected alternate views. Two examples will illustrate the basis of his dismissal of alternate theories. To explain the propagation of disease it has been suggested that effluvia are diffused through the air. Snow claimed if this happened, all who come in contact should contract the disease. For example, it is recognized that all individuals inoculated with smallpox contract the disease unless they have had it and gained immunity. However, in the case of *effluvia*, not everyone contracts the disease. To explain this fact it is postulated that the individual must have a *predisposition*. If, on the other hand, contraction of the disease depends on entrance into the individual of an agent, then it is probable that some individuals will not receive the agent. In that case predisposition has nothing to do with contracting the disease.

Snow cited the "fairly asked" question whether communicable diseases could arise spontaneously. For example, in the neighborhood of a wound there is inflammation, which, Snow acknowledged, probably arises without being communicated. If these infections were caused by the agents specified by Snow, where do they come from if not spontaneously generated? Snow stated, "The material that causes it to be might be as widely diffused as the spores of some fungi." In brief, the infective agent is carried by air.

John Simon provided a contemporary (1858) understanding of Snow's disease theory. At that time Simon had adopted a chemical theory of disease. Simon characterized Snow's theory as a "peculiar doctrine," presumably because it involved a living agent that is excreted by a sick person and accidentally imbibed by another person. The germ of the disease increases in the stomach and bowels where it appears to reproduce and thus has a structure, "most likely that of a cell."

The Discourse About Childbed Fever (Puerperal Sepsis) During the Mid-Nineteenth Century

In March 1847 Ignaz Semmelweis (1818–1865) was appointed first assistant in the maternity ward in the General Hospital in Vienna. It was here that he encountered childbed fever. The disease generally occurred in women on the second to the fourth day after delivery. There was fever and acute pain radiating from the region of the uterus to the whole abdominal area. There were many deaths. The same disease was reported in Scotland and the United States with observations on the mode of contraction of the disease.

In 1795, A. Gordon of Aberdeen wrote a treatise on the fever, summarized by Dr. Oliver Wendell Holmes during the 1840s in the United States. Gordon concluded that the disease was acquired only by women delivered by a practitioner, doctor or nurse, who had previously attended patients who had the disease. Midwives could also transmit the disease if they had been in contact with fever patients. Gordon made the "disagreeable declaration . . . that I myself was the means of carrying the infection to a great number of women." He summed up his view that puerperal fever is a specific contagion or infection, altogether unconnected with the noxious constitution of the atmosphere.

Holmes concurred with Gordon, from his own experience. The disease was not transmitted from patient to patient but by an intermediary, physician or nurse, a very unpopular view among the profession. In a contemporary treatise, on Obstetrics puerperal fever was portrayed as noncontagious, while the publication Philadelphia Practice of Midwifery did not even discuss the disease.

In March 1847 Semmelweis began his second tour at the Clinic and was informed of the death of Professor Jakob Kolletschka, who had been conducting autopsies with students. His finger was nicked by the knife of a student who had already performed an autopsy. Kolletschka developed a range of symptoms that Semmelweis believed to be identical to those of childbed fever. Semmelweis speculated that the cause was the introduction of cadaverous particles into Kolletschka's vascular system. If this was the case, how could infection happen? How could women giving birth come in contact with autopsied material? This was the problem facing Semmelweis. Fortunately, a considerable body of statistics was available about the disease before 1840 and during the years 1841 to 1846 in the Vienna General Hospital.

At all times there were two birthing clinics. Before 1840 student obstetricians and midwives were trained in both clinics. Deaths from childbed fever occurred in both clinics in approximately equal percentages. After 1840 training procedures were changed: all medical students (males) were assigned to clinic 1 and all midwives (females) were assigned to clinic 2. There was a significant change in the mortality rates in the two clinics in the years 1841 to 1846. In clinic 1 there were 20,042 births and 1989 deaths; the average death rate was 9.92 per thousand. In clinic 2 there were 1779 births and 691 deaths; the death rate was 3.38 per thousand. The number of deaths in the first clinic was underestimated since many of the ill mothers were transferred to the general hospital, where they died. Their deaths were not added to those in clinic 1 but assigned to the general hospital.

When women came to the hospital to give birth there was no selectivity in assigning them to the clinics. The first clinic admitted patients four days a week, the second one three days a week. The first clinic therefore had 52 more days of admissions each year. Everyone who came to the clinics in the first 24-hour period was put in clinic 1. In the next 24-hour period everyone was admitted to clinic 2. Once a week, admission to clinic 1 continued for 48 hours. In clinic 1 only obstetricians and medical students delivered babies, while in clinic 2 midwives performed delivery.

Semmelweis considered a number of hypotheses to explain the difference in mortality between the two clinics. A current opinion ascribed the cause to a miasm, some atmospheric influence. Semmelweis rejected this hypothesis, arguing that both clinics existed under the same roof and had the same antechamber. He also considered overcrowding, but the statistics showed that clinic 2 had slightly more patients per day. A commission appointed in 1846 concluded that foreign medical students caused injury when they examined women. The numbers of such students were decreased, and although the mortality rate in clinic 1 declined somewhat, it was clear this was not the answer.

Semmelweis offered a different analysis, similar to Gordon and Holmes, after he learned that physicians and medical students conducted autopsies and from there moved to clinic 1, where they delivered babies. In contrast, midwives did not perform autopsies, and they delivered babies in clinic 2. This was the important evidence for Semmelweis. To explain the occurrence of disease in clinic 1, he proposed that physicians and students were carrying cadaverous material from autopsies and transferring it to women. Although physicians and medical students were washing their hands with soap and water, Semmelweis speculated this did not entirely free their hands from autopsy matter; he insisted they wash their hands in chlorinated lime. When this was done, the mortality rate in clinic 1 fell to that in clinic 2.

In May 1850 Semmelweis gave a lecture and said that puerperal fever is not a contagious or specific disease. It originated when cadaverous particles in the process of putrefying were transferred to the female body by the hands of the examiner. Semmelweis did not suggest what caused "the process of putrefying;" he said the disease was not contagious, presumably because it could not go from one patient to another. Ultimately, what Semmelweis clearly had was a method to prevent the spread of the disease, not an understanding of the underlying cause of the disease. Furthermore, his experience with the disease was known in London, Copenhagen, and Vienna through the many reports and lectures of his original studies.

An attempt was made to create a commission to investigate his work, but it became embroiled in controversy between Semmelweis' superior and young physicians who proposed the commission. It was not appointed, Semmelweis' tenure expired, he was not rehired, and he was slowly "shunted" aside. He left Vienna for Pest, where he ultimately became a professor of obstetrics, and in 1860 he published his book "The Etiology, Concept, and Prophylaxis of Childbed Fever." In this book he reviewed his work and the subsequent studies of others and expressed anger at his detractors, among them Rudolph Virchow (1821–1902) and Carl Braun.

Virchow was the most famous physician-pathologist in Europe and founded a journal that came to be known as *Virchow's Archives*. In 1858 he published a famous text, "Cellular Pathology," which contained the theoretical basis for his disease theory. Virchow adopted the cell theory of Schwann and used the microscope to establish the principle that all tissues in the body were composed of cells. This led him to choose the axiom that every cell comes from another cell: "from it [the cell] emanate all the activities of life both in health and sickness."

Virchow was the first to follow the sequence of pathological effects microscopically and concluded that all disease is due to malfunctioning of existing cells. Cancer was a model example. These pathological conditions spread from cell to cell. What initiated these events, as described by Virchow, were internal failures in cells or extracellular stimuli from within the body or external epidemic conditions. Virchow rejected Semmelweis' theory for the cause of childbed fever.

Carl Braun was Semmelweis' successor at the Vienna clinic and a professor of obstetrics in Vienna. In his medical text he listed 30(!) possible causes of childbed fever; Semmelweis' theory was number 28. In short, there were, according to Braun, better theories.

The Spontaneous Generation Controversy Continued and the Biology of Bacteria

The development of the microscope in the seventeenth century led to the discovery of a world of living creatures invisible to the unaided eye, Leeuwenhoek's animalcules, with their apparently simple structures. It seemed reasonable to place these organisms among the lowest forms of life, which are spontaneously generated. In the mid-eighteenth century the microscopic animalcules discovered by Leeuwenhoek and the ever-present fungi had been observed by Robert Hooke and John Turbeville Needham

(1713–1781), who also concluded they were spontaneously generated from organic material. Furthermore, Needham accepted the presence of these entities in air and in all kinds of environmental materials, since this form of generation was alleged to be a continuous process, and thus the success of his proposed experiments depended on eliminating all these preexisting microscopic organisms in his starting "soups." One of his famous experiments in 1748 used

> "Mutton-gravy hot from the fire and shut up in a phial, closed with a cork so well masticated, that my precautions amounted to as much as if I had sealed my phial hermetically. I thus effectively excluded the exterior air, that it might not be said my moving bodies drew their origin from insects, or eggs floating in the atmosphere. I would not instill any water lest, without giving as intense a degree of heat, it might be thought these productions were conveyed through that element".
>
> JOHN. T. NEEDHAM, "A Summary of Some Late Observations upon the Generation, Composition, and Decomposition of Animal and Vegetable Substances," *Philosophical Transactions of the Royal Society of London*, p. 23, 1748, pp. 615–666, vol. 45, No. 490

Lazzaro Spallanzani (1729–1799) carried out similar experiments in 1765 using hermetically sealed vessels with 11 different kinds of seeds. Here are his important observations: (1) The number of animalcules developed in these various infusions was proportional to the communication with the external air; (2) When more air reached the infusions, there was more microscopic life; (3) There are different animalcules in different seed infusions. He concluded that different kinds of "animalcular eggs" were present "in air, and falling everywhere." Furthermore, the air conveyed germs to the infusions, or assisted the growth of those present.

Spallanzani stated there is no spontaneous generation; all of the microbial life was preexistent, present in the air or already present in the infusions. Probably much of the microbial life observed by Spallanzani were bacteria. Otto Friedrich Muller (1730–1784) reported on "infusoria," which included bacteria; he described them according to their morphology, motion, environment, and group organization.

Therefore, by the close of the eighteenth century and well into the nineteenth century, the scientific community was well aware of the existence of microscopic life, bacteria and fungi (whether one believed they were spontaneously generated or from preexisting life), could see these microorganisms with the microscopes of the time, and knew that they could be present in enormous numbers in the appropriate growth medium. The question, why were they ignored?

"If the work of Bradley, Marten and Goiffon (early 18th century) is considered collectively, they had fully accepted the hypothesis that certain diseases, especially epidemic ones, can only be explained on rational grounds by postulating the presence of a *contagium vivum*; that different diseases are caused by different organisms; that organisms may be conveyed directly from person to person, by air, by food, or by fomite; that once they have entered the body they can be carried by the blood stream to other parts . . .

Why they failed to convince their contemporaries is an interesting problem beyond the scope of this paper".

R. WILLIAMSON, The germ theory of disease. Neglected precursors of Louis Pasteur, Richard Bradley, Benjamin Marten, Jean-Baptiste Goiffon *Annals of Science*, vol. 11, p. 57, 1955, pp. 44–57

"For 150 years from the date of their discovery the bacteria were strangely neglected. Mankind remained inexplicably blind to their importance and almost to their very existence."

CLIFFORD DOBELL, *Antony von Leeuwenhoek and His "Little Animals,"* 1960, p. 381, Dover

. . . "pragmatic clinicians took no note of them. Henle's *On Miasmata and Contagia* (1840) shifted the explanation of disease transmittal from "pestilential emanations" from bogs and swamps to microbial origin, though the designation "microbe" had to wait another 36 years. (p. 566)

OWEN H. WANGANSTEEN, Nineteenth Century Wound Management of the Prurient Uterus and Compound Fracture: The Semmelweis-Lister Priority Controversy *Bulletin of the New York Academy of Medicine*, P. 566, 1970, 46(8):565–596

Why were bacteria ignored or discounted? Why were they not at the very least included as the cause of contagious disease? Are they spontaneously generated, or are they the cause of fermentation? There were of course competing theories for disease causation. There was the chemical theory of fermentation. There was a belief in spontaneous generation, that microorganisms are derived from diseased tissues, which would eliminate microorganisms as the cause of disease. And there was a lack of understanding of the nature or, phrased another way, the biology of bacteria.

Before we clarify and eliminate the dispute about spontaneous generation, it is important to discuss a parallel development covering the biology of bacteria that provides experimental evidence arguing against the theory of spontaneous generation. It is evidence that bacteria are cells that undergo division and grow independently, producing large populations; during these activities they can carry out fermentation and produce contagious diseases.

One can determine the state of understanding of the nature of bacteria in the early decades of the nineteenth century by examining the work surrounding the discovery of a phenomenon that resulted in a red material developing on polenta (cornmeal mush). Peasants near Padua in 1819 reported that the surface of polenta turned a brilliant red; they ascribed the formation of the bloody red color to a supernatural event. Pietro Melo, the Director of the Botanical Garden at Savonara, claimed the red substance was spontaneously generated. Bartolomeo Bizio disagreed. Bizio, a pharmacist residing in Venice, also began his studies in 1819. His paper is entitled *Bartolomeo Bizios's Letter to the Most Eminent Priest, Angelo Bellani, Concerning the Phenomenon of the Red-Colored Polenta.* Bizio believed it was the result of a warm and damp atmosphere. It took about 24 hours for the red color to develop under these conditions.

Bizio investigated the cause of the phenomenon, probably believing that the condition was contagious. He carried out two basic experiments. He placed polenta under a glass dome on a plate about an inch away from polenta containing the red material at 2:30 p.m. At 11 a.m. the next day the fresh polenta had red spots, and by evening it was completely covered with red material. In another such experiment he included "foul vapors," with the same results. To confirm these experiments he engaged a friend in Padua who obtained similar results. Initially, Bizio's theory of its origin was that it was a product of fermentation, which would suggest spontaneous generation: it might be a minute plant or an animal of the "lowest class." To observe the putative entity he mixed the red substance with water and observed the mixture microscopically. He reported no visible filamentous or oscillating objects. He offered no description of what he saw. In another approach he referred to the work of Spallanzani and therefore carried out a series of experiments that relied on "contagion" to transfer the red condition and a series of heating experiments to determine if the red material was sensitive to temperature. He again placed red-colored polenta some distance from fresh polenta under a globe. The uncolored polenta became colored in an interesting way: it acquired spots of red. Bizio described them as small hemispheres; today we would describe them as colonies. Bizio suggested the spots were produced by an organic being. If the red material were a living organism, it must be "a little animal of the class infusoria" or a "very minute plant." To select between these alternatives he used the methods of Spallanzani, who subjected his entities to chemical treatment or to heating; for example, infusoria are sensitive to camphor, turpentine, and tobacco. These materials did not affect the ability of the red matter to grow on polenta. In addition, heat did not "deprive . . .

seeds of vitality." The resistance to heat of the red material "obeys those laws to which Spallanzani found vegetable seeds subject." He concluded that, although the red substance was of the lowest type, it was a vegetable, a living thing. To what class of plant did it belong? Bizio placed it among the fungi, "possessing . . . all the characteristics pertaining to this order."

Bizio concluded it was a living thing. Its transfer from polenta to polenta convinced him it was acquired and not spontaneously generated on non-red polenta. He noted that humid conditions, appropriate temperatures, and foul vapors enhanced the growth of the red matter. It is interesting that microscopic examination of the red matter did not appreciably add to comprehension of the nature of the living being. Bizio named it *Serratia marcescens* to honor his physics teacher Serafino Serrati.

Vincenzo Sette, a physician, followed with a paper, originally read at a meeting in 1820, in which he generally concurred with Bizio. He decided that the red substance was a stemless fungus.

In 1844 Bizio returned to the issue with a short publication whose title stated that the bread was altered by a "cryptogam." The word has its origin in nineteenth-century French, used in botany to categorize a plant that appears to have no sexual stage; these plants included fungi and algae.

A significant contribution to realizing the diversity of morphological types among bacteria was the work of Christian Gottfried Ehrenberg (1795–1876) who published in 1838 "Infusoria-animalcules as Complete Organisms." The descriptions were based on microscopic surveys that revealed the complexity of types such as small and large rod-like shapes, vibrios (probably curved, small rods), and different spiral organisms.

A fortuitous event brought Ehrenberg to investigation of the red matter. In 1848 he came to a home where there was cholera and noticed red spots on a cooked potato. He was aware of Sette's work and had a more powerful microscope than that available to Bizio and Sette. He reported that the red matter lived on moist food, showing blood-red spots. The red substance was composed of oval animalcules that were motile, with a single flagellum. The individual biological units were not quite one micron long, and he calculated that more than 4×10 to the power of 14 of these organisms were contained in a cubic inch. Ehrenberg assigned them the name *Monas prodigiosa*, indicating a miracle, recalling the popular conception of the occurrence of the bloody red spots. He placed them in the animal kingdom.

In an 1853 work C. Montagne, who read Ehrenberg's studies, decided the components of the red matter were algae. He stated that he could not see flagella with his most powerful microscopes and did not believe the bodies were motile. A large number of pigmented microorganisms were discovered between the 1830s and the 1850s. Some were so numerous in water environments that they appeared visible as films on surfaces.

It is clear at the mid-nineteenth century that what were labeled bacteria were distributed widely; however, their origins and their structures remained a mystery. There was clearly a dispute about their origin; were they spontaneously generated? There was the possibility that they carried out fermentations and reproduced during these processes according to the studies of Pasteur on the lactic and butyric acid fermentations; this was disputed by Liebig, although Pasteur's studies inspired the work of Davaine, who initiated work on the cause of anthrax, and Joseph Lister.

"The occasion had not arisen and other work had prevented me from continuing active research when, in February 1861 M. Pasteur published his remarkable work on the butyric ferment, a ferment consisting of small cylindrical rods which possess all the characteristics of vibrios and bacteria".

JEAN THEODORIDES, "Casimir Davaine (1812–1882): A Precursor of Pasteur,"
Medical History, p. 159 1966, 10(2) 155–165,

. . "the philosophic researches of M. Pasteur who has demonstrated . . . minute particles . . ., which are the germs of various low forms of life, [that act like] the yeast plant [that] converts sugar into alcohol and carbonic acid . . . .

JOSEPH LISTER, "On a new method of treating compound fractures,
abscesses, and so forth; with observations on the conditions
of supperations," *Lancet*, 1867, vol. 1, pp. 326–329

In 1851 Ferdinand Cohn (1828–1898) published "Contributions to the Developmental History of Infusoria"; it was the beginning of a systematic program of study over the next 20 years of the characteristics of bacteria. Cohn was in opposition to various investigators who did not believe that bacteria came in different varieties, but rather that environmental conditions determined bacterial morphology. He and colleagues collected information about the structural details of bacteria, their growth in various media, and their organization during cell division. To document the ability of a unique bacterium to reproduce itself in a constant way, his students experimented with cultures of bacteria that produced a distinct pigment. A procedure to isolate distinct pigmented bacteria was developed by J. Schroeter in 1875 and could be used to isolate any single bacterial type. The title of the paper was "Concerning a Few Pigments Generated by Bacteria." The plan was to fix a particular bacterial type at a position on a solid surface. Schroeter used the cut surfaces of potatoes, or solid media made from starch paste, egg albumin, bread, and meat. The stability of a particular type was demonstrated when the pigmented colony maintained its character in successive generations. After comparative microscopic studies of numerous animalcules, which included motile types, Cohn decided to place bacteria in the plant kingdom. Cohn also discovered the characteristics of the "hay bacillus," a large rod-like structure that produced heat-resistant spores. He named the organism *Bacillus subtilis*.

Louis Pasteur (1822–1895)

Louis Pasteur was interested in the cause of fermentation and contagious disease, regarding them as intrinsically related. The demonstration that a living agent causes fermentation would in principle support a living agent theory of disease and would disprove spontaneous generation. We will initially analyze Pasteur's work on fermentation and follow it with his classic studies on spontaneous generation.

Pasteur came to the study of fermentation by an interesting route. He was a student of chemistry and two of his teachers, J. P. Biot (1774–1862) and A. Laurent (1808–1853), were pioneers in the study of the optical activity of organic chemical compounds. A compound that is optically active rotates the plane of polarized light. In 1844 an interesting phenomenon was reported concerning the relationship between crystal structures and "optical activity." It was found that there are two forms of sodium-ammonium tartrate. They both contain the same amounts of carbon, hydrogen, and oxygen. Solutions of one tartrate rotated a plane of polarized light to the right, and it was optically active, while the other did not, and therefore it was optically inactive. Pasteur's studies showed that the crystal forms were not identical. The one that did not rotate the plane of light was separable into two forms that were mirror images of each other. The forms were equivalent to each other as our left and right hands. Each form rotated the plane of polarized light, one to the left and the other to the right. In an equal mixture of the two, each rotation cancelled the other, and therefore the mixture was optically inactive.

The story is more complicated in its development; however, there is no doubt that Pasteur accomplished the remarkable feat of separating physically the two forms of the crystals, the two isomers. But how did this research lead to Pasteur's deep interest in fermentation? Pasteur offered an explanation in his introduction to an 1857 paper on lactic fermentation:

"I think I should indicate in a few words how I have been led to occupy myself with researches on fermentation. Having hitherto applied all my efforts to try to discover the links that exist between the chemical, optical, and crystallographic properties of certain compounds . . . it may perhaps be astonishing to see me approach a subject of physiological chemistry seemingly so distant from my earlier works. **It is, however, very directly connected with them** (my emphasis).

In fact, I ought to admit that my researches have been dominated for a long time by this thought that the constitution of substances—considered from the point of view of their molecular asymmetry or non-asymmetry, all else being equal—plays a considerable role **in the most intimate laws of the organization of living organisms and intervenes in the most hidden of their physiological properties** (my emphasis)".

GERALD L. GEISON, *The Private Science of Louis Pasteur*,
1995, pp. 95–96, Princeton University Press

In a fermentation it was shown that optically active compounds were produced. It was also experimentally demonstrated that a fungus could selectively metabolize one optical isomer while leaving the other intact. Pasteur's axiom was that life alone could make these discriminations. He concluded that only living agents could produce optically active compounds; therefore, the process of fermentation was a vital process.

Pasteur was drawn to the cause of fermentation because there was a theoretical dispute between a chemical and a biological cause (that is, between a mechanistic and a vitalistic interpretation of the phenomenon), represented prominently on the chemical side by J. Liebig and on the biological side by Schwann, Cagniard de la Tour, and Kutzing. Each side relied on experimental facts in support of its theories. Liebig cited the known action of ferments (currently they are termed enzymes), such as pepsin and diastase, and noted that fermentation could be carried out by water extracts of yeast cells and by animal tissues, processes that did not involve intact cells. The others (Schwann, Cagniard de la Tour, and Kutzing) had carried out extensive experiments evidently demonstrating that intact, growing yeast cells carry out fermentation.

Pasteur was a proponent of the biological cause espoused by the trio of Schwann, Cagniard de la Tour, and Kutzing and began his studies on fermentation studying the production of lactic acid in 1857. Pasteur had to contend with the work and reputation of Liebig and his hostility to a biological explanation of fermentation and contagious disease. We reproduced Liebig's quotes at the head of the chapter; they fundamentally opposed a living agent theory of disease and fermentation. His charge was essentially that there was no experimental evidence to support such a theory:

"A theory of the cause of fermentation and putrefaction, which is utterly fallacious in its fundamental principles, has hitherto furnished the chief support of the parasitic theory of contagion. There is no opinion so destitute of a scientific foundation as that which admits, that miasms and contagions are living beings, parasites, fungi or infusoria, which are developed in the healthy body, are there propagated and multiplied, and thus increase the diseased action, and ultimately cause death".

J. LIEBIG, *Animal chemistry in its application to physiology and pathology*,
ed. by W. Gregory, 1852, p. 127, 3rd edn, John Wiley, New York

Pasteur introduced his work by presenting the elements of Liebig's methodology, argument, and conclusion:

"The facts appear therefore to be very favorable for the ideas of M. Liebig. In his eyes, the ferment is a substance which is highly alterable which decomposes and in so doing induces the fermentation because of the alteration which it experiences itself, communicating this agitation to the molecular groups of the fermentable material and in this way bringing about its decomposition. According to M. Liebig, this is the principal cause of all fermentations and the origin of the majority of contagious diseases".

Milestones in Microbiology: 1546–1940, 1961 tr. and ed. by
Thomas D. Brck, p. 41, ASM Press

It was already known to chemists that the production of lactic acid could be carried out by "animal membranes" without the appearance of independent microscopic organisms. Pasteur, obviously, did not accept this experiment, although he did not explain how the lactic acid was produced using animal membranes (tissues) or how carbon dioxide and alcohol production was carried out by the contents of yeast contained in the wash water. He accepted as a fact that "lactic yeast" (which are bacteria) were the cause of fermentation and made the analogy between a lactic fermentation, where sugar is converted to lactic acid by a "lactic yeast," and fermentation carried out by beer yeast, producing alcohol and carbon dioxide.

In the lactic fermentation Pasteur noted a "grey substance" that on microscopic examination was found to be composed of tiny globules, smaller than yeast, arranged singly or in groups and non-motile. He added a trace of this substance to a nutrient medium containing sugar. The next day there ensued a vigorous fermentation; the sugar was converted to lactic acid. Pasteur wrote, "only a small amount of this `yeast' [a bacterium] is needed to convert a large amount of sugar."

This work, however, presented some difficulties. Lactic acid was not the only end product due to the presence of more than one kind of "organized body" (that is, more than one kind of bacterium). Furthermore, the medium contained the complex organic matter that Liebig maintained, with regard to alcoholic fermentation, was responsible for the fermentation. Nevertheless, Pasteur was committed to the principle that fermentations were the consequences of the life of living organisms and proceeded almost immediately to study the role of yeast in alcoholic fermentation in 1858. These experiments were designed to demonstrate that yeast had an independent existence and were the cause of fermentation.

In a careful quantitative experiment Pasteur showed that yeast grow and assimilate a nitrogen source from the medium while carrying out fermentation. This was accomplished by growing yeast in a semi-defined medium that supplied all the nutrients the organism needed to reproduce itself. It consisted of a solution of sugar; ash (from 1 g of yeast), which is a source of inorganic salts (minerals, such as phosphate and sulfate); and ammonium tartrate (a source of nitrogen). He added to this medium an amount of fresh beer yeast that "could fit on the head of a pin." It was important to start with a trace of yeast to clearly demonstrate that the yeast multiplied, the sugar was fermented and the ammonia disappeared from the medium. He included a very important "control" in these experiments. He added a pin-head of yeast into the same sugar solution with ash but no nitrogen source, and observed "there is hardly any sign of fermentation." The conclusion was that the growth of the yeast cell depended on a nitrogen source (the ammonium tartrate) and growth was essential for the coincident production of alcohol and carbon dioxide.

Pasteur also carried out a microscopic study of yeast cells. He reported that yeast reproduced by developing bud-like structures, which confirmed the studies of Leeuwenhoek in 1680 and those of Theodore Schwann and Cagniard de la Tour in the 1830s. He concluded with a statement of his results: whatever amount of yeast he introduced into a medium with a nitrogen source, fermentation and growth occurred.

Pasteur carried out one experiment that Liebig criticized. Pasteur demonstrated that the addition of a substantial amount of yeast in a sugar solution, without other components such as a nitrogen source, caused fermentation. Liebig questioned how yeast could carry out fermentation without the materials necessary for growth when fermentation is alleged to be the result of reproduction. Pasteur did not address the criticism but simply reaffirmed that yeast could ferment under either condition.

Pasteur could not adequately explain why there was fermentation without growth of the yeast cells. Today we can. The crucial issue is the amount of yeast used in the experiment in which the necessary requirements for growth were missing. Recall Pasteur wrote that he used "a significant or substantial amount" of yeast. Without cell multiplication, the existing large amount of yeast cells would convert the sugar to alcohol and carbon dioxide.

Pasteur carried out additional experiments whose goal was to prove that yeast cells had the metabolic capabilities that would enable them to grow as an independent entity. For example, Pasteur studied the effect of oxygen on the fermentative powers of yeast. In the presence of air and under conditions where the yeast were not submerged in the fluid, yeast grew rapidly without the production of alcohol. During this process the yeast absorbed some of the oxygen from the air. Yeast grown in the absence or presence of air (oxygen), Pasteur said, "has not changed its nature." What had changed is that yeast did not produce alcohol and carbon dioxide in the presence of air. Pasteur could not explain this phenomenon but remained completely convinced and confident that he had demonstrated that yeast could grow with or without air. Pasteur was correct: yeast can grow with or without air. We now know that under aerobic conditions (with air-oxygen), yeast does not accumulate alcohol and carbon dioxide; it metabolizes a metabolic precursor of alcohol to water and carbon dioxide.

Pasteur and Spontaneous Generation

In 1858 Felix Pouchet (1800–1872), a distinguished French biologist, published a short paper on the spontaneous generation of microorganisms in boiled hay infusions incubated under a cover of mercury. One year later, Pouchet published "Heterogenesis: A Treatise on Spontaneous Generation." The term *heterogenesis* specifies that living organisms arise from non-living organic matter.

In the infusions Pouchet claimed there was no preexisting microscopic life at the start of the experiments, and thus living microorganisms arose anew. The paper was almost immediately attacked as methodologically flawed and philosophically insufficient. It was charged that there was no proof that boiling killed all microorganisms present in ambient air. In addition, mercury could contain numerous living bodies that could also be the source of the microbial life growing in the infusions.

Pouchet also had to contend with a most serious charge, especially in France, that he encouraged a materialist and consequently an atheist philosophy that was linked to spontaneous generation, all part of the conflict between the forces of republicanism and the monarchy allied to the church. In the Second Empire under the control of Napoleon III (Louis Napoleon) the Catholic Church gained control of education and grew in size and strength, and its conservatism (Ultramontanism) was enhanced by Rome with the promulgation by Pius IX of the Syllabus of Errors in 1864 condemning liberal ideas.

The conflict was exacerbated by the publication of Charles Darwin's "Origin of the Species," in a French translation in 1862 by Clemence Royer, characterized as "materialist, republican, and atheist." In the preface Royer included an extended denunciation of the Catholic Church. Darwinism, evolution, became part of the political and theological conflict in France. It was recognized that Darwinism required that at some time in the past life emerged spontaneously. This was, of course, anathema to church doctrine, which ascribed creation to a deity. The attack on the Church was viewed as an attack on the state.

Pouchet, sensitive to the view that equated spontaneous generation with atheism, argued that the Deity had used it in the beginning to create life and had established laws and forces that would enable these events to occur again. Pouchet denied abiogenesis—that is, living things coming from inorganic matter rather than arising from organic matter. Nevertheless, there was the charge of heresy and a vigorous attack by Louis Pasteur on his methodology.

Pouchet had used heat to ensure the destruction of preexisting microscopic life. This method had been used by Needham some 100 years before and more recently by Theodore Schwann in 1836–37. He had heated to 100 degrees C a "small amount of an infusion of meat" in a closed glass sphere that remained free of putrefaction and free of any "infusoria." Schwann was aware of the theory that the development of microscopic life depended on the presence of air-oxygen and therefore devised a way to introduce heated air into the infusion. No putrefaction or microbial life appeared. If unheated air were introduced, microbial life developed. Schwann concluded that there is no spontaneous generation; growth of microorganisms was simply the introduction of air containing such life. There was an unaddressed possibility, that heating the air destroyed some postulated vital force. Pasteur later dealt with this problem.

Pouchet reported that he had repeated these experiments and consistently obtained growth of "germs" in his infusions. He continued to hold that the heating process killed all preexisting germs. Pouchet made another claim, that the microbial life in the infusions was different from those germs found in the air. No evidence was presented for this contention.

In 1859, when Pouchet published his work *Heterogenesis*, the Paris-based Academy of Sciences "announced a prize which would throw new light on the question of spontaneous generation." Pasteur entered the contest in the midst of experiments on the lactic and alcoholic fermentations that convinced him that all fermentations were caused by the growth and reproduction of living microscopic organisms.

On experimental grounds Pasteur was at odds with Pouchet. But there were, possibly, other factors, external, non-experimental factors, that determined Pasteur's position. There is a fascinating disagreement among scholars over whether these political and religious components influenced Pasteur's position on spontaneous generation. We will discuss this disagreement shortly—but first Pasteur's elegant experiments on spontaneous generation.

Pasteur, in his own words, "wanted to arrive at an opinion of spontaneous generation." Why? He "would perhaps be able to uncover a powerful argument in favor of my ideas on the fermentation themselves"; they were due to the presence and multiplication of living microorganisms.

Pasteur: Microbial Life is Present in the Air

Pasteur initially addressed the question of whether microorganisms are present in the air. This is an important and appropriate starting point for Pasteur, for he was laying the groundwork for his contention that the growth in heated infusions was the result of the introduction of air containing microscopic life, an issue already addressed by Schwann.

Pasteur used round flasks, with a long narrow neck, filled with broth to about 40 percent of their volume. The broth was boiled, and while the fluid boiled the neck was sealed with a flame. No growth occurred in these flasks. If the seal was broken and air entered the flask, growth of microbial life was visible in a few days. This result was understood to indicate that air contained microscopic life.

To demonstrate directly the presence of microscopic life in air, Pasteur drew air through a cotton filter and observed microscopic agents trapped in the cotton. Inoculation of broths with this material yielded growth. Following these results he tested whether heated air contained microorganisms. He once again started with a growth medium in a flask with an extended neck. The medium was boiled and cooled, but the cool air reentering the flask through the neck was heated, and finally the neck was sealed in a flame. The flask was placed at 30 degrees C. Pasteur stated that nothing grew in the flask even after 18 months. If, however, the flask were filled with unheated air, a number of different microorganisms flourished. Pasteur concluded that microorganisms entered from the air. However, one argument remained unanswered, that in heating air a postulated "vital force" was destroyed. To counter this theory Pasteur carried out the same experiment using the same round-bottomed flasks, except that the necks of various flasks were very long and differently curved downward, but most importantly, remained open to the air. Pasteur boiled the broth until steam issued from the flasks through the curved necks. The broths were allowed to cool. During the cooling process, outside, unheated cool air was drawn into the flask. Pasteur stated, "In no case is there the development of organized bodies in the liquid." This is how Pasteur explained these results. Although unheated air returned to the broths, there was no growth because the microscopic bodies in the air were trapped in the moisture, coating the walls of the glass curved necks. What would happen if unheated air entered the flask without passing through the long, curved glass passageway? When the long, curved neck of the flask was removed with a file and the broth came directly in contact with air that did not have to pass through the long, curved, humid passageway, numerous organisms grew in 24, 36, or 48 hours.

Pasteur continued to pursue the presence of microorganisms in the air by exposing broths to air at various elevations throughout the country. He opened 20 flasks with broth to the air at the foot of a mountain, 20 at the top of a mountain 850 meters high (2,762 feet), and 20 on a glacier at an altitude of 2,000 meters. Microorganisms grew in eight of the flasks at ground level, five at the top of the mountain, and one at 2,000 meters. At all levels there were microorganisms in the air, although at the higher altitudes there appeared to be fewer organisms.

Pouchet responded with experiments in 1860–61 and continued to claim the presence of microorganisms in heated, hermetically sealed flasks. Nevertheless, Pasteur was awarded the prize of the Academy after Pouchet withdrew from the competition. Pouchet returned in 1863 to conduct experiments in the mountains and glacier areas. He reported that he opened four broth flasks and obtained growth in two from each area. Pasteur objected to his methodology and results.

For one moment Pouchet and collaborators proposed to continue the experimentation and asked the Academy to create another Commission to judge the conflicting results. There were disagreements about the way the experiments were to be performed and Pouchet withdrew. In 1865 the Commission supported Pasteur's claims.

The Pasteur–Pouchet controversy may be summarized this way. Pasteur, under specifically defined conditions, asserted that he never obtained growth in his various broths; Pouchet, under his set of conditions, almost always obtained growth in his infusions. Pasteur concluded there was no spontaneous generation; Pouchet said there was. Pasteur stated that all of the microorganisms contained in Pouchet's broths came from microbial life present in the air or preexisting in the broths. Pouchet claimed there were no bodies in the air to intrude into his flasks; he stated that any microorganisms preexisting in the broths were killed by the heating procedure he used. Pasteur claimed that his heating procedures might have been inadequate to kill all microbial life, for there may be those resistant to heat.

These are all arguments about scientific questions, questions that can presumably be resolved by experiments. But, according to John Farley and Gerald L. Geison, there is more to this dispute, explicitly expressed in the title of Chapter 5 of Geison's book The Private Science of Louis Pasteur: "Creating Life in Nineteenth-Century France: Science, Politics, and Religion in the Pasteur–Pouchet Debate over Spontaneous Generation." Geison opened the chapter in a dramatic way:

"On the evening of 7 April 1864, Pasteur took the stage at the large amphitheater of the Sorbonne to give a wide-ranging public lecture on spontaneous generation and its religio-philosophical implications. It was the second in a glittering new series of "scientific soirée" at the Sorbonne, and *tout Paris* was there, including the writers Alexander Dumas and George Sand, the minister of public instruction Victor Duruy and Princess Mathilde Bonaparte".

Pasteur opened the talk with a review of the great problems confronting the minds of people:

"The unity or multiplicity of human races; the creation of man several thousand years or several thousand centuries ago; the fixity of the species or the slow and progressive transformation of one species into another; . . . the notion of a useless God".

What is obvious, reflected in these opening remarks, is the impact that Darwin's *Origin of Species* made on Pasteur and the whole society. Pasteur then reviewed the controversy over spontaneous generation, labeling it an example of a false idea. After that, as Geison phrased it, "Pasteur struck to the heart of the matter":

> "Very animated controversies arose between scientists, then [in the late eighteenth century] as now—controversies the more lively and passionate because they have their counterpart in public opinion, divided always, as you know, between two great intellectual currents, as old as the world, which in our day are called materialism and spiritualism. What a triumph, gentlemen, it would be for materialism if it could affirm that it rests on the established fact of matter organizing itself, taking on life of itself; matter which already has in it all known forces! . . . Ah! If we could add to it this other force which is called life . . . what would be more natural than to deify such matter? What good then would it be to resort to the idea of a primordial creation, before which mystery it is necessary to bow? Of what use then would be the idea of Creator-God? . . . Thus, gentlemen, admit the doctrine of spontaneous generation, and the history of creation and the origin of the organic world is no more complicated than this. Take a drop of sea water . . . that contains some nitrogenous material, some sea mucus, some "fertile jelly" as it is called, and in the midst of this inanimate matter, the first beings of creation take birth spontaneously, then little by little are transformed and climb from rung to rung—for example, to insects in 10,000 years and no doubt to monkeys and man at the end of 100,000 years. Do you now understand the link that exists between the question of spontaneous generations and those great problems I listed at the outset"?
>
> GERALD L. GEISON, *The Private Science of Louis Pasteur*, 1995,
> pp. 110–111, Princeton University Press

It appears clear that Pasteur was opposed to spontaneous generation on religious grounds. However, this position did not interfere in the quality of his experiments to counter this theory. Pasteur had demonstrated to his satisfaction, by way of experiments that were creative and methodologically sound, that there was no spontaneous generation and fermentations were the result of the growth of a particular microorganism in the appropriate medium. If there was no spontaneous generation and if fermentation was caused by a microscopic agent, then the analogy between microbe-causing fermentation and microbe-causing contagious diseases represented a viable, biological theory of disease. Nevertheless, the controversies continued, as we have indicated, well into the 1870s. Why?

Although Pasteur's experiments were brilliantly conceived, other experimenters continued to obtain growth in various media despite precautions to exclude contamination from the air. The demise of the issue of spontaneous generation began with Pasteur's work, and it was almost completely abandoned after the biological studies of Ferdinand Cohn on bacteria in Germany and John Tyndall (1820–1893) in England, who demonstrated that certain bacteria remain viable after immersion in boiling water.

Ferdinand Cohn and the Formation of Spores

Cohn found there were microbes in hay infusions still able to grow after hours of boiling. There appeared to be no possibility that organisms were introduced from the air after boiling because he used cotton plugs, in some experiments, that prevented introduction of microbes; in other experiments he used the curved flasks devised by Pasteur to exclude external microscopic entities, confirming Pasteur's results. Cohn also confirmed the experiments of Pouchet, who had claimed that he obtained growth in flasks that were exposed to boiling temperatures for more than one hour. Microscopic studies of boiled hay infusions revealed the presence of rod-like bacteria, which Cohn named *Bacillus subtilis*. Cohn discovered another form of the same bacillus that he called a spore. This was a smaller structure than the rod-like structure and formed under conditions under which the bacillus could not grow. These spores could withstand boiling for considerable periods of time. When these spores germinated, they produced the characteristic rod-like bacilli, which were sensitive to boiling.

The Spontaneous Generation Controversy in England

In England there was also a conflict over spontaneous generation. The controversy revolved around the quality of experimentation. An idealized version of the state of the controversy in England and how it would be resolved is provided by John Fiske, a Professor of Philosophy at Harvard who had visited Thomas H. Huxley and H. Charlton Bastian in 1873. Huxley and Bastian held opposing views.

According to Fiske there were no "external" issues that intruded into the controversy in England. He believes that this is an argument about scientific hypothesis and it should be ultimately resolved by a crucial experiment.

To trace the development of the dispute in England we shall discuss the work and writings of three individuals who, of course, were not the only persons involved but were influential and represented some of the positions taken by members of the scientific and medical community. The three are Thomas H. Huxley (1825–1895), John Tyndall (1820–1893), and H. Charlton Bastian (1837–1915). Huxley was a Professor in the Government School of Mines and at different times was Secretary and President of the Royal Society. He was an outstanding naturalist and a vigorous supporter of Darwin's evolutionary theory. Tyndall was a Professor of Physics at the Royal Institution and at the Government School of Mines. He was an opponent of spontaneous generation and a proponent of a germ theory of disease. He devised a procedure to sterilize broths that contained the heat-resistant spores discovered by Cohn. Bastian was a physician who published books and numerous papers claiming to have demonstrated spontaneous generation.

Thomas H. Huxley: His Influence

It may be that Huxley was drawn to the theory of spontaneous generation not because he did not believe Pasteur's experiments, but rather because he was a convinced evolutionist, which logically would require a belief that at some point in the distant past (but possibly more recently), life was being generated from non-life. This may explain Huxley's interest in a sample from a deep-sea sediment, characterized as "a gelatinous substance with microscopic bodies scattered throughout," which Huxley thought was "a primitive undifferentiated form of protoplasm" and which Ernst Haeckel (1834–1919) thought was an example of an organism being spontaneously generated on the ocean floor. Did Huxley believe in Haeckel's theory? He never stated that he did. In 1875 the material was found to be an artifact, and by that time Huxley was a vigorous opponent of spontaneous generation.

On the evening of November 18, 1868, Huxley delivered an address in Edinburgh with the title "On the Physical Basis or the Matter of Life". The ideas contained in this seminal essay were a powerful argument against a theory of spontaneous generation. He argued for the interconnectedness of all living creatures, all made from the same chemicals. He wrote "that there is some one kind of matter which is common to all living beings and that their endless diversities are bound together by a physical, as well as an ideal unity." He suggested that this view does not fit a common-sense observation of what is present in the living world:

> "What, truly, can seem to be more obviously different from one another, in faculty, in form, and in substance, than the various kinds of living beings? . . . the brightly colored lichen . . . resembles a . . . mineral incrustation . . . the microscopic fungus-a mere infinitesimal ovoid particle . . . the wealth of foliage, the luxuriance of flower and fruit . . . the giant pine of California . . . the finner whale . . . eighty or ninety feet of bone, muscle, and blubber . . . : and contrast him with the invisible **animalcules** (my emphasis)-mere gelatinous specks, multitudes of which could, in fact, dance upon the point of a needle . . . what hidden bond can connect the flower which a girl wears in her hair and the blood which courses through her youthful veins: . . . what is there in common between the dense and resisting mass of the oak . . . and . . . disks of glassy jelly . . . pulsating through the waters of a calm sea . . . ? . . . I propose to demonstrate to you that, notwithstanding these apparent difficulties, a threefold unity—namely a unity of power . . . a unity of form, and a unity of substantial composition—does pervade the whole living world".

Huxley continued,

> . . . "the matter of life depends on the pre-existence of certain compounds; namely carbonic acid, water and certain nitrogenous bodies. Withdraw any one of these three from the world, and all vital phenomena come to an end. They are as necessary to the protoplasm of the plant, as the protoplasm of the plant is to that of the animal. Carbon, hydrogen, oxygen, and nitrogen are all lifeless bodies. Of these carbon and oxygen unite in certain proportions and under certain conditions, to give rise to carbonic acid; hydrogen and oxygen produce water; nitrogen and other elements give rise to nitrogenous salts. These new compounds, like the elementary bodies of which they are composed, are lifeless. But when they are brought together, under certain conditions, they give rise to the still more complex body, protoplasm, and this protoplasm exhibits the phenomena of life".

THOMAS H. HUXLEY, *Select works of Thomas H. Huxley*, 1886, p. 452, J.B. Alden, New York

Huxley appeared to have argued for continuity between organic and inorganic matter and between living and non-living matter. But Huxley clearly stated that although protoplasm of plants and animals had fundamental similarities, there were differences in the synthetic powers of the two groups. Whereas plants could fabricate their protoplasm from mineral elements and carbonic acid, animals required preformed materials that could be provided only by plants. And, of course, only living organisms had the power to assimilate these chemical compounds and turn them into protoplasm. It is important to recognize that Huxley included animalcules, certainly bacteria, among living organisms that are made of the same materials that make up the protoplasm of the cells of animals and plants and have the capacity to assimilate nutrients and reproduce themselves.

Bastian was a physician who believed in evolution and spontaneous generation. Over a period of years he produced experimental papers and books purporting to demonstrate spontaneous generation. These prolific writings brought him to the attention of Huxley. Bastian had read Darwin but appeared to be more immediately influenced by Huxley's lecture of 1868. Bastian adopted the doctrine that there was continuity between inorganic and organic matter as a theoretical support for spontaneous generation. In a series of experiments beginning in 1870 Bastian claimed that organisms were present in fluids exposed to temperatures of 146 and 150 degrees C for four hours. While reviewing Bastian's papers Huxley emerged from a position of a neutral, perhaps sympathetic, reader to a severe critic of Bastian's knowledge and methodology. Huxley noted that in Bastian's hermetically sealed, boiled fluids a structure resembling a leaf of the moss *Sphagnum* was present, in addition to some spiral fibers. After conducting some of his own experiments Huxley wrote to Joseph D. Hooker, Director of the Royal Botanic Garden, that Bastian was an incompetent experimenter.

Huxley, a tenacious champion of Darwinian evolution, did believe that at some time in the distant past life emerged from non-life, but it was not happening in Bastian's experiments nor anywhere else.

John Tyndall and Spontaneous Generation

In 1870 John Tyndall gave an address to the Royal Institution entitled "Dust and Disease". Tyndall contended that the organic nature of dust strongly suggested that Pasteur was correct about the presence of microorganisms in the air. From there they could enter boiled infusions and thus account for growth. This would be a refutation of a theory of spontaneous generation. The airborne microorganisms could also account for the origin of contagious disease, a position that had been taken by Pasteur. The experiment carried out by Tyndall to support his contention that air contained microorganisms involved an enclosed cabinet with glass panels on two sides so that light could be directed through the box, and it was designed so that boiled broth could be introduced. The insides of the box were coated with glycerin to trap dust particles, whose presence could be visualized by a beam of light passed through the glass panels. When the light was no longer scattered because the dust particles had settled and adhered to the walls, broth was introduced and remained free of growth.

Tyndall continued to work to disprove spontaneous generation for although he was convinced by Pasteur's experiments, he was having some difficulty with his boiled hay infusions. Furthermore, Bastian continued to publish papers in which he claimed to have demonstrated spontaneous generation.

Tyndall pointed out that most infusions could be sterilized with short boiling periods but other infusions could not, even with extended boiling periods. Since he was convinced that spontaneous generation did not occur, he envisioned a resistant-to-heat phase of the bacterium (equivalent to Cohn's spores) that could convert to a sensitive-to-heat bacterium (the rod-like structures). In 1877 he published a paper that depended on the cycling of bacteria from the heat-resistant phase to the heat-sensitive state. A unique sequence of heating procedures would eliminate all sensitive-to-heat bacteria. Since all bacteria in the population would at some point cycle into the heat-sensitive phase, all bacteria would ultimately be killed, and the broth culture would remain free of microorganisms after that.

This is how Tyndall carried out the experiment. Two sets of tubes were prepared containing hay infusion. The A infusions were boiled for one minute and the B infusions were boiled for 10 minutes; both sets were subsequently incubated at the appropriate growth temperature. Twelve hours later both sets of infusions were completely clear, indicating there had been no visible growth during that period. The A infusions were again subjected to one minute of boiling; then they were allowed to remain for 12 hours at incubation temperatures and boiled for 30 seconds. This procedure was repeated three more times for set A. Consequently the set A infusions were boiled for a total of four minutes, although the boiling was carried out intermittently.

What were the results? The B infusions were completely turbid after two-and-a-half days and contained what Cohn had called *Bacilli*. The tubes boiled intermittently remained clear and free of *Bacilli* for two months.

Tyndall explained these results in the following way. Boiling for one or 10 minutes killed all the "sensitive-to-heat" cells in the A and B sets but not the heat-resistant form of the bacterium. After 12 hours the heat-resistant form developed into the sensitive state; therefore, they were killed by the next one-minute boiling procedure applied to set A. The B set was initially boiled for 10 minutes and was not subject to any further heating. In set A there may have been some heat-resistant forms remaining or developed again; thus, another round of heating was carried out 12 hours later, after these heat-resistant forms had converted to the sensitive form. Several rounds of this procedure killed all the sensitive cells; during the rounds of heating, all heat-resistant cells were converted to heat-sensitive cells and killed by heating. There were no heat-resistant forms remaining in the population, nor any sensitive cells. Meanwhile the tubes in set B, which had been boiled for 10 minutes, had some heat-resistant forms remaining after the first and only heating, and they developed and continued to grow to produce the turbid suspension that Tyndall observed. His conclusion was that there was no spontaneous generation. After these studies, there were few proponents of spontaneous generation.

If spontaneous generation of microorganism did not exist, it was reasonable to attribute the cause of fermentations and contagious disease to the presence of these microscopic living entities. That, however, was not the only alternative. It was conceivable that microorganisms were present fortuitously in fermentations or in diseased hosts.

At mid-century and later into the early 1870s, a living agent cause of contagious disease was one among many causes. It was speculated that contagious diseases such as cholera and smallpox were spread by entities that behaved like living agents. There was a chemical theory of disease, according to Liebig, a zymotic theory, that compared the disease entity to a ferment that causes a chemical change. There was Rudolph Virchow's *Cellular Pathology* view of disease. Recall that Carl Braun in Vienna listed 30 causes of childbed fever, while a later text in England identified "ten types of ferments, parasites, septic products" as the causes of disease. The issue of causality of specific contagious diseases remained to be resolved after the middle of the nineteenth century.

Experimental Evidence That External Living Agents, Bacteria, Are the Causes of Contagious Diseases of Humans and Domestic Animals

An impediment to identifying the cause of human contagious disease was the impossibility of studying the transmission of the disease to uninfected hosts under controlled conditions. These kinds of procedures were rou-tinely carried out with material conferring plant diseases since the work of Tillet and Prevost, culminating in the experiments of de Bary.

Anthrax

Anthrax is a disease that can occur in horses, cattle, sheep, and, at times, humans. It leads to a blood infection, hemorrhage, and death. Humans contract the disease by handling animal parts and the wool of infected sheep. An important development in the early nineteenth century was the attempt, in an 1823 experiment, to transfer anthrax to healthy animals using blood from infected animals. Anthrax was not always transferred. Nevertheless, a number of investigators were led to examine the blood of infected animals and found unique microscopic entities. In 1850 Casimir Davaine (1812–1882) was involved in an investigation of anthrax with P. F. O. Reyer (1783–1867), who was head of a hospital in Paris and was an expert on skin diseases and was also interested in animal diseases. Reyer discovered that the blood of a sheep that died of the disease contained ab-errant clumps of corpuscles and something unusual, "filiform bodies . . . about twice as long as a blood corpuscle." The role, if any, of these bodies in the death of the animal was not discussed. By the mid-1850s other in-vestigators, among them A. Pollender (1823–1879), who appeared to be looking for a disease agent, found similar bodies but concluded he could not answer the question of whether they caused the disease. The methodol-ogy required to provide some answers was not available. Shortly thereafter these filiform bodies were found in the blood of a human who had died of the disease.

Davaine returned to the problem in 1863:

> "I thought at that time [1850] that I should be able . . . to check on the exis-tence of those filiform . . . bodies found in the blood of sheep which have died of anthrax and to find out if the development of these microscopic beings (rather like algae) was not the cause of deterioration in the blood and afterwards the death of the animal".
>
> JEAN THEODORIDES, "Casimir Davaine (1812–1882): A Precursor of Pasteur." Medical History, 1966, .10(2):155–165 (citation on p. 159)

He explained the delay of his return and what inspired him to adopt the possibility that the filiform bodies were living agents and that they were the cause of the disease. He wrote that after 1850

"The occasion had not arisen and other work had prevented me from continuing active research when, in February 1861 M. Pasteur published his remarkable work on the butyric ferment, a ferment consisting of small cylindrical rods which possess all the characteristics of vibrios and bacteria. The filiform corpuscles that I have seen in the blood of anthracic sheep are much like the vibrios in shape and I was led to try and discover if this kind of corpuscle (or others of the same nature as those which determine butyric fermentation) when introduced in the blood of the animal would not act as a ferment".

J. THEODORIDES, "Casimir Davaine (1812–1882): A Precursor of Pasteur." *Medical History*, 1966, 10(2), p. 159

Davaine was convinced by Pasteur's work. The cause of fermentation was the result of the activity of a living agent and, reasoning by analogy, Davaine accepted that the cause of anthrax was a living agent present in the blood of infected animals in the form of filiform bodies. That is why he proceeded with the following studies. He used blood of infected sheep to transmit the disease to rabbits and a rat. He described the *bacterides* (the name he created for the microscopic entity) in infected blood. He discovered that dried anthrax blood could transmit the disease, a phenomenon only understood 20 years later with the work of Robert Koch. Davaine recognized that transferring whole blood from one animal to the other to cause the disease did not exclude the possibility that there was something other than the filiform body in the blood that caused the disease. Davaine concluded that he had to separate bacteria from the blood to carry out this experiment, but he had no method to carry out this procedure. There were methods available for cultivating fungi, including growth on solid surfaces, procedures used a century before by Bradley and Miceli. There appeared to be no sharing of information between investigators studying plant and human diseases—or rather individuals studying human diseases did not pay attention to work on plant diseases.

Tuberculosis and Puerperal Sepsis

The visible manifestations of tuberculosis (consumption) were striking: bloody coughing and wasting of the body. Postmortem examinations revealed nodules (tubercles) on lungs that hardened with interior cheese-like deposits. The Virchow school proposed that these nodules were secondary responses to internal tissue inflammation. Another cause, an external cause, became a possibility in the mid-1860s when Jean-Antoine Villemin (1827–1892) claimed transmission of the disease by introducing tuberculous tissue into susceptible animals.

In 1865 it was reported that puerperal fever (childbed fever) could be experimentally transferred to healthy rabbits. These results indicated that these contagious diseases could be initiated by some external source. A number of the investigators referred to Jacob Henle in their studies, which suggested they were proposing a living agent cause of contagious diseases.

The Cattle Plague

In June 1865 a disease of cattle appeared in England and by 1866 it had spread to Wales and Scotland. A Royal Commission was appointed to study the causes of the disease and measures to prevent it. The Commission recognized that the disease was contagious, as it moved from animal to animal and place to place. It is important to examine the Commission's reports, and particularly the testimony of John Simon, on the causes of the disease since they illustrate the state of contemporary views on the origin of contagious disease, human or animal, in England.

The Commission issued three reports, the last one in May 1866, which contained the results of seven scientific papers founded on current causal theory and sanitary practices. These reports contained laboratory investigations, an important decision because it endorsed this sort of methodology to understand the cause of contagious disease. An important concept was stressed in the report, the connection between animal and human diseases. John Simon compared cattle disease to various human epidemic diseases such as smallpox, scarlet fever, typhus, and typhoid fever. These fevers were "zymotic diseases." Simon sounded like a Liebigean, but in fact he objected to Liebig's interpretation of disease. Simon's view, like that of Schwann and Cagniard de la Tour, was that yeast is an "organized vegetable production" that indeed causes fermentation, but such a living entity was not a good analogy for the infectious disease process. Simon "conjectured that the disease process could be classed as a new species of catalytic action," but like Liebig he was unable to explain the increase in the disease-causing material. The report recognized that each disease had its specific cause and its generation was a regular phenomenon; cattle plague always caused cattle plague just as dogs always produced dogs and cats always made cats. The report exclaimed, "What is there like it in chemistry or in physics?" The contemporary answer would be this: There is nothing like it in chemistry or physics.

Is it biology?

The investigations of J. Burden Sanderson (1828–1905) provided the laboratory evidence in the Commission's findings. He found that blood of a diseased animal contained an agent that could produce the disease in another animal. The disease agent, the "poison," must multiply in the blood of an infected animal in a short time, possibly less than 48 hours, since:

> "a minute portion of the mucous discharge from the eyes and mouth of an animal ill with the cattle plague (when introduced into the blood of a healthy animal) the whole mass of blood, weighing many pounds, is infected, and every small particle of that blood contains enough poison to give the disease to another animal".
>
> TERRIE M. ROMANO, *Making Medicine Scientific: John Burden Sanderson and the Culture of Victorian Science*, 2002, p. 68, Johns Hopkins University Press

It is clear that there was an entity that caused the cattle plague, but its nature was not defined. The Commission noted the problem by citing the different terms used to refer to the probable entity, such as poison, particles of poison, seeds of disease, a contagion (a physical substance), and a zymotic element (that is, a ferment).

It was clear that John Simon was struggling to arrive at a definition of the cause of various contagious diseases using concepts that described biological entities, such as "breeding true" and "increasing of morbid material," while he continued to refer to cattle plague as a "zymotic disease," which suggested a chemical agent. Thus, for about 15 years Simon was contending with the "chaotic phenomena of contagion" (multiple causes) and was caught somewhere between a biological and a chemical theory to explain its cause. Simon and the members of the Commission recommended laboratory investigations of the disease, procedures that would also be applicable to human infectious diseases. Included were microscopic studies by Lionel S. Beale (1828–1906) and studies by J. Burdon Sanderson on the general characteristics of the disease. Beale examined "different tissues and fluids of the body at different stages of the disease" and claimed he saw small, living particles in blood—but the Committee found no definitive proof. Sanderson found that the disease could be transferred from an infected animal to another via blood. What could one make of such evidence? The conclusion was that the disease agent was a "poison" that could increase. To complicate the issue, the *British Medical Journal* did not mention these laboratory studies since the establishment was more interested in clinical symptoms and ignored laboratory work.

The work on the cattle plague illustrated the difficulties confronting investigators concerning the cause of contagious diseases. Nevertheless, there were individuals, such as Davaine, who committed themselves to a living agent theory of disease armed with the most limited evidence that such a phenomenon existed. Another such individual was Joseph Lister (1827–1912), who unreservedly adopted a living agent cause of contagious disease founded on the researches of Pasteur. With Pasteur's fermentation studies before him, Lister wrote the following in 1867:

"Turning now to the question how the atmosphere produces decomposition of organic substances, we find that a flood of light has been thrown upon this most important subject by the philosophic researches of M. Pasteur, who has demonstrated by thoroughly convincing evidence that it is not to its oxygen or any of its gaseous constituents that the air owes this property, but to the minute particles suspended in it, which are the germs of various low forms of life, long since revealed by the microscope, and regarded as merely accidental concomitants of putrescence, but now shown by Pasteur to be its essential cause, resolving the complex organic compounds into substances of simpler chemical constitution, just as the yeast plant converts sugar into alcohol and carbonic acid".

Joseph Lister, in THOMAS D. BROCK, *Milestones in Microbiology 1546 to 1940*, 1961, p. 84, Prentice Hall (ASM press, Washington, DC.)

Here we have all the elements of a living agent theory of disease, with the same analogy used by J. Henle in 1840, but now buttressed by the experiments of Pasteur.

Joseph Lister

On August 9, 1867, Lister, a British surgeon, presented a lecture," On the Antiseptic Principle and the Practice of Surgery," at the British Medical Association meeting in Dublin. The contents were subsequently published in the *Lancet* under the title "On a New Method of Treating Compound Fracture, Abscess, and so Forth; with Observations on the Conditions of Suppuration."

To understand his philosophical position, Lister's logic, we shall quote some essential portions of one of his papers. Lister compared simple fractures (where there is no break in the skin) with compound fractures (in which the broken end of a bone protrudes through the skin, leaving an open wound). He inquired why are there "disastrous consequences" from compound fractures "contrasted with the complete immunity from danger to life and limb in simple fracture." As a result of a compound fracture, a serious infection can occur that can be life-threatening. Among the possible causes Lister chose one: the open wound has access to the atmosphere. "Turning now to the question how the atmosphere produces [this effect] we find that a flood of light has been thrown upon this most important subject by M. Pasteur."

Pasteur had not demonstrated that human diseases were caused by "germs of various low forms of life" but had established, to Lister's satisfaction, that microorganisms carry out various forms of fermentation and are present in the air. Lister rejected spontaneous generation of these microorganisms and rejected anti-contagionism. He accepted as valid the analogy between fermentation caused by living microscopic agents and contagious diseases caused by living microscopic agents. I believe that Lister was not arguing simply from similarity but consistently believed the two processes were caused by similar agents, living microscopic organisms.

So how does Lister recommend treating compound fractures? The rationale is given in this way. Since the disease arises from the presence of atmospheric "particles,"

> ... "it appears that all that is requisite is to dress the wound with some material capable of killing these septic germs provided that any substance can be found reliable for this purpose, yet not too potent as a caustic".
>
> Joseph Lister, in Thomas D. Brock, Milestones in Microbiology,
> p. 84, ASM Press

Lister used dressings soaked in weak carbolic acid (phenol). This chemical was suggested to Lister in 1864 when it was observed that sewage treated with carbolic acid did not emit any odor when used as fertilizer on irrigated land. Carbolic acid was also found to kill living agents that infected cattle feed.

Additional Evidence for an External Living Agent as the Cause of Contagious Diseases: Rules for Establishing Causality

Edwin Klebs (1834–1913) was a student of Virchow, whose pathological theory contended that diseases were due to internal inflammatory causes;

external possibilities were ignored. Klebs went off in a different direction, believing that external microorganisms were the causes of disease, frequently referring to Jacob Henle's conjecture that microorganisms cause contagious diseases. In 1871 he was studying wound infections and growing bacteria in special media, including hen's eggs. Why was he growing bacteria in culture? In 1872, in a paper on gunshot wounds, he explained why:

> . . . **"tracing the invasion and the course of micro-organism can make causality probable, but the crucial experiment is to isolate the efficient cause and allow it to operate on the organism** (my emphasis).

To obtain the decisive evidence, it is necessary to

> . . . **"isolate substances from the body and use them to induce further cases of infection** (my emphasis)".

K. CODELL CARTER, "Koch's Postulates in Relation to the Work of
Jacob Henle and Edwin Klebs," *Medical History*, 1985, 29:353–374
(citation from p. 365)

Klebs continued to be more detailed in his requirements to establish causality. He considered it necessary to obtain "matter-fluid" from a host carrying a disease, separate it into its components, and determine which constituent can transmit the infection. The basic assumption in Klebs' discourse was that the causative agent was a living entity, a microorganism. He contended that to prove cause, a number of criteria needed to be established: (1) The agent is present in a host when the disease is present; (2) The agent can be isolated from the host and grown in a pure state; (3) The pure agent is subsequently used to infect a new host.

A Pure Culture

In 1877–78 Joseph Lister (1827–1912) published a paper entitled "On the lactic fermentation and its bearing on pathology." This was not a surprising combination, fermentation and pathology, in view of his writings on the treatment of fractures, where he supported a living agent as the cause of wound infections based on the role of microorganisms in the alcoholic and butyric acid fermentations carried out by Louis Pasteur. Lister accepted as one of the principles for demonstrating a living agent cause of contagious disease or fermentation, the necessity to isolate a unique organism and produce the phenomenon using this microorganism.

Lister's intent "was to obtain, if possible," **absolute proof** that a pure culture of a unique bacterium, which he labeled *Bacterium lactis*, was responsible for the production of lactic acid and the consequent curdling of milk. By demonstrating that a pure culture of a microscopic, living entity could cause fermentation, he would be able to argue "that other organisms may exist . . . smaller than the *B. lactis* and not readily visible in diseased human tissues, that could be the cause of infectious disease of humans."

Lister was searching for a simple experimental system, and the souring/curdling of milk fulfilled his requirements. The effect on milk is visible: the milk solidifies quickly. The process is unique to the lactic acid bacterium, and consequently it is unlikely that accidental contamination will interfere with the experiment.

The experiment depends on obtaining a pure population, but he did not have one because microscopic examination revealed the presence of a much smaller number of another, morphologically different microscopic organism in his culture of the lactic bacterium. To obtain a pure culture Lister adopted the strategy that simply diluting the sample would eliminate the minority organism and at the highest dilution could yield a single *B. lactis* bacterium in a specific volume. He diluted the sample to a point where there is "calculated . . . to contain on the average a single *Bacterium lactis* in the volume he will use to initiate his experiment."

Lister prepared 16 tubes of milk preheated to 210 degrees F that eliminated all existent bacteria. He added to tubes 1 through 10 a volume containing, statistically, one bacterium. To tubes 11 through 15 he added twice the volume, two bacteria per tube. For the last tube, number 16, he used twice again the volume, which resulted in four bacteria in this tube. The milk of tube 16 was curdled in three-and-one-half days, and the same for tubes 11 through 15. At three-and-one-half days tubes 1 through 10 remained fluid, which indicated no fermentation, but during the next day five tubes became curdled at different times. Five tubes remained permanently fluid, without fermentation, even four months later, and did not contain any bacteria. Thus, whenever at least one bacterium was introduced into a milk sample, a lactic fermentation occurred.

The general applicability of this procedure to isolate and characterize microbial agents was demonstrated by Lister using highly diluted tap water to inoculate boiled milk. Some tubes remained free of fermentation while others showed "different kinds of fermentation" (not lactic fermentation) that demonstrated again the particulate nature of the ferments and, importantly, showed that water "contains several different kinds of ferments, which though generally confused through being mixed up together, declare their individual peculiarities when isolated by this method of separation."

Lister's strategy for obtaining a pure culture, although it was successful and was used to make a strong argument for causality, was superseded by a method devised by Koch and collaborators that made the identification of purity more reliable. I will consider Koch's work below.

Did these studies influence contagious disease theories of individuals who occupied important positions in the medical establishment in England, for example? From this perspective it is of interest to examine the views of John Simon, since they are about causality and reveal the changing foundational assumptions from 1850 to 1878.

Simon was indeed an important figure in the British medical establishment. He was the first Medical Officer of Health for the City of London and was later Head of the Medical Department of the Privy Council. He was Head at the time that a Royal Commission was appointed to study the cattle plague in 1865. Simon was influential in choosing the members of the Commission, some of whom were laboratory scientists, since Simon viewed these methodologies as necessary for establishing causality. In 1850 Simon supported a chemical explanation for the cause of disease. During the 1860s he was impressed with the animal experiments of Villemin in his work on tuberculosis and J. Burden Sanderson's work on the cattle plague, in which each investigator could transfer the disease from an infected animal to a healthy animal..

In a review in the British Medical Journal of December 20, 1879, more than a decade after the cattle plague reports, Simon revisited the issue with *An Essay on Contagion: Its Nature and Mode of Action*. He opened the essay by asking the crucial question: What is the nature of the "contagious matter"? On this date, if this question had been put to Pasteur, Lister, Klebs, Davaine, and Koch, they would have designated the cause as a living, microscopic agent, with different kinds of bacteria causing different diseases, although there was no definitive experimental work at this time, and experimental work would be required to establish causality.

Simon, like theorists before him, compared the cause of disease to the cause of fermentation. However, in his philosophy disease was not simply a catalytic process leading to breakdown of organic matter. It was instead an "anaplastic" or "constructive" process. It was analogous to the development of a fertilized egg, when a sperm and egg combine. In the case of tuberculosis there was continuous generation of tubercles as the disease passed from one host to another. It "seems more and more tending to show that the true unit of each metabolic contagion must either be, or must essentially include, a specific living organism, able to multiply its kind."

Simon also considered fermentation a constructive process. The agents that cause fermentation, he wrote, "multiply themselves . . . it seems to be established beyond reasonable doubt . . . the self-multiplication of each of them . . . the infinite multiplication of a specific microphyte."

"Microphyte," a term proposed by Burdon Sanderson, a colleague of Simon, designates a self-multiplying organic form specific for each disease. Simon was convinced by the experiments on fermentation and the transfer of disease, using blood or diseased material, from one host to another that the disease agent was a living microscopic entity.

Robert Koch (1843–1910): Anthrax, Tuberculosis, Pure Cultures, and Specific Criteria That Must Be Satisfied to Prove the Cause of Human Contagious Diseases

In the latter part of the 1870s there was general but not universal agreement that there was no spontaneous generation, and living microorganisms, yeast and bacteria, cause fermentation. The analogy was made between fermentation and disease, both caused by microorganisms. Thus, there appeared to be no theoretical barrier to the role of bacteria, for example, in contagious disease, but it remained to be demonstrated that living microscopic organisms (bacteria) were the **necessary** cause of human contagious diseases. If it is the case that living agents cause disease, does a specific microorganism cause a specific disease? Finally, is there an experimental methodology that can provide unambiguous answers to these questions?

Jacob Henle and Edwin Klebs had proposed an experimental method that would establish a cause-and-effect relationship between a living agent and a disease. The method included the isolation in pure culture of the putative causal agent. Anton de Bary had carried out such studies using spore preparations of fungi to transmit specific diseases to plants. Plant studies presented fewer difficulties for investigators. The spores were larger than the microscopic microorganisms viewed by Pasteur and Lister and thus easier to manipulate and isolate. And infection of plants posed none of the difficulties associated with human disease transmission until the discovery of animal hosts that could be infected by human disease-causing agents. Thus it was not surprising that the first "proof" of a cause-and-effect relationship between a bacterium and a human infectious disease involved an agent that could also infect an experimental animal like the mouse. This was the work of Robert Koch, who demonstrated that a microorganism named *Bacillus anthracis*, a spore-forming rod-like bacterium, was probably the cause of anthrax in mice. It was not possible for Koch to state definitively that *B. anthracis* caused anthrax because he did not have a pure culture.

How could one purify one specific bacterium, for example, from a sample of material that may have many different kinds of bacteria? There was no way one could "reach" into a liquid culture and pick out one bacterial cell among millions. There were approaches available, however. Unknown and certainly unmentioned was the technique fortuitously revealed by Richard Bradley and deliberately by Pier Micheli, in the eighteenth century, who isolated a fungus on the solid surface of the interior of a melon, a surface presumably free of all microorganisms. In the nineteenth century there was the work of Prevost and A. de Bary, who worked on isolating fungi. Ferdinand Cohn, who studied microorganisms, algae, fungi, and bacteria, classifying them on the basis of morphology, did grow them on solid surfaces like potato slices. It was Koch, however, who applied these methods to the isolation of bacteria that were presumptive agents of human contagious diseases.

Koch was a physician who had some experiences with disease. He was an army physician in the Franco-German war of 1870, where there were battlefield wounds and the ever-present diseases of war like typhoid fever. When he returned to life as a civilian doctor in the town of Wallstein, east of the Oder River, now in Poland, there was an anthrax epidemic in 1873 that affected sheep, cattle, and humans. Koch initiated a research program in his house under the most primitive laboratory conditions. He examined microscopically the blood of sheep and saw rod-shaped bodies, which Koch acknowledged were revealed by the work of C. Davaine, who categorized them as bacteria and had infected healthy animals with fresh and dried blood containing these rod-like structures. Koch, a country doctor with no organized research equipment except for a microscope that he had used in his natural history interests, arrived at some positions in contrast to Davaine and set out to clarify the issue of contagion and to understand the growth characteristics of the anthrax bacterium, which he regarded as the causative agent of anthrax. In addition to Davaine, an important influence on Koch was Ferdinand Cohn, who discovered the spore stage of the family of rod-like bacteria, which he named *Bacilli*. This structure, a developmental stage in the life of the bacteria, became an important element in Koch's work on the anthrax disease. Koch wrote that the bacteria that cause anthrax form spores that possess the ability to form bacteria after a long or short resting period.

A summary of the studies before Koch demonstrated that anthrax was a disease that struck many domestic animals as well as humans. In 1849, A. Pollender observed rod-shaped microorganisms in the blood of cows that had died of anthrax. Subsequently, the French scientist C. J. Devaine transferred the infection from a diseased animal to a healthy one by the transfer of blood containing the microscopic agent. While it was not unusual to examine microscopically material from animals or humans, the impulse to do so derived from a theory of disease that involved a microscopic agent. The presence of the agent in an infected host was not enough to prove the cause of the infection. However, the work of Devaine constituted a major advance, for he demonstrated the transfer of the deadly disease from one animal to the other. The microorganism present in blood, already indicated by Pollender, was a large, rod-shaped bacterium that was easily visible, especially with the advanced microscopes of the day, and grew readily in various nutrients.

When Koch initiated his studies he pointed out a number of difficulties regarding the origin, the cause of the disease revealed by previous studies. Koch cited experiments, other than Devaine's, where the bacteria were not detected in the blood of diseased animals but nevertheless caused the disease when the blood was transferred to recipient animals. There were instances when the disease would apparently arise spontaneously in animals when there were no instances of the disease in the environment. How was one to explain the origin of disease without the apparent presence of bacteria? Koch's experiments resolved these problems by clarifying the life cycle of the bacillus that was the necessary cause of anthrax, a life cycle identical to that revealed by Cohn for the "hay bacillus" that Cohn had named *Bacillus subtilis*.

Koch initially experimented with various ways to grow the bacterium. He placed a fragment of spleen from an animal with anthrax on a slide, in a growth chamber containing fresh serum or corneal fluid. In this way he could follow microscopically the development and growth of the bacillus. After about 20 hours at 35 degrees C the fluid surrounding the spleen fragment contained the entire range of structures of the organism, including individual bacilli, strings of bacilli, strings of bacilli with "grains" (developing spores), and free spores. He was able to observe the development of rods and filaments from spore masses and the formation of spores from rods, the entire life cycle of the bacterium.

Understanding the life cycle of the bacillus explained why the disease could be transmitted using either the blood of infected animals that contained no rods, or dried blood. Davaine used dried blood that transmitted disease. Blood without rods still contains spores that are smaller than the bacilli and can be missed on microscopic examination. Furthermore, spores are not only heat-resistant but can withstand drying. The number of spores in ordinary blood or dried blood need not be large. Once transmitted to the blood of a new host, the spores germinate and produce a large population of the vegetative state. The "spontaneous" occurrence of anthrax in animals is explained by the fact that spores can remain in a pasture, for example, for long periods and can at times cause infection.

Koch demonstrated that mice inoculated with spore masses died of anthrax. Koch also carried out a series of experiments with mice in which he sequentially inoculated 20 mice, reproducing anthrax in each mouse. Koch reasonably concluded that the bacillus had multiplied each time in each mouse to cause the disease. Koch carried out a number of experiments to determine whether the disease was caused by this specific bacillus. He introduced into animals the organism that was present in hay infusions, which was morphologically similar to the bacillus involved in anthrax. There were no infections. These experiments constituted for Koch a powerful argument that the bacterium known as *B. anthracis* causes anthrax. The disease could only be induced by a unique bacterium, and the presence of this bacterium was necessary for the disease to occur. His philosophical position was formalized in a second paper on anthrax: "anthrax never occurs without viable anthrax bacilli or spores. . . In my opinion no more conclusive proof can be given that anthrax bacilli are the true and only cause of anthrax."

Koch realized that this statement was not supported by proper evidence; he knew that only the use of a pure culture of bacteria to produce a disease would constitute proof, and he did not have a pure culture. Indeed, no investigator, except perhaps Lister, up to this period, had such a pure culture—not Pasteur, not Tyndall, and not Koch.

To be sure that one unique microorganism caused one human disease, it was incumbent to isolate one kind of microorganism spatially from all others, and this could be done on a solid nutrient surface that was sterile. The procedures used to isolate a single bacterial type were devised by J. Schroeter in 1875, while he was a student in Cohn's laboratory. The title of the paper "Uber Einige Durch Bacterien Gebildete Pigmente" (Concerning a Few Pigments Generated by Bacteria) does not do justice to the importance of the method used by Schroeter to characterize different bacteria according to their pigment-producing capabilities. Colonies of different colors could be visualized by having each colony fixed at a position on a solid surface separated from a colony of another color growing on the same surface. Each colony displayed a pigment unique to itself, which was maintained on subsequent subcultures. Schroeter initially used the cut surfaces of potatoes to culture the bacteria, and he also used solid media made from starch paste, egg albumin, bread, and meat. Koch, in 1881, although apparently not citing Schroeter, discussed in great detail how growing bacteria on the surface of potatoes, "is a very simple method for the production of perfect pure cultures." However, Koch recognized that these cut potato surfaces would not support the growth of many pathogenic bacteria. Thus, he had to create another solid surface containing nutrient materials that would do so. He incorporated various broths in gelatin. These preparations were useful, but the solid gelatin surface could not be maintained at higher temperatures and were particularly subject to heat and melting during the summer months. At that time a polysaccharide product derived from *Rhodophyta* (a seaweed that yielded agar) was used commercially as a gelling agent, and associates of Koch, Fannie and Walter Hesse, used it as a solidifying agent for a variety of rich nutrient broths. In early experiments the solid media were layered on flat glass plates that were covered by a bell jar, a rather cumbersome apparatus. In 1887 Richard J. Petri devised a simple flat dish that was 11 cm in diameter and about 1.5 cm deep. This was covered by another glass dish with only a slightly larger diameter. The double dishes could be sterilized separately and cooled and the hot liquid media could be added and allowed to solidify. (These Petri plates are still used today, even though they may be made of plastic: not many tools of biological research have had such a long and distinguished life in the laboratory!)

To isolate individual bacteria from a source presumed to contain a mixed population, it was necessary to dilute the sample of bacteria until there were a limited number of individual cells in a small volume of liquid that could be spread on the surface of an appropriate solid growth medium. At each position on the solid surface, where one bacterium had been deposited, the bacterium would undergo division and in time a visible colony would be generated. All the cells in the colony would be derived from a single cell. Ideally, no colony would be near another colony to rule out cross-contamination. Several sequential repetitions of this procedure could be carried out to ensure clonal purity.

In the late 1870s Koch turned to investigations on wound infections and began to codify criteria for disease causality, standards similar to those proposed by Edwin Klebs. First, the microorganism must be present in all cases of the disease; it is necessary for the disease. Second, for each different disease, a morphologically distinguishable microorganism must be identified. Koch initiated a research program on tuberculosis based on the criteria he had developed.

Tuberculosis

Koch began his work on the cause of tuberculosis in the summer of 1881. The number of deaths due to TB was declining in the nineteenth century, but it was still the major cause of death in Western Europe. There was longstanding evidence that the disease was contagious among humans, and the French physician Villemin had demonstrated that it could be transmitted to experimental animals. Koch was convinced from his reading of the literature that a "parasite" was the causative agent of tuberculosis. He presented his results to the Physiological Society of Berlin on March 24, 1882, and published the work one month later. The lecture was entitled "The Etiology of Tuberculosis."

By a variety of staining procedures developed in his laboratory over the years, Koch was able to visualize microscopically bacilli that were the suspected cause of tuberculosis. He stated categorically that even though they may not have occurred in large numbers in infected tissues,

> "On the basis of my extensive observation I consider it proven that in all tuberculosis conditions of man and animals there exists a characteristic bacterium which I have designated as the tubercle bacillus.
>
> ROBERT KOCH in THOMAS D. BROCK, *Milestones in Microbiology 1546–1940*,
> 1961, p. 111, ASM Press)

To prove that this bacterium caused TB, it would have to be isolated from the body as a pure culture and would have to cause the disease when inoculated into a healthy uninfected animal. To isolate the bacterium he removed small tubercles from the lung tissue of infected guinea pigs and placed them on the surface of a solidified nutrient medium. The growth of the bacteria, when successful, took about two weeks at 37 degrees C. To maintain growth, the culture was transferred to a fresh surface. Finally, Koch used these cultures to inoculate a large number of animals in many different ways and demonstrated the production of tubercles and tuberculosis. His results allowed him to establish a series of principles coupled to a methodology to be followed to prove causality in all case of contagious diseases caused by an external living agent:

> "First it is necessary to determine whether the diseased organs contain elements that are not constituents of the body or composed of such constituents. If such alien structures can be exhibited, it is necessary to determine whether they are organized and show signs of independent life . . .

Such considerations enable one to conclude that there is probably a causal connection between the structures and the disease. Facts gained in these ways can provide so much evidence that only the most extreme sceptic would still object that the organism may not be the cause, but only a concomitant of the disease. Often this objection has certain justice, and, therefore establishing the coincidence of the disease and the parasite is not conclusive.

In addition, one requires direct proof that the parasite is the actual cause. This can only be achieved completely [by] separating parasites from the diseased organism and from all products of the disease that could be causally significant.

If the isolated parasites are then introduced into healthy animals they must cause the disease with all its characteristics".

K. CODELL CARTER, "Koch's Postulates in Relation to the Work of
Jacob Henle and Edwin Klebs," *Medical History* 1985, 29:253–274
(citation from p. 361)

Finally, from these infected animals the identical agent initially introduced to cause the disease must be isolated from the infected diseased animal. This experimental methodology establishes the necessary role of the specific bacterium in the disease process.

Two hundred years after Leeuwenhoek revealed the presence of bacteria, Koch provided unambiguous evidence that a pure culture of a specific bacterium was the necessary cause of TB. Over the following 20 years it would be shown that numerous contagious diseases are caused by specific bacteria using the same methods employed by Koch.

Summary

Clifford Dobell and others contended that,

"from the date of their discovery … bacteria were strangely neglected. Mankind remained inexplicably blind to their importance and almost to their very existence".

A microscopic living agent theory of the cause of contagious diseases was obliterated by the power of an alternate theory at the end of the seventeenth century and the early decades of the eighteenth century, a theory founded on a version of the mechanical philosophy that proposed that all matter is composed of corpuscles that, according to Robert Boyle, for example, have the physical properties of size, shape, and motion and originate in various environments to bring disease to humans through the air or ingested with food and water. The conflict between Richard Bradley and Richard Mead, described in detail in Chapter 10, illustrates the triumph of one philosophical position over another.

Another blow to a microscopic living agent theory of contagious disease was a theory that persisted into the mid-nineteenth century that contended that these very small living entities were spontaneously generated. Because of their minute size and the inability to characterize their composition, they occupied a territory somewhere between living and non-living matter and remained even into the nineteenth century suspended in a category that was neither plant nor animal.

In the nineteenth century a cell theory was proposed to account for the construction and the metabolic activity of all animals and plants. The inclusion of microscopic entities as living cells created the possibility that these cells could carry out independent metabolic and pathological activities, a theory opposed by a chemical theory for fermentation and disease causation. Consequently, a third important impediment to considering bacteria as a possible cause of contagious disease was for a long period the impossibility of establishing the nature of these microscopic agents. To accept the possibility that they could be spontaneously generated differentiated them profoundly from all other living beings that came from preexisting living animals or plants. What contributed to the demise of this divide between bacteria and fungi and "higher" plants and animals was the inclusion of these organisms as cells, the basic building blocks of all plants and animals, and the understanding that they have the chemical capabilities that enable them to reproduce more of their kind.

Finally, and most importantly, a methodology to enable the isolation of bacteria in pure cultures was established that made it possible to associate the production of one unique disease with one unique bacterial agent.

CHAPTER 13 | Filterable Agents, Designated as Viruses, Cause Contagious Diseases of Plants, Animals, Humans, and Bacteria

"The central problem in the study of the intracellular behavior of viruses is that of their reproductive mechanisms. On the one hand, clarification of these mechanisms would give us a better understanding of the nature of viruses . . . On the other hand, information on the mechanism of viral reproduction may throw light on the central problem of biology, that of the reproduction of individual specific biological elements, such as genes. Viruses carry specific genetic material. Viruses may indeed be the material of choice for the investigation of biological reproduction".

S.E., General Virology, 1953 LURIA, pp. 157–158, John Wiley & Sons

"It seems likely that the nucleic acids may themselves be, in part or exclusively, the carriers of specific biological configurations".

S. E. LURIA, *General Virology*, 1953, p. 100

ON APRIL 1, 1717, Lady Mary Montagu, the wife of the British Ambassador to Turkey, wrote to one Miss Sarah Chiswell, from Adrianople, about the plague; in this instance the term specified smallpox. She appeared to be delighted to report that the disease "does so little mischief":

"Apropos of distempers, I am going to tell you a thing that I am sure will make you wish yourself here. The small-pox so fatal, and so general amongst us, is here entirely harmless by the invention of ingrafting which is the term they give to it. There is a set of old women who make it their business to perform the operation every autumn, in the month of September when the great heat is abated".

Groups of 15 or 16 people gather and

. . . "the old woman comes with a nut-shell full of the matter of the best sort of small-pox, and asks what veins you please to have opened. She immediately rips open that you offer to her with a large needle (which gives you no more pain than a common scratch), and puts into the vein as much venom as can lie upon the head of her needle, and after binds up the little wound . . . The children or young patients play together all the rest of the day, and are in perfect health to the eighth. Then the fever begins to seize them, and they keep their beds two days, very seldom three. Every year thousands undergo the operation . . . There is no example of anyone who has died of it; and you may believe I am very well satisfied of the safety of this experiment, since I intend to try it on my dear little son".

A final remark by the lady includes a nasty swipe at doctors in England:

"I am patriot enough to take pains to bring this useful invention into fashion in England, and I should not fail to write to some of our doctors very particularly about it, if I knew any of them that I thought had virtue enough to destroy such a considerable branch of their revenue for the good of mankind".

LADY MARY WORTLEY MONTAGU, Letters of the Right Honorable Lady,
letter 36, 1796, vol. 1, pp. 167–169, Anthony Henrici

What Lady Montagu observed was a procedure, also used in Asia and Africa, that exposed persons to an entity that caused a mild disease to protect them from contracting a serious, generally fatal case of the same disease should they encounter it at some future date. Montagu referred to the entity as the matter of smallpox. The question not addressed at this time was: What is the nature of the entity introduced on the head of the needle? An insight, however tentative, into the essence of the entity was revealed in 1798 in a paper by Edward Jenner (1749–1823). Jenner used material from pustules surrounding the nipples of cows to achieve the same immune result as the Turkish women. Jenner stated, ". . . what renders the Cow-pox virus so singular, is, that the person who has been thus affected is for ever after secure from the infection of the Small-Pox."

Jenner referred to the entity as a virus, but his understanding was not the characterization 100 years later. During the eighteenth and nineteenth centuries virus could signify a poisonous liquid, some morbid principle, that produced a disease in a body, and that disease may induce the disease in another body. It came under the category of "contagions."

Louis Pasteur was not looking for the cause of smallpox but had been intensely pursuing the cause of rabies since the mid-1870s in attempts to produce a way to protect individuals from the horrible disease, in the same way that Jenner protected humans against smallpox. On December 10, 1880, a surgeon contacted Pasteur and said he had a patient, a five-year-old boy, who had been bitten by a rabid dog one month previously. He died the next morning. Four hours later Pasteur collected mucus from the boy's mouth, mixed it with water, and injected it into two rabbits, who died in 36 hours. Blood obtained from the rabbits produced death in other rabbits and dogs. It was obvious that the blood contained a lethal principle. Pasteur attempted to culture the "microbe" on culture media, media that would support the growth of bacteria, without success. Nevertheless, he persisted and looked for some "rabid" material in spinal fluid, and attempted to culture some agent from this fluid, again without success.

Pasteur was pursuing an established methodology for observing, cultivating, and testing the disease-causing ability of bacteria that had been developed in the final decades of the nineteenth century (Chapter 12). Koch had already demonstrated that the agent that causes tuberculosis is a pure culture of a unique bacterium. It appeared impossible to apply this methodology to the entity that is presumed to cause rabies, however. Pasteur reported that a colleague suggested there might not be a rabies microbe. However, Pasteur was confident he could diagnose a rabid brain by microscopic examination. In his writings one can detect his frustration after years of inability to culture the causative agent of rabies. It is clear that there was a "morbid principle" that could not be cultured on the usual growth media that support the growth of many bacteria; however, in the appropriate host it increased in potency so that sequential transfers of fluid from infected animals could transfer the disease to new hosts.

The Term "Virus" Takes on a New Identity

Plant diseases are caused by fungi; animal diseases and human diseases are caused by bacteria. They are living microscopic cells that can reproduce in different ways and are responsible for disease. These were the facts in the 1880s (Chapter 12). This secure pattern of disease causation began to be upset in this decade by the investigations of Adolph Mayer, an agricultural research chemist and Director of the Agricultural Experiment Station at Wageningen, Holland, who was studying a disease of tobacco in 1879. In 1886 he published a paper, "Concerning the Mosaic Disease of Tobacco":

> "About 3–5 weeks after the young plant has been planted in the field . . . and has begun to grow vigorously . . . (there appeared) . . . . a mosaic-like coloring of light and dark green of the leaf surfaces".
>
> ALFRED MAYER, *Concerning the Mosaic Disease of Tobacco, Phytopathological Classics*, 1942, Number 7, p. 11, trans. by James Johnson, Published by The American Phytopathological Soc., The Cayuga Press, Ithaca, NY

Plant growth is retarded, there is curling and brittleness of the leaves, and ultimately large holes are created in the leaves as the texture disintegrates. Mayer, of course, had resources to study this disease, means that were developed in studying fungi and bacterial diseases. In experimenting with methods to transfer the disease,

> "I suddenly made the discovery that the juice from a diseased plant obtained by grinding [leaves] contained a certain infectious substance for healthy plants".
>
> ALFRED MAYER, *Concerning the Mosaic Disease of Tobacco*, Phytopathological Classics, 1942, p. 20

The juice, a "thick green emulsion," was injected, using fine capillary glass tubes, into the large veins of a leaf, and about 10 to 12 days later the "very young leaves" of the plant displayed the disease. What was in the sap derived from the diseased leaf that transmitted the disease? Mayer cultured the fluid, expecting to uncover some unique bacterium, but none of the "bacterial vegetation" that grew caused infection. He attempted to cause the disease using extracts of the various fertilizers and the variety of bacteria available in pure cultures in his collection. Nothing worked. He speculated that there might be a "ferment" that caused the disease and thus filtered the sap and used the clarified juice to transmit the disease. The experiment did not work, and Mayer concluded that the disease was caused by a bacterium that could not be cultured.

In 1892, Dimitri Ivanowski read a paper before the Academy of Science, St. Petersburg, with a title almost identical to the one used by Mayer: "Concerning the Mosaic Disease of the Tobacco Plant". He confirmed

Mayer's finding that the sap was infectious for healthy plants. However, he stated,

> "I must contradict most emphatically the author's statement that the sap of leaves attacked by the mosaic disease loses all its infectious qualities after filtration through double filter paper. According to my experiments the filtered extract introduced into healthy plants produces the symptoms of the disease just as surely as does the unfiltered sap".

DMITRII IVANOWSKI, *Concerning the Mosaic Disease of the Tobacco Plant,*
Phytopathological Classics, 1942, Number 7, p. 29, trans. by James Johnson,
Published by The American Phytopathological Soc.,
The Cayuga Press, Ithaca, NY

Ivanowski followed with the most important experimental result:

> "Yet I have found that the sap of leaves attacked by the mosaic disease retains its infectious qualities even after filtration through filter-candles".

DMITRII IVANOWSKI, *Concerning the Mosaic Disease*
of the Tobacco Plant, Phytopathological Classics, p. 30

The "filter-candles" were invented by Charles Chamberland, working with Louis Pasteur, as a way to prepare water free of bacterial contaminants. Their efficacy was established over many years of use: bacteria would not readily pass through these filters. Ivanowski confirmed these results in his experiments. The question is this: What passed through these filters? Ivanowski chose, from among the available options, a soluble ferment, a toxin secreted by a bacterium—in effect, a chemical agent.

The story was continued by Martinius William Beijerinck (1851–1931) with his paper published in 1898, "Concerning a Contagium Vivum Fluidum as Cause of the Spot Disease of Tobacco Leaves". He called it a "contagious living fluid." We will get to an elaboration of this phrase. Beijerinck was trained in chemical engineering and the biological sciences and was at one time a student of Anton de Bary (see Chapter 11). At the time of publication of the paper he was Professor of Microbiology in the Technical School of Delft. During the previous decade he had lectured at Wageningen, where Mayer worked, and it was there he learned of Mayer's work on the tobacco disease. What led Beijerinck to the conclusion contained in the title were these facts:

1. The sap from diseased plants caused infection after being processed through porcelain filters (thus the "*Contagium*").

2. The material in the sap that caused disease diffused in an agar prep-aration. Agar is like gelatin and forms a firm gel. Because it diffused some small distance from where it was placed, Beijerinck concluded that the disease agent behaved like a fluid or an enzyme, a soluble entity (thus the *"Fluidum"*).

3. In the infectious process only those parts of the plant that are actively growing, "in which the division of cells is still in full progress," could be infected. Recall Mayer reported that only young leaves dis-played the disease.

4. The clarified extract of infected leaves contained a highly infectious quantity of the disease-causing agent. The increase in quantity of in-fectious agent led to the term *"Vivum."*

5. Finally, Beijerinck discovered when a certain high concentration of alcohol (not specified) was added to the sap that contained the infec-tive principle, a precipitate was formed consisting of the filterable agent that caused the disease. This agent could be dried at 40 degrees C and retained its ability to cause the disease of tobacco.

Certain facts indicate that the agent was not a bacterium or a fungus. These cells could not ordinarily survive after treatment with alcohol. It could not be cultured on media that support the growth of many bacteria but in-creased in quantity in living, rapidly growing plant tissue. It could be pre-cipitated and dried; this suggests it is a chemical compound, but there is no instance of a chemical compound that is reproduced when in contact with a living cell.

And therefore, in the last decade of the nineteenth century, there was uncovered a disease of plants whose cause was unknown.

A Filterable Agent That Causes An Animal Disease

Friedrich Loeffler (1852–1915) and Paul Frosch (1860–1928) reported, a few months before Beijerinck's paper on the tobacco disease in 1898, on a disease of cows named *foot and mouth disease*. The disease manifests itself in the formation of vesicles, blisters in the mouth area, udders, and hooves of cows. Loeffler and Frosch were part of a commission to investi-gate the disease. Their approach was to mobilize all the methodology that had been developed during the previous decades that revealed the presence of bacteria as causes of diseases such as anthrax and tuberculosis. To ac-complish this goal they obtained material from the interior of the vesicles and attempted to culture the presumptive bacterial agent. The surfaces of vesicles were initially treated with alcohol to sterilize the exterior, while the interior contents were removed with a sterilized glass capillary tube and examined microscopically in various ways, including staining techniques that had been developed to view bacteria. Vesicle fluid was also placed on a variety of growth media under aerobic and anaerobic conditions (that is, with and without oxygen) in the event that an organism sensitive to oxygen or inhibited by oxygen turned out to be the culprit. No bacteria were observed microscopically and no growth of bacteria occurred in any of the media even after many weeks of incubation. Nevertheless, the vesicular material, when used to inoculate the upper and lower lips of calves and heifers, transmitted the disease. It was concluded that the causative agent was not a toxin, since healthy animals housed in the same place with sick animals contracted the disease. The disease was contagious; it passed from one animal to another. It was a devastating illness: all animals had to be destroyed and disposed of in such a way to prevent the disease agent from remaining active.

Following this work the investigators carried out experiments hoping to find some way to create immunity in animals and discovered that the material that caused the disease could pass through a filter that held back bacteria. To demonstrate this experimentally, lymph from infected animals was used:

. . . "the lymph was diluted with 39 parts of water, then inoculated with an easily culturable and identifiable bacterial species which had been cultured from lymph—*Bacillus fluorescens*—and then filtered two to three times through a sterilized Kieselguhr candle".

The investigators deliberately added an identifiable bacterium to the lymph to ensure the efficacy of the filtration process:

"The addition of the bacteria served to demonstrate that the filtrate was really bacterial-free since large inoculums of this were placed on nutrient media and examined for growth after incubation. If colonies of this bacterium did not appear on the seeded media, then it was assumed that the filtration had succeeded, and that all the bacteria that had been previously present in the lymph were retained by the filter candle. Filtrates tested in this manner were always bacterial free".

A series of calves were inoculated intravenously with measured amounts of these filtrates corresponding to one tenth to one fortieth of pure lymph, in order to ascertain if there was a dissolved substance in the lymph which would aid in the production of immunity.

The results of the injections were surprising (my emphasis). The animals inoculated with the filtrates died in the same time as had the control animals which had received corresponding amounts of lymph. Their deaths occurred with all of the typical symptoms of the disease.

Obviously, there was a lethal substance present in filtered lymph from which all ordinary-sized bacteria had been eliminated. In a complex series of calculations the authors decided that if it was a toxin in the filtrates, its potency was greater than any previously found, particularly more powerful than tetanus toxin. From these considerations they concluded that

"It therefore seems more appropriate to conclude that the activity of the filtrate is not due to the presence in it of a soluble substance, but due to the presence of a causal agent capable of reproducing. This agent must then be obviously so small that the pores of a filter which hold back the smallest bacterium will still allow it to pass".

FRIEDRICH LOEFFLER and PAUL FROSCH, "The Report of the Commission
for Research on the Foot-and-Mouth Disease, 1898," In Thomas D. Brock,
Milestones in Microbiology: 1546 to 1940, 1961, pp. 150–152, ASM Press

Loeffler and Frosch believed that the disease-causing agent of foot and mouth disease was a smaller version of the smallest previously discovered bacterium known as the Pfeiffer bacillus, suggested to be the cause of influenza. They proposed that miniscule bacteria may be responsible for other diseases, such as smallpox, cowpox, scarlet fever, measles, typhus, cattle plague, and so forth. In 1898 the causes of these diseases were unknown. And thus, in the last decade of the nineteenth century, there was uncovered a disease of animals whose cause was physically smaller than most bacteria and was invisible even microscopically. In short, the cause was unknown.

A Human Contagious Disease Caused by a Filterable Entity, Transmitted to Humans via a Mosquito

In 1900 Walter Reed (1851–1902), a physician, curator of the Army Medical Museum, and Professor of Bacteriology at the Army Medical College, was appointed to head a commission to study, primarily, yellow fever on the island of Cuba. The conventional wisdom was that the disease was caused by a bacterium, spread by mosquitoes, according to the Cuban physician Carlos Finlay. The blood of patients with the disease did not reveal any bacteria, nor did the bodies of individuals who died of the disease. Reed was allowed to carry out human experimentation, exposing volunteers to mosquitoes that had fed on the blood of individuals with yellow fever, and they contracted the disease. On microscopic examination of blood, however, no bacteria were visible. Reed was informed by William H. Welch, a professor at Johns Hopkins, of the studies by Loeffler and Frosch on foot and mouth disease, and this encouraged Reed to repeat their filtration experiments.

At 11 a.m. on October 15, 1901, Reed carried out one such experiment. A volunteer "was given a subcutaneous injection of 3 c.c. [about a teaspoon] of the diluted and filtered serum from a patient with the disease." Reed used a newly developed Berkefeld laboratory filter, which was tested to determine whether it prevented the bacterium *Staphylococcus pyogenes,* a common skin resident, from traversing the filter. The filter did indeed eliminate bacteria. In four days after the subcutaneous introduction of filtered serum into a volunteer, the individual developed fever, which varied over the next four days and included the usual symptoms of yellow fever. Reed, like everyone else performing this kind of procedure, had to explain why a liquid containing no visible microorganisms or any living entity that could grow in a rich broth could cause disease. Reed rejected the presence of a toxin, noting the inactivation of the disease-causing entity by a level of heat that did not inactivate the known toxins, such as tetanus toxin. Reed concluded, "The important questions which naturally arise from the foregoing experiments must be left for the future observations to determine." The cause of yellow fever was unknown.

So, at the opening of the twentieth century, there were diseases of humans, domestic animals, and plants whose causes could not be cultivated using media that support the growth of bacteria and fungi and were microscopically invisible. This new disease phenomenon upset the prevailing model of the bacteria or fungal cause of contagious diseases of plants, domestic animals, and humans. Bacteria and fungi are living agents, are microscopically visible, and can be grown on various nutrient media. This paradigm was not discarded but appeared to be too restrictive to explain the nature of all contagious disease agents. The difficulties in developing a more comprehensive model of the causes of contagious disease can be illustrated by the difficulties encountered in identifying the cause of human influenza well into the twentieth century.

Influenza: A Human Disease Caused by an Entity That Passes Through Filters

Influenza epidemics were known in the nineteenth century, and it was during the epidemic of 1889–90 in Germany that Richard Pfeiffer, once associated with Robert Koch and now head of the Berlin Institute for Infectious Diseases, isolated *Bacillus influenza* from the respiratory tract of people with the disease. It became known as the Pfeiffer bacillus. A 1919 textbook of bacteriology stated that "the relationship between the clinical disease known as influenza or grippe and the Pfeiffer bacillus has been definitely established by numerous investigators."

Various studies did not clarify the issue of the Pfeiffer bacillus as a cause of influenza. In New York and London the bacillus was isolated from patients but the isolates were not identical strains. That indicated they did not come from a possible central disease source but rather were present as "normal" inhabitants of the respiratory tract.

In Germany, in 1914, there was a report that the symptoms of a cold could be reproduced by inserting the filtered nasal contents from an individual with a cold into the respiratory tract of a healthy person. This finding suggested that influenza was caused by a filterable agent.

Influenza reemerged as a lethal disease of catastrophic proportions in 1918, the last year of World War I, the result of a unique biological transformation. The disease broke out in the spring of 1918 as a contagious but not lethal disease, but in the fall it turned into a devastating killer all over the world. Many individuals died quickly due to the destruction of the lungs, while a majority died from secondary infections such as bacterial pneumonia. In the United States various studies were conducted, but they produced confusing results. In 1918, in the Boston area, where the disease struck hard on Army and Navy personnel, an investigation to reveal the cause was carried out on Gallups Island, a quarantine station in Boston Harbor, by members of the U.S. Public Health Service and the Navy. Human volunteers were recruited and were injected with sputum and blood samples and lung tissue from individuals who died of the infection. Investigators considered the idea that the causative agent might be filterable, and they used such material to infect volunteers. No one contracted the disease. Similar results were obtained with volunteers in the Angel Island quarantine station in California.

These results differed from those of Olitsky and Gates at the Rockefeller Institute, who published a series of papers in 1921. Nasopharyngeal washings from patients with early stages of epidemic influenza were used to infect rabbits, which exhibited clinical and pathological conditions in the blood and lungs similar to influenza. Washings from many patients free of the disease did not induce a disease in rabbits. They recommended "further investigation of the inciting agent of epidemic influenza."

At this point the investigators claimed they had a system to study the disease since they believed they had transmitted influenza to rabbits. The nasopharyngeal washes from humans and material contained in the lungs of infected rabbits contained a filterable agent that passed through two versions of the Berkefeld filter:

> "From the filtered nasopharyngeal washings of patients in the first 36 hours of uncomplicated epidemic influenza [recall the concerns with secondary infections such as bacterial pneumonia] . . . we have cultivated a minute bacilloid body, *Bacterium pneumosintes*, 0.15 to 0.3 microns in length [0.1 micron is 0.00003 of an inch] . . . capable of indefinite propagation on artificial media . . . was also recovered in pure culture from the unfiltered and filtered lung tissue of rabbits and guinea pigs inoculated with nasopharyngeal washings of early influenza case . . . The organism grows only under strictly anaerobic conditions, passes Berkefeld V and N filters".
>
> PETER K. OLITSKY AND FREDERICK L. GATES, "Experimental Studies of the Nasopharyngeal Secretions from Influenza Patients, IV," 1921, pp. 727–728, *J. Exp. Med.*, 33:713–729

They concluded that the cause of influenza was a filterable, unusually small bacterium that could pass through a "coarse" (V) and "normal" (N) filter; the organism grows on artificial culture medium only under conditions where there is no oxygen.

The formal rules for establishing causality, Koch's postulates, require the isolation of the putative causal agent in pure culture and the subsequent infection of a susceptible host to reproduce the identical disease. It turned out that the influenza disease could not be transmitted to rabbits.

A major step toward accomplishing the ideal of obtaining a proper test animal began in the fall of 1918 at the height of the human influenza epidemic. J. S. Koen, a veterinarian and inspector with the Federal Bureau of Animal Husbandry, was attending a swine breeders' convention in Cedar Rapids, Iowa, in the first week of October when he noticed that many pigs were ill with a contagious respiratory disease. The phenomenon was not pursued until 1929, when there was an outbreak of a similar disease among pigs in Iowa. One N. McBryde obtained "filtered" mucus from the respiratory tract of infected pigs and placed this material into noses of healthy pigs. The disease was transmitted. The symptoms displayed by the pigs were so similar to those displayed by human influenza victims that the disease of pigs was called swine flu. The disease of pigs captured the attention of Richard Shope (1901–1966) of the Rockefeller Institute. He was born in Iowa, spent his youth on a farm, and received a medical degree at the University of Iowa Medical School. It is not surprising that his knowledge of farm animals and his medical training brought him to the study of swine flu. At Rockefeller he began his studies with Paul Lewis, an expert on virology, who subsequently died of a yellow fever infection.

From pigs with swine flu, Shope isolated from respiratory samples a bacterium that was not present in uninfected pigs. It grew well in an artificial medium containing blood and was designated *Hemophilus influenzae* (variety *suis*). Attempts to infect pigs with this bacterium were unsuccessful. Shope resorted to filtration experiments. Berkefeld filtrates of material from pigs with an active case of swine influenza transmitted into uninfected animals a mild version of the disease. When this filtrate, containing the agent that gave this version of the disease, was mixed with the bacterium *H. influenzae,* an authentic case, clinically and pathologically, of swine influenza occurred. The bacterium that was present in the bronchial fluids of infected pigs did not itself produce the disease. Certainly, the filterable agent made the key difference. Perhaps these results could be relevant for the cause of human influenza.

Within two years there was a definitive answer from three investigators in England, C. H. Andrewes (1896–1988), Wilson Smith (1897–1965), and P. P. Laidlaw (1881–1940), during an epidemic of influenza. The title of their paper was A *"Virus Obtained from Influenza Patients"*, published in *The Lancet* of July 8, 1933, 15 years after the great pandemic of 1918. The paper is a classic, and we will quote parts of the narrative to enjoy its substance because there is something substantial to relate.

The key hypothesis, stated in the opening paragraph, was that the causative agent was a virus. It is important to remember that the term "virus," at this point in the history of contagious diseases, meant that there was a disease-causing entity that was filterable and consequently smaller than any bacterium. It was impossible to conceive of their size since these agents had never been seen. They appeared to increase in number, but only when they were in contact with their respective hosts. The British investigators wrote,

"The epidemic of influenza at the beginning of 1933 afforded an opportunity of making an experimental study of this disease . . . On the assumption that the etiological agent of influenza was probably a filterable virus the throat-washings were filtered before use through a membrane impermeable to bacteria. The filtrates, proved to be bacteriologically sterile, **were used in attempts to infect many different species. All such attempts were entirely unsuccessful until the ferret was used** (my emphasis)".

The ferret had been used by British investigators to study a disease known as distemper:

"The initial successful experiment was made with two ferrets, both of which received a filtrate of human throat-washings, both subcutaneously and by intranasal instillation. Both animals became obviously ill on the third day after infection and exhibited symptoms of the characteristic disease . . . It was found that the disease could be transmitted either by contact or by direct transference of nasal washings from a sick ferret to a healthy ferret. . . .

Coincidentally with the primary rise of temperature the ferret looks ill, is quiet and lethargic, often refuses food, and may show signs of muscular weakness. The catarrhal symptoms usually begin on the third day. The eyes become watery and there is a variable amount of watery discharge from the nose . . . The animal sneezes frequently, yawns repeatedly, and in many cases breathes partly through the mouth with wheezy . . . sounds which clearly indicate a considerable degree of nasal obstruction. The disease has frequently been transmitted by placing a normal ferret in the same cage as a sick one for 24 hours. The majority of virus passage, however, have been made by the following technique".

The authors described processing nasal material and inserting it into the nostrils of another ferret:

"In this way 26 serial passages of one strain of virus have been made and every animal of the series has shown . . . symptoms of the disease. A hundredfold dilution of the usual preparation has also been found to be regularly infective.

Throat-washings from four human subjects not suffering from influenza were non-infective . . . The nasal secretions of a man suffering from a severe common cold were also non-infective.

Most of the human throat-washings were filtered before use through membranes having an average pore size of 0.6 micron. Invariably filtrates of an emulsion of the nasal mucosa from a sick ferret through membranes having an average pore size of 0.6 micron were found to produce the typical disease. A tight membrane (0.25 micron) was used on one occasion only; the resultant filtrate was infective . . . . .

The infectivity of the filtrates, coupled with the fact that we failed to grow anything from the filtrate on a variety of media under aerobic or anaerobic conditions, has convinced us that we are dealing with a true virus".

C. H. ANDREWES, WILSON SMITH, and P. P. LAIDLAW, "A Virus Obtained from Influenza Patients," Lancet, 1933, 222, pp. 66–68

It is interesting to reflect on what the three English experts on contagious disease meant when they concluded that the agent that caused influenza was a "true virus." Part of the definition is what it is not: it is not a bacterium. What it is, is filterable, and it infects certain hosts, where it is reproduced. Alone, in some artificial culture medium, it is inert. They compared it to the cause of a disease of pigs:

"A disease of swine, which arose spontaneously at the time of an influenza epidemic in America, has been described by Shope. We are indebted to him for samples of the swine influenza virus, and also for cultures of *Haemophilus influenzae (suis)*, an organism which plays an important role in the serious and fatal cases of the swine disease. The virus when inoculated intranasally into ferrets gave rise to a disease . . . indistinguishable from the ferret disease caused by virus of human origin . . . Cross-immunity tests have shown that this swine influenza virus bears a close antigenic (structural) relationship to the virus strain of human origin which has been chiefly used in our work".

A Filterable Contagious Agent That "Eats" Bacteria

"So, naturalists observe, a Flea
Hath smaller fleas that on him prey,
And these have smaller Fleas to bite 'em,
And so proceed ad infinitum".

Jonathan Swift, Miscellanies in Prose and Verse. vol. 5, p. 172, 1735, London

In December 1871, the Brown Institution was founded in London "for the purpose of investigating, and if possible, endeavoring to cure the diseases of animals useful to man." The institution and its work were mildly satirized by a cartoon in the periodical *The Graphic* of April 1875 entitled *OUR PATIENTS—Monday Morning*, which shows horses and a variety of other animals waiting to be treated. The serious nature of this institution, however, was not in doubt, as revealed by the comment in the *Journal of the Agricultural Society of England* in 1876 that foot and mouth disease was being studied there with a "number of experiments on different animals, but it would be premature to make any statement of the results."

Frederick Twort (1877–1950) was a member of this institute studying "filter-passing viruses." He observed an unusual phenomenon: colonies of micrococci growing on agar took on a "glassy appearance" and in time were replaced "by fine granules." In fact, the bacteria dissolved. Fresh colonies touched with a sample of the glassy material were killed. The effect was most lethal to young cultures and did not affect inactive or dead bacteria. The glassy (transparent) material that killed bacteria passed through "the finest porcelin (sic) filters" (Pasteur-Chamberland F and B). The killing capacity could be serially transmitted to fresh cultures indefinitely, but it could not be cultured without bacteria. Twort speculated that the entity might be a minute bacterium, an ultramicroscopic virus, or an enzyme with the power of growth.

Twort suggested that it might be an infectious disease of micrococci.

He had discovered a lethal, contagious disease of bacteria. After this work in 1917 Twort participated in wartime activities and carried out work on influenza. He never did return to studies of the "lytic principle."

Felix d'Herelle (1873–1949), of French-Canadian heritage, received his medical training in Montreal. From there he went to Guatemala as a professor of bacteriology and later as bacteriologist to the Mexican government, where he studied a bacterial infection of locusts. The usual procedure in these studies was to culture putative contagious bacteria on agar plates. At times he observed that bacterial colonies growing on agar contained "clear spots." Microscopic examination of material from these clear areas revealed that bacteria were absent. In 1915 d'Herelle was in Tunisia intending to eliminate a locust population using disease-causing bacteria when he encountered the same clear spot phenomenon; however, he did not conduct further studies on this event. On arrival at the Pasteur Institute he was asked to study an epidemic of dysentery and, again, he encountered this clearing phenomenon. This time he followed the story. He reported, from stools of patients recovering from dysentery, there appeared an "invisible microbe" that is "antagonistic" to the bacterium that causes dysentery. He determined that the invisible entity increased with the disappearance of the *Shiga* bacillus (which causes dysentery) as the patient regained health. This entity was not recovered from healthy individuals. d'Herelle believed that such an agent might be of therapeutic value. The anti-dysentery bacillus material was isolated in the following way:

1. Stools of patients with dysentery were incubated in broth for 18 hours at body temperature.
2. The contents of the tube were filtered through a particular Chamberland filter candle.
3. A small drop of the filtrate was added to a fresh culture of the *Shiga* bacillus and incubated at body temperature. Depending on the amount of filtrate and the number of bacteria, lysis (that is, killing of the bacterial population) was complete in a period ranging from hours to days.

He performed this experiment 50 times, and each time the same phenomenon occurred. At this point he stated that the active antagonist was a "living germ."

d'Herelle developed a procedure for isolating and counting the number of these "living germs." He added a highly diluted portion of a filtrate containing the "living germs" to a fresh bacterial culture, mixed the contents thoroughly, spread a small volume of this mixture on a nutrient-containing agar plate, and incubated it at 37 degrees C, human body temperature. In this procedure he obtained a thin lawn of bacteria that covered the entire surface of the agar except for "a certain number of sterile circles of about 1 mm in diameter where the bacilli did not grow." d'Herelle theorized that these isolated spots were the result of the deposit of a bacterium, infected with the invisible killing agent, at that position on the agar. The bacterium was killed, releasing "killing agents" that lysed other bacteria. The end effect was a sterile circle. By counting the number of such loci, it was possible to quantitate the number of entities that destroy bacteria. When d'Herelle carried out such experiments, he discovered that there were five to six billion such "germs" per cubic centimeter of filtrate. The increase in numbers of this filterable agent occurred only in the presence of actively growing *Shiga* bacilli. d'Herelle concluded in 1917 that the anti-*Shiga* microbe was an "obligate bacteriophage," an entity that eats (devours) bacteria. The designation "bacteriophage" continues to be used today.

Twort and d'Herelle agreed that there was something, however they named it, that was filterable and killed bacteria. From there on, however, it became a battleground between the two, on two issues. The first was priority, a very important issue in science: Who discovered this phenomenon? The second issue is important for our discussion: What is the nature of the filterable agent that kills bacteria? The argument about priority will not concern us.

A grand debate about the nature of the filterable agent occurred at the ninetieth annual meeting of the British Medical Association, which was held in Glasgow in July 1922, under the rubric of the Section of Microbiology (Including Bacteriology). Both Twort and d'Herelle spoke, followed by various commentators.

d'Herelle presented four hypotheses concerning the cause of the killing of bacteria. They included the two principal theories, that enzymes cause lysis and that lysis is due to the presence of an ultramicroscopic entity. d'Herelle presented a series of quantitative experiments that unambiguously, according to him, demonstrated that a "particulate" entity caused the destruction of bacteria. His methodology was so simple and instrumental that it was still in use 35 years later when I attended the famous Phage Course at Cold Spring Harbor, taught at that time by Salvador Luria, a Nobel Laureate, and George Streisinger.

Initially, d'Herelle provided evidence that the lytic principle is reproduced:

> "After more than a thousand passages, the thousandth bacteriolysed culture contains a bacteriophagic principle as active as, and generally much more active than, that of the primitive (original) filtrate". (p. 290)

In the following series of experiments d'Herelle demonstrated that bacteriophage can be counted in the same way we count bacteria. For example, a broth culture may have tens to hundreds of millions of bacteria per cubic centimeter. Obviously one cannot count these by microscopic observation. It is done by removing a known amount from the culture, making a known amount of dilutions of that quantity, and depositing on the surface of nutrient agar a known amount of the highest dilutions. In time colonies grow, each derived from a single bacterium. By simply multiplying the number of colonies by the dilution factor, the number of viable bacteria in the original culture is known. d'Herelle carried out such an experiment with his filtrate containing bacteriophage:

> "The enumeration of the bacteriophage corpuscles contained in a filtrate is estimated in exactly the same way. But as the bacteriophage only grows at the expense of living bacteria, we must make dilutions of the filtrate in a bacterial emulsion. For this we pipette over agar the bacterial emulsion containing a given quantity of filtrate: after incubation we obtain a bacterial

layer strewn with circular bare spaces, each of these spaces being a colony of bacteriophage issued from one corpuscle [one bacteriophage]. The number of bare spaces multiplied by the titre of the dilution gives the number of ultramicroscopic bacteriophage corpuscles contained in 1 c.cm. of the primitive filtrate. (p. 290)

This experiment shows that the behavior of the bacteriophage is exactly the same as that of any ordinary microbe. [He is of course thinking of bacteria.] But this last [bacteria] develops at the expense of the nutritive substances contained in the medium; the bacteriophage develops at the expense of the bacterial bodies which constitute its nutritive medium. The bare spaces represent places cleared up by the growth of the ultramicroscopic bacteriophage corpuscles". (p. 291)

FELIX D'HERELLE, "Discussion On The Bacteriophage (Bacteriolysin)," *British Medical Journal, Section of Microbiology*, 1922, pp. 289–297, Aug. 19.

d'Herelle's logic and the methodology of his experiments manifest the way he conceived of, and probably always comprehended, the bacteriophage identity. It was an ultramicroscopic particulate entity; this concept was reinforced by his use of the term "corpuscle" to identify it.

The experiment that provided him with the critical evidence of the particulate nature of the bacteriophage was presented at the 1922 meeting. In principle it was identical to the experiment carried out in 1877–78 by Joseph Lister to show that a single bacterium can cause the fermentation of milk (Chapter 12):

"Now dilute a filtrate so as to obtain a dilution such that 1c.cm. contains one bacteriophage corpuscle. Dilute this 1c.cm. with 9 c.cm. of sterile water, [call this tube X] and inoculate ten tubes of bacterial emulsion each with 1 c.cm. [from tube X]. It will be obvious that only one of the ten tubes will contain the generator of a bare space; the nine others will not contain any. Place the ten tubes in the incubator at 37°C for twenty-four to forty-eight hours; it will be seen that only one of the ten microbial emulsions shows bacteriolysis; the nine others will remain unchanged, the bacteria remain living, normal, and subculturable.

The lytic action, therefore, is complete when only **one** generator of a bare space is introduced into a bacterial emulsion; . . . This experiment can only be explained on the supposition that the bacteriophagic principle, the source of the lytic enzymes, is a corpuscle; [d'Herelle subscribed to the formula that enzymes cause the lysis of the bacterial cell; they are produced as the result of the invasion of the cell by the particle] and that each corpuscle deposited on the agar in the midst of the bacteria gives rise to a colony of these ultramicroscopic corpuscles, such a colony being represented by a bare space". (p. 291)

The Particulate Nature of Filterable Agents

Is it the case that other filterable agents that cause diseases of plants, farm animals, and humans have a particulate character? The answer is yes for the agent that causes the tobacco mosaic disease. The method for supporting this claim was similar to the procedure used to observe the appearance of "bare spaces," the sites of destruction of bacteria by bacteriophage, on a "lawn" of bacteria on a plate of nutrient agar. The experiments were performed by Francis Holmes.

A source of the agent of tobacco disease was rubbed on the upper surface of many young tobacco leaves. The leaves were rinsed with water to wash off excess agent. In four or five days there appeared 300 to 600 lesions on the leaves. The procedure was thus shown to cause infection. The same experiment was carried out with a series of dilutions of the disease agent that was rubbed on leaves. In time, lesions developed, proportional to the extent of dilution: the higher dilutions yielded fewer lesions. The results may be explained in this way: wherever a particle was deposited on the leaf, a lesion developed due to the propagation of the agent at the place where the particle entered the leaf tissue.

The Chemical Nature of the Agent That Causes Tobacco Mosaic Disease

In 1935, in the journal *Science*, Wendell M. Stanley (1904–1971), a chemist at the Rockefeller Institute for Medical Research in Princeton, New Jersey, whose specialty was purifying proteins, published a two-page paper, "Isolation of a Crystalline Protein Possessing the Properties of Tobacco-Mosaic Virus."

Stanley had infected tobacco plants with the agent that causes tobacco mosaic disease, crushed the diseased leaves to obtain a liquid extract, and by a large number of chemical procedures obtained about 10 grams of a crystalline material that had the power to infect plants with the mosaic disease. In fact, one cubic centimeter of a 1:1 billion dilution of the solubilized crystals generally proved infectious. The crystalline material was a protein by many chemical tests. The molecule was estimated to be about two million in molecular weight. In solution it passed through a Berkefeld "W" filter. Treatment of the material by heat or high acidity destroyed infectivity. Stanley concluded,

"Tobacco-mosaic virus is regarded as an autocatalytic [self-generating] protein which, for the present, may be assumed to require the presence of living cells for multiplication."

WENDELL M. STANLEY, "Isolation of a Crystalline Protein Possessing the Properties of Tobacco-Mosaic Virus," *Science*, 1935, 81:644–645
[citation is on p. 645]

Such an entity was heretofore unknown. The crystallization of an alleged living, pathological agent was a stunning event. The material could be treated like table salt, stored in a dry state, and at some point rubbed onto tobacco leaves to produce a disease. The work made headlines in *The New York Times* and other papers. In the *Times*, Stanley's name appeared in the sub-headline with his affiliation to the Rockefeller Institute. Ten years later he was among three individuals who received the Nobel Prize in Chemistry.

In June 1937 Stanley presented a detailed paper at the Fourteenth Colloid Symposium on the physical properties of the protein that causes the tobacco disease. Although he had used the term "autocatalyst," Stanley indicated that he had done so simply to indicate that the end effect was to produce more of itself. How this happened was unknown. Stanley suggested it might be the method that geneticists proposed for the replication of genetic material.

Stanley then pointed out "another amazing phenomenon," that during the production of the disease-producing entity, some disease-producing progeny differed from the majority. These new viruses contained a protein different from the original virus and thus became a new strain of virus. Stanley commented that although the virus protein was a protein with the chemical characteristics of other proteins, it also had the ability to produce more of itself **and** to change its chemical composition (mutate), which was unique among this class of molecules. This "protein represents an entity unfamiliar to us."

Stanley envisioned the protein reproducing itself, a process he could not explain. Very soon after Stanley's 1935 paper appeared, two investigators in England carried out experiments to confirm his work: F. C. Bawden (1908–1972) of the Rothamsted Experimental Station, dedicated to agricultural research, and N. W. Pirie (1907–1997) of the Biochemical Laboratory of

Cambridge University. In their work, published in 1937, they purified three strains of tobacco mosaic virus from infected leaves by methods similar to those of Stanley. They obtained data different from Stanley's, however. To understand what they found, it is important to discuss a bit of biochemistry. Stanley had reported that the crystals that cause the tobacco disease were protein. It is essential to understand that proteins contain carbon, hydrogen, oxygen, and nitrogen, very little sulfur, and no phosphorus and no carbohydrate (that is, sugar). Bawden and Pirie's preparations, however, always contained a significant, constant amount of phosphorus and carbohydrate. Obviously, tobacco mosaic virus consists of protein and something else. This other constituent is nucleic acid containing a five-carbon sugar characterized as ribose. The nucleic acid component came to be known as ribonucleic acid or RNA and constituted 6 percent of the weight of the virus. These results led one historian to write, "Stanley's work was flawed by technical errors and misconceptions." The major error was missing the presence of RNA, and it follows that the major misconception was to regard the virus as *simply* a protein, equivalent to an enzyme catalyst. Clearly this conclusion was at the very least premature and additional studies were required before a belief in a pathological protein was established. Nevertheless, Stanley continued to treat the agent that causes tobacco disease as a protein, even after the results of the English group were made known that the various strains contained RNA. In addition, Stanley ignored the results of Frances Holmes that indicated the particulate nature of the tobacco mosaic agent.

A question was proposed by a historian writing about the history of these discoveries:

"How can we explain these technical and conceptual errors in the light of the international recognition bestowed on the young Stanley and his leadership in science"?

LILY E. KAY, "W.M. Stanley's Crystallization of the Tobacco Mosaic Virus, 1930–1940," *Isis*, 198677(3):450–472 [citation on p. 451]

My question is this: Why was Stanley satisfied with the concept that the virus was a pure protein? To provide some context for a possible answer, we shall trace some history of the Rockefeller Institute before Stanley's appointment in 1931 and the era of his tenure there.

In 1931 the Institute was under the leadership of Simon Flexner, who had reformed medical education in the United States. Flexner believed that a chemical approach to the study of viruses would help solve the issue of their identity (for example, whether they were living entities) and how they caused disease. Stanley was to provide additional chemical expertise to study this problem, although he had no exposure to biological research and knew little about viruses. At this date there were hundreds of virus entities whose major characteristics were that they caused diseases of humans, animals, plants, and, according to Twort and d'Herelle, bacteria. They could not be grown independent of a host; however, in the case of human and animal viruses, they could be cultured on animal tissues or on the membranes of embryonic eggs. Initially, they were all grouped as filterable agents, but now were considered to be of different sizes. It was believed they could reproduce and undergo change, mutations, which are properties of living agents. In short, they were still a mysterious collection of disease-causing entities some 40 years after the discovery of filterable agents that cause a disease of tobacco.

In 1931 there was a satellite of the Rockefeller Institute in Princeton, New Jersey, including a department of animal pathology. There was added a unit of plant pathology. The research facilities were generous and state of the art for doing work on pathogenic microorganisms and viruses. The identity of viruses and their effects had been an interest of Flexner, and when Stanley arrived at the Institute the distinguished biochemist John Northrup was studying bacteriophage, which like all viruses appeared to be somewhere between living and non-living entities. What occupied Northrup were protein molecules, which he believed were responsible for all physiological properties of living entities. Northrup was a leading light at the Institute and had made his deserved reputation as an expert at purifying enzymes, which are proteins, producing them in their crystal form so that their chemical and physical properties could be understood. Northrup was trained differently from Stanley, with a background in biology obtained at Columbia University with E. B. Wilson (1856–1939) and the great geneticist Thomas Hunt Morgan (1866–1945). However, he had turned to an approach to biological problems with the tools of chemistry and physics. Such a philosophy was valued at the Institute. It was also encouraged in the larger scientific community. In the *American Naturalist* of June 1917 Leonard Thompson Troland, a professor at Harvard University, extolled the great strides taken by physicists with their modern theories of matter, and exhorted biologists to abandon vitalism and look to chemistry and physics to explain the properties of living things, particularly those issues of heredity and evolution "which make up the heart of the biological mystery." He contended that the success of Mendelism (the science of genetics, where the results of "crossing experiments" had revealed the phenomena of heredity) had not revealed the underlying assumed mechanical aspects of heredity. The most far-reaching claim he made was that the Mendelian factors, genes, were enzymes, and therefore proteins.

In his desire to spread the message of this modern version of the mechanical philosophy among biologists he quoted a lecture presented in 1913 by W. Bateson, a pioneer geneticist who stated that the solutions to problems of inheritance "must presuppose knowledge of the chemistry and physics of living things." Support for a functional approach to the identity of genes was strongly endorsed by H. J. Muller (1890–1967), another pioneer geneticist, who demonstrated that radiation could cause mutations. Muller was inspired by d'Herelle's work on the entities that destroyed bacteria, bacteriophage, and wrote in 1922,

> . . . "if these d'Herelle bodies were really genes, fundamentally like our chromosome genes, they would give us an utterly new angle from which to attack the gene problem. They are filterable, to some extent isolable, can be handled in test tubes, and their properties, as shown by their effect on the bacteria, can be studied after treatment. It would be very rash to call these bodies genes, and yet at present we must confess that there is no distinction known between genes and them. Hence we cannot categorically deny that perhaps we may be able to grind genes in a mortar [this was being done to extract enzymes from cells and tissues] and cook them in a beaker after all. **Must we geneticists become bacteriologists, physiological chemists, and physicists, simultaneously with being zoologists and botanists? Let us hope so** (my emphasis).
>
> H. J. MULLER, 1922, *Variation Due to Change in the Individual Gene,*
> *The American Naturalist*, 1922, vol. 56, pp. 48–49

In Northrup's philosophy enzymes (proteins) were the key to understanding the function of living organisms, and he proceeded to develop methods to purify these components, although he was not the first to do so. James Sumner in 1926 was the first to crystallize a protein, the enzyme urease, a protein obtained from jack beans, which converted urea to ammonia and carbon dioxide. By 1938 10 enzymes had been obtained in crystalline form, four by Northrup and collaborators. Two of the proteins he prepared were precursor proteins. They were inactive in this form but when acted on by the active enzyme, which originally came from the precursor, they were transformed into the active enzyme. This was Northrup's model for an autocatalytic process that he invoked for the formation of viruses.

In this scientific, technically sophisticated environment that placed proteins at the center of biological activity, Stanley carried out his purification and crystallization of the agent that caused tobacco mosaic disease. (For convenience we will call it TMV.) In 1941 Stanley and collaborators published three papers that examined the properties of TMV proteins. Why did they do this? Because he judged, reasonably, that the structure of its protein was uniquely responsible for its biological activity since the virus consisted solely of protein. Including the RNA component in the discussion appeared useless since nothing could be said of its role in the biology of the disease, although in the early 1940s all viruses were shown to contain nucleic acid.

In Stanley's experiments, chemical derivatives were made of a solution of the TMV protein. Under these conditions a majority of the constituent amino acids were modified and the altered TMV was used to infect young plants. The altered TMV protein, alleged to represent the disease-causing activity, did indeed cause a disease, identical to that caused by untreated TMV protein. The TMV protein that was produced in the plants infected with modified TMV protein was the same as the TMV protein derived from plants infected with normal TMV protein. **Modification of the protein had no effect on progeny TMV**. What were the possible explanations for these results? One hypothesis was that the protein had nothing to do with the character of the progeny. In Stanley's view such a conclusion was impossible since it was only the protein that was involved in infection. To rely on the essential role of the protein, Stanley speculated that the virus molecules (proteins) were possibly inexact autocatalysts, since it appeared that the proteins could be altered without affecting the ability of the virus to cause disease.

The Impact of Stanley's Work

The initial fundamental research by Stanley that demonstrated that a crystalline material caused a disease stimulated Max Delbruck (1906–1981) to speculate about the nature of viruses and the problems of heredity. To understand the roots of this philosophical enterprise it is useful to consult a memorandum prepared by Delbruck in 1937, immediately before he left Germany for the United States to study genetics at the California Institute of Technology, the center of fruit fly genetics. During the mid-1930s Delbruck, a physicist and assistant to Lisa Meitner in Berlin, met informally with colleagues to discuss theoretical physics, but soon they turned their attention to biology. Their teacher was a geneticist, N. Timofeev-Ressovsky (1900–1981), doing research on the effect of radiation, a field initiated by H. J. Muller. A group that included Muller, Ressovsky, and Delbruck travelled to Copenhagen for a meeting arranged by Niels Bohr. All except Ressovsky and Delbruck were Nobel Prize recipients. As Delbruck phrased it, "These discussions occurred very much under the impact of the findings of W. M. Stanley reporting the crystallization of tobacco mosaic virus."

The memorandum that was prepared contained the speculations of theoreticians of physics who wondered how the study of genetics and the study of viruses intersect. Delbruck listed the following kinds of questions and speculations:

1. Viruses are large molecules with defined structures.
2. Are they living organisms; can they multiply within a living cell?
3. Is it the host that provides the ability to produce the virus, or does the host provide the right environment for reproduction to take place?
4. It seems unlikely that the host can produce a complicated molecule "which is unknown to the host." The host, of course, can produce itself. It has all the heredity material, genes, that would make this possible. This indicates that the cell does not know how to produce a viral component because it does not have the hereditary material to do it. If this is the case,
5. Virus replication is an autonomous accomplishment of the virus.
6. From the point of view of genetics, what kind of replication takes place?

"we want to look upon the replication of viruses as a particular form of a primitive replication of genes, the segregation of which from the nourishment supplied by the host should in principle be possible. In this sense one should view replication . . . as a particular trick of organic chemistry."

M. DELBRUCK, "A Physicist's Renewed Look at Biology: Twenty Years Later," *Science*, 1970, 168:1312–1315 [citation from p. 1315]

The question left unaddressed is this: Is a protein molecule capable of controlling and carrying out the complex reactions needed to produce a virus? Delbruck referred to complicated molecules but did not further qualify the components of the apparatus (chromosomes) that was revealed by "modern cytology" and Mendelian genetics to be involved in inheritance. The apparatus of heredity consists of chromosomes that undergo mitosis (division) and meiosis (the production of half the number of chromosomes contained in germ cells contributed by each parent during sexual reproduction). In the late 1930s it was recognized that chromosomes contain DNA, but this was regarded as merely a structural component. In 1937 Delbruck did not refer to the presence of RNA in TMV. **We must stress that nucleic acids were not part of the discourse about heredity.**

Delbruck came to the California Institute of Technology to do research in genetics and there he came into contact with Emory Ellis (1925–1993), a physical chemist, who was studying the production of bacteriophage using the knowledge and methodology obtained by d'Herelle. This apparently simple biological system, a bacterium and a virus, seemed, to Ellis and Delbruck, to be an ideal model system to study the nature of the genetic material. The two devised a procedure, a classic experiment known as *the one-step growth experiment*, later described by Delbruck as follows:

"Bacteria first are grown in a test tube of liquid meat broth. Enough virus of one type are added to the test tube so that at least one virus is attached to each bacterium. [Attachment was observed later with electron micrograph pictures, but at this time an accurate description would be that the ratio of virus to bacteria was one to one.] After a certain period (between 13 and 40 minutes, depending on the virus, but strictly on the dot for any particular type), the bacterium bursts, liberating large numbers of viruses. At the moment when the bacteria are destroyed, the test tube, which was cloudy while the bacteria were growing, becomes limpid. Observed under the microscope, the bacteria suddenly fade out".

HORACE FREELAND JUDSON, The Eighth Day of Creation, 1979, p. 51,
Simon & Schuster

In this experiment it appeared that one viral particle entered the bacterial cell, where some form of reproduction occurred. The mode of propagation was unknown, but that it took place was obvious since at a certain time a large number of viruses were suddenly released. During the following two years electron micrograph pictures of phage interacting with bacteria were obtained. The virus has a unique structure, consisting of a head and a tail. These images provided the visual evidence that the virus was a particle, the original contention of d'Herelle. The bacteriophage attached to bacteria via the tail. During infection neither the head nor the tail appeared to enter the bacterium. This raised the question: What element of the virus got into the cell, and once that happened, what were the internal events that led to more viruses? Did the virus instruct the cell to make more virus, or did the cell already "know" how to make virus, and therefore the entrance of a viral component stimulates the cell to produce viruses?

Chemical analyses revealed that the virus used by Ellis and Delbruck was made of protein and DNA, a compound present in chromosomes of all cells. It turns out that all viruses, whether they infect plants, humans, a variety of animals, or bacteria, contain one kind of nucleic acid, either DNA or RNA. Since 1937, it had been known that tobacco mosaic virus contained RNA. The bacterial virus was composed of proteins and DNA. If the virus instructed the cell to make virus the hereditary apparatus must be one of these components.

The answer to the nature of the hereditary molecule began to emerge in 1944 in studies concerning an apparently unrelated phenomenon known as "transformation," where one bacterial characteristic was changed permanently to another after treatment with a preparation of DNA. To tell this momentous story we need to discuss a bit of bacteriology.

The Genetic Material Is DNA

There are bacteria that cause pneumonia in humans and mice. There are three major varieties of these bacteria, 1, 2, and 3, differentiated by their immunological properties (that is, their reaction with antibodies), since each kind contains at its surface a capsular substance made of a complex sugar molecule identified as a polysaccharide. When these various strains are grown on agar medium, they appear as *smooth* colonies and are labeled S. S bacteria are lethal to mice. At times these bacterial strains permanently lose the ability to make this capsule, and when grown on agar medium they present colonies that are not smooth but are characterized as *rough*, R. The R strain derived from an S strain is not lethal to mice. Making the capsule is an inherited trait just as eye color in fruit flies and humans is an inherited trait.

An English bacteriologist, Frederick Griffith (1879–1941), performed an experiment in the 1920s to study this interchangeability, S to R, and achieved a startling result. He introduced into mice, simultaneously, two different *pneumococci,* living type 1 R (which do not kill mice) and dead (heat-killed) type 2 S (which certainly do not kill mice). No mice should have died in these experiments. However, some did, and when bacteria were examined from the organs of the dead mice there were found living type 2 S, not living type 1. Type 2 S continued to propagate as type 2 S.

This subject interested Oswald Avery (1877–1955) of the Rockefeller Institute, who was skeptical about the results. However, these results were confirmed in Germany and at the Rockefeller, where it was demonstrated that the phenomenon, conversion of one cell type to another, could be reproduced without mice by placing the two strains together in culture. Later they could reproduce the event using an extract of S bacteria that had been filtered to ensure there were no live S cells that could transform R cells to the S type. For a while work was suspended until Colin MacLeod (1909–1972) came to the Institute in 1934 and worked for a number of years developing a susceptible R strain that could be transformed. In 1940 the research was reinitiated, with great emphasis placed on purifying the material that caused transformation. Avery and collaborators understood they were studying a phenomenon with important biological consequences: it appeared possible to confer on an organism a new biological trait using a chemical compound, a pure, unique compound that caused the transformation in a specific way. The work was carried out over a period of four years. The paper was published in 1944, and all the evidence indicated it was DNA. Nonetheless, this fact was very guardedly presented: the agent responsible for the permanent acquisition of a new characteristic was referred to as the "transforming principle," the "inducing substance," or the "active principle." In a summary, the fact that the principle may be desoxynucleate is stated and there is a cautious statement, that such a molecule may have biological activity:

"The inducing substance, on the basis of its chemical and physical properties, appears to be a highly polymerized and viscous form of sodium desoxynucleate . . . assuming that the sodium desoxynucleate and the active principle are one and the same substance, then the transformation described represents a change that is chemically induced and specifically directed by a known chemical compound. It is of course possible that the biological activity of the substance described is not an inherent property of the nucleic acid but is due to minute amounts of some other substance absorbed to it or so intimately associated with it as to escape detection . . . if the results of the present study . . . are confirmed, then nucleic acids must be regarded as possessing biological specificity the chemical basis of which is as yet undetermined".

OSWALD T. AVERY, COLIN M. MACLEOD, and MACLYN MCCARTY,
"Studies on the Chemical Nature of the Substance Inducing
Transformation of Pneumococcal Types: Induction of Transformation by a
Desoxyribonucleic Acid Fraction Isolated from Pneumococcus Type 3." *Journal of
Experimental Medicine*, 1944, 79:137–159 [citation from p. 155]

Avery, in a letter in 1943 to his brother, who was a bacteriologist at Vanderbilt University, was more speculative and expansive:

"If we are right . . . then it means that nucleic acids are not merely structurally important but functionally active substances in determining the biochemical activities and specific characteristics of cells . . . This is something that has long been the dream of geneticists . . . sounds like a **virus** (my emphasis) . . . may be a **gene** (my emphasis)".

HORACE FREELAND JUDSON, *The Eighth Day of Creation*,
1979, p. 39, Simon & Schuster

Avery is definitely not engaging in hyperbole. Let us reexamine what has been accomplished. A chemically defined material, an active principle, has affected a living cell (we don't know how this has occurred) to produce a very large, complicated molecule, a capsule, that it could not produce previously. This large molecule must be constructed in many steps, and as the biochemists of the twentieth century amply demonstrated, steps that are catalyzed by enzymes that are proteins. It is now possible to contemplate that the active principle is in some way responsible for the appearance of these enzymes.

When Avery's brother received the letter at Vanderbilt, Delbruck was at the university and was shown the letter. The contents disturbed Delbruck. Why? On the one hand Delbruck believed DNA to be a "stupid" molecule constructed of repetitive sequences that could not specify the complex structures, for example, of proteins. Proteins were more likely to contain hereditary information. This was a view that persisted until the early 1950s. Therefore, various investigators continued to insist that the preparations of Avery must contain traces of protein components that were the actual genes; thus the transforming principle could not be DNA. If, however, it was DNA, then the material was not "stupid," and it was clear that much work needed to be done to resolve the relationships between nucleic acids and proteins.

As a background to this work, four years previously in 1940, two investigators, one a biochemist, Edmund Tatum (1909–1875), and the other a geneticist, George Beadle (1903–1989), had carried out experiments with a certain fungus, *Neurospora crassa,* that established that a genetic change in the organism manifested itself as a change in one enzyme. There was no doubt about the nature of the enzyme: it was a protein. These results established a connection between a gene and a protein but did not reveal the chemical composition of the gene.

The Genes of the Virus That Infects *E. coli* are Composed of DNA

Research over the next decade clarified the nature of the genetic material of the bacterial virus. Three important developments revealed the fundamental properties of bacterial viruses:

1. Certain strains of bacteria have the inherited capacity to produce viruses. All the genetic information for the production of a certain bacterial virus is present on the bacterial chromosome; all the viral genes are present on the bacterial chromosome. During normal reproduction of the bacteria, the genes controlling the character of the bacterial virus are reproduced with the genes of the bacteria. This is labeled the *lysogenic state*. Infrequently, viruses are produced because of some failure in the control mechanism that prevents viral production. In this phenomenon, labeled *lysogeny*, under certain conditions the genetic material of the virus is expressed, resulting in the destruction of the bacterial cell.
2. The bacterial viruses that always kill *E. coli* have genes, like animal or plant cells, that undergo mutation, and in crossing experiments genetic exchanges lead to viruses with new characteristics.
3. The DNA of the viruses that infect *E. coli* is the material that enters the cell when infection occurs. It is sufficient for infection that this event takes place.

Lysogeny

During the 1920s it was recognized that certain strains of bacteria produced, at times, bacteriophage (virus) in the absence of free viruses. They were lysogenic. How was it determined that a low level of viral production had occurred? Because there are strains that **are** sensitive to these viruses and are lysed. These results raised important issues about the nature and origin of bacterial viruses since they appeared to arise from a bacterium and not from an external infection. The discoverer of this process, Jules Bordet, commented in 1931 that the virus of d'Herelle did not exist. Lysis represented a pathological function that was part of the physiology of the bacterium.

F. Macfarlane Burnet (1899–1985) was impressed with the production of virus from bacteria under conditions where no external virus was present. He considered this ability a hereditary property and speculated that the ability to produce virus was present in every bacterium. He looked for some evidence of the virus by breaking open bacteria that were lysogenic but found no virus. Convinced that something existed in the cell, he commented in 1934, "One is almost forced to postulate that each lysogenic bacterium carries in intimate symbiosis one or more phage particle which multiply by binary fission concurrently with the bacterium." If Burnet was correct, each cell had the hereditary property to produce virus. In what form was this property contained in the cell?

After World War II the work was reinitiated by Andre Lwoff (1902–1994) at the Pasteur Institute. Using a microscope it was possible to observe the reproduction of a lysogenic bacillus into two bacilli, and so on, until the organism had divided 19 times. Each isolated bacterium, obtained with a micro-manipulator, when cultured, gave rise to lysogenic clones, described in this way:

"When microcolonies of lysogenic *B. megaterium* growing in microdrops are carefully and constantly watched for a reasonable length of time, the lysis of some bacteria may be observed. A bacterium is there, and suddenly it disappears. When this happens, phage [virus] is found in the droplet, around 100 phages per lysed bacterium".

ANDRE LWOFF, "Lysogeny," *Bacteriological Reviews*, 1953, 17:269–337 [citation from p. 280]

The entity carrying this hereditary power Lwoff labeled prophage. It was now clear why a reasonable definition of lysogeny is the heredity power of bacteria to produce virus: each bacterium was carrying the genetic information to produce virus, although that information was not expressed. Now the question became: What can prophage be? Since all viruses were composed of proteins and nucleic acid, it must be one of these components. Since the work of Avery and associates in 1944 indicated that the transforming principle that conferred permanent hereditary changes to pneumococci was DNA, this was suggested to be the nature of prophage. Two developments in the fields of the genetics of bacterial viruses and the genetics of bacteria established that prophage was a gene situated at a particular position on the bacterial chromosome. In 1951 Joshua (1925–2008) and Esther Lederberg discovered that the lysogenic character was present in a strain of *E. coli* that could exchange genes with other members of this group. They established that the lysogenic trait could be passed to bacterial progeny in the same way that other bacterial genes were transmitted.

The gene for lysogeny "is located at a specific chromosomal site; it is replicated in coordination with bacterial reproduction."

These data demonstrate that the genetic material of the virus becomes part of the genetic constitution of the host cell. The bacteria produce virus under certain conditions because they are carrying the genes for virus production, not because of some aberrant physiological function of the bacterial cell. The proof that every bacterial cell in a lysogenic population is carrying prophage was provided by Lwoff when he found conditions that would induce every bacterial cell in a lysogenic population to undergo lysis. The nature of prophage will become clear shortly.

The T series of viruses that infect *E. coli* (and kill them) are composed of proteins and DNA. DNA constitutes about 40 percent of the weight of the virus. The virus particle is made of a "head" and "tail" structure. When infection takes place the virus attaches to the bacterial surface via the tail, which remains with the head outside the bacterium during infection. All of these initial events were revealed by electron microscope studies. Conclusive support for this model of infection, that no intact viral particles are found in newly infected bacteria, was revealed when these bacteria were artificially disrupted to discover there were no viruses. This was the identical finding when lysogenic bacteria were artificially broken open: no virus particles were found. Why no intact viral particles were present at certain points in the developmental cycle of phage was revealed by the studies of Alfred Hershey (1908–1997) in 1952. Here are the essential elements of that experiment.

Again, to remind us, bacterial viruses are made of proteins and DNA. Proteins are made of amino acids. There are 20 amino acids. Two of them contain the element sulfur. DNA does not contain sulfur but contains lots of the element phosphorus, while proteins do not contain phosphorus.

Bacterial viruses were produced from bacteria grown in two separate media. In one, the growth media contained radioactive phosphorus, which was incorporated into the DNA of the mature virus particles. In another flask the medium contained radioactive sulfur so that the viral particles produced contained radioactive sulfur in their proteins. Each population of radioactive particles was used to infect, separately, bacterial cells growing in a non-radioactive medium. Very shortly after infection the bacteria were separated from the "heads" and "tails" that remain outside the infected bacterium. The bacteria were recovered and the radioactivity present in the bacterial cells was recorded. Only the radioactive phosphorus had entered the bacteria; thus, only the DNA of the viral particles had entered the bacteria. These bacteria were allowed to grow and produced a population of viruses. The DNA that entered the bacterium contained all the necessary information for the generation of bacterial viruses. DNA is the genetic material of the virus. In the establishment of the lysogenic state, the DNA entered and somehow became part of the bacterial chromosome. It became prophage. In the case of the T viruses that kill *E. coli*, DNA entered and a cascade of events was triggered that led to more virus and cell lysis. This sequence of events could also be induced in a bacterium carrying the prophage.

All these events were recognized in 1953 in the first textbook, *General Virology*, written by S. E. Luria (1912–1991). Luria was very careful in evaluating the role of the nucleic acid and protein components of bacterial viruses. Although Luria included the structural study on DNA of Watson and Crick, there was, as yet, no understanding of the relationship of this structure to the characteristics of any organism except for the work of Avery and collaborators on the transforming principle of pneumococci, and the work of Hershey using radioactive-labeled phage. Luria offered two possibilities for the role of DNA in the life of the virus:

"It may supply the rigid framework for the maintenance of the configuration of a protein in the unfolded two-dimensional state, in which a protein molecule, acting as a model for production of similar molecules, might have to arrange itself in order to make possible an identical point by point replication".

S. E. LURIA, *General Virology*, 1953, p. 100, John Wiley, New York

In short, Luria reported the likelihood that protein was the genetic material. However, he did allow that "it seems likely that the nuclei acids may themselves be, in part or exclusively, the carriers of specific biological configurations."

That is the story that would emerge conclusively after Watson and Crick's proposal for the structure of DNA.

The answer to the general question, "What are viruses?" can be derived from the significant work on bacterial viruses. A virus is an entity composed of proteins and DNA. DNA contains the genetic code, established after the elucidation of the structure of DNA by James Watson and Francis Crick in 1953 (a great story in itself), which encodes the information for the primary structure of proteins that perform the essential functions leading to growth and cell reproduction. DNA enters the host cell to commandeer the preexisting machinery of the host cell to make the new components that will assemble to constitute the virus particles that break out of the host cell. These viruses go on to infect other cells, a contagious process, until the entire population may be lysed. At this point, as Delbruck pointed out, the turbid culture in a test tube containing as many as millions of bacteria per milliliter becomes clear due to the destruction of all the cells.

To provide a climax to this story, let us return to the first filterable agent to be identified in the last decade of the nineteenth century, which caused a disease of tobacco. It was initially characterized as a protein in 1935. It was identified to contain RNA about a year later. In 1956 Heinz Fraenkel-Conrat (1910–1999), a biochemist in the Virus Laboratory at the University of California, Berkeley, headed by Wendell Stanley, separated the protein and RNA portion of the tobacco mosaic virus. It had already been demonstrated that the virus contained 2,130 copies of the same protein component assembled in such a way as to provide a hole in the interior where the RNA part was sequestered. The reconstituted virus, resulting from the combination of purified RNA and protein, had the same shape as the original virus, viewed in electron microscope pictures, and caused an infection of tobacco plants. Fraenkel-Conrat then carried out an experiment in which he reconstituted the virus, reassembled it, from the protein of one strain, for example A, and the RNA from another strain that contained protein B. He used the hybrid virus to infect plants; the progeny virus contained protein of strain B. Thus the genetic information for the new progeny was contained in the RNA of strain B. In the same year in Germany, G. Schramm (1910–1969) was able to infect plants by using pure RNA! Tobacco mosaic viral heredity was determined by its nucleic acid component, in this case RNA.

These data explain the results obtained by Stanley in 1941 when he used protein modified TMV to infect leaves, which led to the production of viruses without modified protein because the RNA of the virus, the hereditary material, was unchanged.

All Viruses Are Composed of Proteins and Nucleic Acid

All viruses are composed of protein and DNA or RNA. In every case it is the nucleic acid that is the genetic material. In every case it is the introduction of the genetic material into the host cell that initiates the production of viral particles. Viruses depend on the synthetic machinery of the host cell to make new viral components; the information for these components is contained in the primary structure of the nucleic acid.

Proof of Causality

Viruses are the causative agents of diseases of plants, humans, other animals, and bacteria. The proof of this principle is demonstrated by the fact that one can obtain a pure preparation of viral particles that are specific for a particular host, infect that host, and recover from that host the identical viral particles that were used initially to infect the host.

APPENDIX

THE DIFFICULTY of evaluating theories of the cause of contagious diseases by writers of the past is demonstrated by the numerous interpretations offered for Girolamo Fracastoro's use of the word "semina," seeds, in the treatise "On Contagion, Contagious Diseases, and their Treatment,"

In the mid-sixteenth century, Giambattista da Monte, a contemporary of Fracastoro, accused him of adopting Epicurean philosophy, where seeds and atoms are interchangeable. In the twentieth century Fracastoro's treatise became part of the discourse on disease causation, after the discovery in the last quarter of the nineteenth century that microorganisms cause contagious diseases.. Here are the analyses in chronological order.

In 1913 his writing was described by Fielding H. Garrison as an early, unique account of a living agent theory of disease:

"His medical fame rests. . . . his treatise "De contagione" (1546) in which he states, with wonderful clairvoyance, the modern theory of infection by microorganisms (seminaria contagionum)"

Not so, wrote Charles Singer in "The Scientific Position of Girolamo Fracastoro." He admired Fracastoro's doctrine of contagion via seeds since it "formed the basis of the best work on the subject in the centuries that followed." Singer opposed a vitalist (a living agent) interpretation of Fracastoro's use of the term "seed": "It is idle . . . to discuss whether he regarded these germs, seeds, or semina as living or non-living since the distinction would not have appeared important to him." To support this contention Singer pointed out that Fracastoro "believed that infectious diseases could be originated anew"; simply stated, they could be spontaneously generated from "within us."

In 1928 G. B. Stones connected Fracastoro's seeds to the seeds of Lucretius in "De rerum natura". In the Lucretian version seeds were linked to the atomistic theory of Epicurus. Stones contended that Fracastoro's philosophical base was expressed in *Sympathy and Antipathy,* where, Stones alleged, Fracastoro adopted a corpuscular theory to explain chemical changes. In this interpretation, Fracastoro had a physical theory of disease. Seeds were equivalent to the atoms of Epicurus and Lucretius.

In the English version of "On Contagion", published in 1930, Wilmer Cave Wright consistently translated "seminaria" as "germs," commenting that it was the nearest approximation in English to the meaning of the Latin. In the twentieth-century discussions of disease, "germs" signify a living organism. Wright was obviously taking the position that Fracastoro intended for "seminaria" to have that meaning.

In 1935 a translation of Fracastoro's poem "Syphilis or the French Disease" was published by H. Wynne-Finch. In it he commented on Fracastoro's prose work "On Contagione": "he proposed his theory of minute reproductive germs, anticipating the discoveries of bacteriology in our own day." According to Wynne-Finch, Fracastoro concluded that contagion was caused by a living agent.

In 1943 C. E. A. Winslow, in "The Conquest of Epidemic Disease," wrote glowingly of Fracastoro. His

"analysis was a truly marvelous triumph of close observation and clear reasoning. It is the first really philosophical statement of the contagionistic theory of disease—a mountain peak in the history of etiology, perhaps unequalled by any writer between Hippocrates and Pasteur".

Winslow appropriated throughout his evaluation the term "germ," described as a transmissible, self-propagating entity. He contended that the term was used in the sense of a germ of an idea. Germs were not living organisms but rather chemical substances.

In 1948 Walter Pagel expressed high regard for Fracastoro's theories. He found them to contain the modern concept of specificity, that the cause of the disease and the disease are specific entities. Consequently, Pagel concurred in Fracastoro's differentiating between poisons and contagions. Pagel associated "seminaria" with *life* in contrast to occult qualities or simple miasms. The ability of seeds to act at a distance made it "atomistic" in Pagel's evaluation.

In 1961 Thomas D. Brock reproduced small sections of the 1930 translation by Wright in a volume titled "Milestones in Microbiology", *1546–1940* (1546 was the publication date of "On Contagion"), which contained the word "germ" for the agent of infection. Brock, agreeing with Wright, maintained that it was the English word that best explains the entity that causes disease. In a short, nuanced analysis of the paper Brock said that Fracastoro's discourse was important "because it comes as close as it does to hitting the nail on the head." This sentiment does not reveal exactly how Brock reads "germs of disease"; perhaps he is suggesting that Fracastoro was close to the idea of a living agent theory of disease.

In 1965, in a text by H. A. Lechevalier and M. Solotorovsky, "Three Centuries of Microbiology", Fracastoro appears on the first page. He is characterized as "the father of the germ theory of disease." Later in the text the authors identified an individual who had postulated in the mid-nineteenth century that a number of diseases were caused by living parasites. They commented, "For the first time, the contagion theory of Fracastoro . . . was receiving solid scientific support." Their construal was that Fracastoro had proposed that the particles, seeds, were living agents.

In 1990 V. Nutton wrote an extended analysis of Fracastoro's disease theory called "The Reception of Fracastoro's Theory of Contagion: The Seed that fell among Thorns"?

He concluded that Fracastoro was using seeds as a metaphor. Nutton suggested that the analogy was to a physical particle, which is not a living agent.

In 2011, in a chapter titled "Lucretius and the History of Science" in the Cambridge Companion to Lucretius, the authors stated,

"Among the first of the scientists in the modern era to use Lucretius' text ("De rerun natura") was the humanist physician Girolamo Fracastoro (1478–1553). In "On Sympathy and Antipathy of Things" (1545) he developed a theory of contagious disease, proposing that some sicknesses are the products of exhalations of seeds or tiny living bodies".

M. R. JOHNSON and CATHERINE WILSON, Chapter 8, "Lucretius and the History of Science," in *Cambridge Companion to Lucretius*, 2011, pp. 131–148, ed. by Stuart Gillespie and Philip Hardie, Cambridge University Press.

My analysis is contained in Chapter 7.

SELECTED READINGS

General

C. Burnett, Ed. *Adelard of Bath: An English Scientist and Arabist of the Early 12th Century*. pp. 182, 188, 1987, The Warburg Institute, of London, 1987.

Charles S. F. Burnett and Danielle Jacquart, Eds. *Constantine the African and 'Alt Ibn Al–'Abbas Al Magust: The Pantegni and Related Texts*. Brill, 1995, Leiden. An authoritative source of information about Constantine, who came to Salerno, then under control by the Normans, where the first school for medical education was established. Constantine translated numerous works from Arabic into Latin, including *Pantegni (Pantechne), The Total Art*, by the Muslim physician Haly Abbas. He also translated works of Hippocrates and Galen, thus making these writings available in Western Europe.

H. Floris Cohen. *The Scientific Revolution: A Historiographical Inquiry*. 1994, University of Chicago Press. There are many versions of what constitutes the "revolution" of the seventeenth century by numerous philosophers and historians. Cohen presents their analyses.

Lawrence I. Conrad, Michael Neve, Vivian Nutton, Roy Porter, and Andrew Wear. *The Western Medical Tradition: 800 BC to AD 1800*, 1995, Cambridge University Press.

Jared Diamond. *Guns, Germs, and Steel*, 1999, W. W. Norton and Co., New York, Traces the origin of a number of contagious diseases from livestock to humans in Chapter 11, "Lethal Gift of Livestock."

Roger French and Andrew Wear, Eds. *The Medical Revolution of the Seventeenth Century*. 1989, Cambridge University Press.

Marcus Hellyer, Ed. *The Scientific Revolution: The Essential Readings*, 2003, Blackwell.

Carl Hempel. *Philosophy of Natural Science, 1966*, Prentice Hall. How science is done. To illustrate the invention of a hypothesis Hempel uses the explanation by Semmelweis of the differential occurrence of childbed fever in adjacent clinics of the Vienna General Hospital.

Jonathan I. Israel. Chapter 25, "The Collapse of Cartesianism," in Radical Enlightenment, Philosophy and the Making of Modernity 1659–1750, pp. 477–501, 1993. Oxford University Press.

Thomas Kuhn. The Structure of Scientific Revolutions. 1962, University of Chicago Press. How models of scientific theory change over time.

Herbert A. Lechevalier and Morris Solotorovsky. Three Centuries of Microbiology, 1965, McGraw–Hill, New York. An historical overview beginning with G. Fracastoro.

G. E. R. Lloyd, Ed. Hippocratic Writings. 1978. Penguin Books, London. The book to read to gain an introduction to the Hippocratic Corpus. There is a clear and instructive introduction by G. E. R. Lloyd. Treatises such as Tradition in Medicine (On Ancient Medicine), Air Waters Places, The Sacred Disease, and The Nature of Man give the flavor of the different points of view in the writings, including, of course, the premise that diseases have natural causes.

Gerrad Naddaf. The Greek Concept of Nature. 2005, SUNY Press. A book devoted to understanding a theory of nature revealed in the writings of the pre-Socratics.

Vivian Nutton. Ancient Medicine, 2004, Routledge, London, 2004. A history covering the period before the Hippocratic writings to Galen and medicine in the Roman Empire. In Nutton's words, "a history of medicine not only as a system of ideas but also as a network of practices rooted within a particular society, overlapping, competing and changing with time. It is this historicity as well as that diversity that this study aims to convey."

Charles E. Rosenberg. The Cholera Years: The United States in 1832, 1849, 1866, 1962. University of Chicago Press, The book charts the cholera epidemics of the nineteenth century in the United States, which parallel the epidemics in Europe.

P. Slack. The Impact of Plague in Tudor and Stuart England. Routledge and Kegan Paul, 1985, London. The history of the plague in Britain from 1485–1665, its effect in different localities in 1665, and the contemporary responses by governmental authorities and peoples.

Andrew Wear. Knowledge and Practice in English Medicine, 1550–1680. 2000, Cambridge University Press.

C. E. A. Winslow. The Conquest of Epidemic Disease, 1943, Princeton University Press, The heroes conquering contagious diseases.

Books for the General Reader

Erwin A. Ackerknecht. A Short History of Medicine, 1982, Johns Hopkins University Press.

John M. Barry. The Great Influenza: The Epic Story of the Deadliest Plague in History, 2005, Penguin Books, New York, The flu of 1918.

Paul De Kruif. Microbe Hunters, 1939, Harcourt, Brace and Co., New York, This book was initially published in 1926 and is the most popular account of the 11 heroes who, according to De Kruif, discovered microorganisms and worked to uncover their role in causing disease.

Stephen Jay Gould. "Syphilis and the Shepherd of Atlantis," Natural History, 109.8, pp. 38–42. Gould compares Fracastoro's discussion of the cause of syphilis, contained in a poem written in 1530, and what is currently known about the bacterium that causes the disease.

Sandra Hempel. The Strange Case of the Broad Street Pump, John Snow and the Mystery of Cholera, 2007, University of California Press, . The cholera epidemics, the studies of John Snow, and the sanitary movement in England. Also a brief survey of a living agent theory of disease in the nineteenth century.

Gina Kolata. Flu: The Story of the Great Influenza Pandemic of 1918 and the Search for the Virus That Caused It. 1999, Farrar, Straus and Giroux, New York.

S. E. Luria. A Slot Machine, A Broken Test Tube: An Autobiography. 1984, Harper and Row, New York, 1984. An autobiography by one of the pioneers in the study of viruses that infect bacteria (bacteriophage).

William McNeill. Plagues and Peoples.1976, Doubleday, Garden City, New York.

McNeill states that his aim is to bring the history of infectious disease into the realm of historical explanation by showing how various patterns of disease circulation have affected human affairs in ancient as well as modern times.

A. Lloyd Moote and Dorethy C. Moote. The Great Plague; The Story of London's Most Deadly Year. 2004, Johns Hopkins University Press. This is the classic plague that struck England in 1665. It was the last time this disease occurred in England.

William Shakespeare. The Tragedy of King Lear. Ed. Barbara A. Mowat and Paul Werstine. 1994, Washington Square Press, New York, pp. 55–56. Act I, Scene 2, contains Shakespeare's view(?) of astrology about the year 1600 CE.

Thucydides, History of the Pelopennesian War, Tr. Charles F. Smith. Book 2, pp. 337–355, 1919, Harvard University Press. A powerful description of the "plague" in Athens.

Barbara Tuchman. "This is the End of the World," in A Distant Mirror: The Calamitious 14th Century, pp. 92–125 1978. Random House, The Black Death.

Cecil Woodham-Smith. The Great Hunger, Ireland 1845–1849. 1962. Hamish-Hamilton, London. A brilliantly written book on the fungal disease that destroyed the potato crop and in concert with the policies of the British government permanently altered the lives of the Irish people and Ireland.

Philip Ziegler. *The Black Death*. 1969, Penguin,

Hans Zinsser. *Rats, Lice, and History*.1935, Little Brown, Boston. Mostly about typhus fever.

Readings by Chapter

CHAPTER 1

Herodotus. The Histories. Tr. Aubrey de Selincourt, rev. by A. R. Burn.1954, Penguin Books.

Hesiod. The Poems of Hesiod. Tr. R. M. Frazer, 1983, University of Oklahoma Press.

Hesiod. Theogony and Works and Days. Tr. M. L. West. 1988, Oxford University Press.

Homer. The Odyssey. Tr. Robert Fitzgerald. Anchor Books, 1963, Doubleday, New York.

Homer. The Iliad. Tr. Robert Fitzgerald. Anchor Press, 1975, Doubleday, New York.

JPS Torah Commentary, 1989, Jewish Publication Society of America, Philadelphia.

The Hebrew Writings; The Holy Scriptures, 1955, Jewish Publication Society of America, Philadelphia.

Sophocles. Oedipus The King. Tr. David Grene, pp. 11–76, 1954, University of Chicago Press.

CHAPTER 2

Aristotle. Poetics. Tr. Benjamin Jowatt 1942, The Modern Library.

Diogenes Laertius. Lives of Eminent Philosophers. Tr. R. F. D. Hicks. Loeb Classical Library, Book X, Epicurus (341–271 BCE), 1925. William Heinemann, London.

James Longrigg. Greek Rational Medicine: Philosophy and Medicine from Alcmaeon to the Alexandrians.1993, Routledge, London.

Lucretius. *De rerum natura*. Tr. and commentary by Cyril Bailey, 1947. Clarendon Press, Oxford. A poem, in Latin, composed of six books containing physical and biological theories of everything in the world, based on the writings of Epicurus, ending with the cause of diseases.

The Medical Writings of Anonymous Londinensis, Tr. W. H. S. Jones, 1947, Cambridge University Press. In 1892 the *Classical Review* announced that in the British Museum was a papyrus of more than 1,900 lines containing ancient Greek medical and philosophical ideas. In one portion 20 medical authorities were cited on the causes of disease. The work appears to have been written in the second century CE.

Plato. *The Collected Dialogues*. Ed. Edith Hamilton and Huntington Cairns. 1971, Princeton University Press.

Robert Waterfield. The First Philosophers, 2000, Oxford University Press. An authoritative introduction to the philosophers designated as pre-Socratics.

CHAPTER 3

Hippocrates. Tr. W. H. S. Jones. 1923 William Heinemann, London. Volume 1 contains sections on Ancient Medicine, p. 1; Airs Water Places, pp. 71–81; Epidemics, p. 155.

Hippocrates. Tr. W. H. S. Jones, 1923. William Heinemann, London. Volume 2 contains sections on The Sacred Disease, p. 127, and The Law, p. 255.

Hippocrates. Tr. W. H. S. Jones, 1931. William Heinemann, London. Volume 4 contains sections on Nature of Man, p. 1; Aphorisms, p. 97; Regimen 1, 2, 3, pp. 222–367.

Hippocrates. Tr. Paul Potter, 1995, Harvard University Press. Volume 8 contains a section on Fleshes, p. 127.

Jacques Jouanna. Hippocrates, 1992, Johns Hopkins University Press.

G. E. R. Lloyd, Ed. Hippocratic Writings, 1978, Penguin Books, London, 1979, Harvard University Press.

G. E. R. Lloyd. In the Grip of Disease: Studies in the Greek Imagination, 2003, Oxford University Press.

CHAPTER 4

Vivian Nutton. "The Seeds of Disease: An Explanation of Contagion and Infection from the Greeks to the Renaissance." *Medical History*, 27:1–34, 1983. Emphasis on Galen's use of the term "Seeds" and what he signified when using this term to describe the cause of disease.

CHAPTER 5

A. C. Crombie. *Augustine to Galileo*, Vol. 1, 1979, Harvard University Press.

CHAPTER 6

Rosemary Horrox, Tr. and Ed. *The Black Death*. Manchester Medieval Sources Series, 1994, Manchester University Press. The book is an invaluable source of translations of contemporary narratives of the plague in continental Europe and the British Isles. The first two parts deal with religious explanations of the causes, followed by the contemporary "scientific" causes and human causes of the disease.

Anna Montgomery Campbell. The Black Death and Men of Learning, 1931, Columbia University Press. This book is also a rich sources of contemporary writings on the cause of plague. There is a translation of a "masterly description of the nature and course of the disease" written 15 years after 1348 by Guy de Chauliac, physician to Pope Clement VI. The historical writings about this disease are enormous. Campbell provides about 600 references.

CHAPTER 7

Brian Copenhaver, Ch. 22 Magic, in The Cambridge History of Science, Ed. Katherine Park and Lorraine Daston, 2006, Cambridge University Press.

Allen G. Debus. Man and Nature in the Renaissance, 1978, Cambridge University Press,

G. Fracastoro. On Contagion, Contagious Diseases, and their Treatment, Tr. and notes by Wilmer Cave Wright, 1930, G. P. Putnam & Sons, New York. Fracastoro presents a theory of causality consistent with his Epicurean philosophy, transmitted by Lucretius, with its physics that holds that all matter is made of physical semina, which are the vehicles of contagion, passing from one individual to another in various ways.

G. Fracastoro. Syphilis or the French Disease A Poem in Latin Hexameter. Tr. Heneage Wynne-Finch, 1935. William Heinemann Medical Books, London (Elsevier).

Hieronymos Fracastorius of Verona. Concerning Sympathy and Antipathy. Hand-written English translation obtained from Deakin University Library, Australia. No translator is indicated. The text contains philosophical principles that influence his contagious disease theory contained in the treatise On Contagion.

Daniel Garber and Michael Ayers, Eds. The Cambridge History of Seventeenth Century Philosophy, Vol. 1, 1998. Cambridge University Press. See Chapter 2 by Daniel Garber, "Physics and Foundations," p. 26; Chapter 17 by Steven Nadler, "Doctrines of Explanation in Late Scholasticism and in the Mechanical Philosophy"; and Chapter 21 by William R. Newman, "From Alchemy to 'Chymistry,'" p. 497.

Walter Pagel. Paracelsus: An Introduction to Philosophical Medicine in the Era of the Renaissance, 1958, Karger, Basel.

Katharine Park and Lorrine Daston, Eds. The Cambridge History of Science, Vol. 3: Early Modern Science, 2006. Cambridge University Press.

Spencer Pearce. "Intellect and Organism in Fracastoro's Turrius." In The Cultural Heritage of the Italian Renaissance: Essays in Honour of T. G. Griffith, Eds. C. Griffiths and R. Hastings. Edwin Mellen Press, Lewiston/Queenston/Lampeter, 1993, p. 236.

Spencer Pearce. "Nature and Supernature in the Dialogues of Girolamo Fracastoro." Sixteenth Century Journal, 27:111–132, 1996.

Charles Singer. "The Scientific Position of Girolamo Fracastoro." Annals of Medical History, 1–34, 1917.

CHAPTER 8

Peter R. Ansley. The Philosophy of Robert Boyle, 2000. Routledge, London.

Marie Boas Hall. Henry Oldenburg: Shaping the Royal Society, 2002. Oxford University Press, Oxford.

Theodore M. Browne. The Mechanical Philosophy and the "Animal Oeconomy": A Study of the Development of English Physiology in the Seventeenth and Early Eighteenth Century, 1968, University Microfilms, Inc., Ann Arbor, MI.

Walter Charleton. Physiologia-Epicuro-Gassendo-Charltoniana: A Fabric of Science Natural Upon the Hypothesis of Atoms, Founded by Epicurus, Repaired by Petrus Gassendus, Augmented by Walter Charleton, 1654, London, Tho. Newcomb for Thomas Heath.

Allen G. Debus. The Chemical Philosophy, Vol. 1, 1977. Neale Watson Academic Publications, Inc., New York. Chemical theories of Paracelsus and others.

René Descartes, Treatise on Man. Tr. and commentary by Thomas Steele Hall, 1972. Harvard University Press.

René Descartes. The Philosophical Works of Rene Descartes, Vol. 1. Tr. Elizabeth S. Haldane and G. R. T. Ross, 1979, Cambridge University Press.

Denis Des Chene. "Mechanism of Life in the Seventeenth Century: Borelli, Perrault, Regis." Studies in the History and Philosophy of Biological and Biomedical Sciences, 30(2):245–260, 2005.

Saul Fisher. Pierre Gassendi's Philosophy and Science: Atomism for Empiricists, 2005. Brill, Leiden.

William Harvey. The Movement of the Heart and Blood in Animals. Tr. Kenneth J. Franklin. 1957, Blackwell Scientific Publishers, Oxford. Herman Boerhaave, commenting on the "immortal Harvey," wrote that he "founded Physic upon a new and certain basis . . . he demonstrates the human body to be an engine, all whose offices depend on the Circulation of the Blood" (from Dr. Boerhaaves Academical lectures on the theory of physic [Translation]. 1742–1746, Vol. 1, pp. 41 and 42, London.

William Harvey. The Works of William Harvey, Tr. Robert Willis. In The Sources of Science, No. 13, pp. 610–611, 1965, Johnson Reprint Corp., New York.

Robert Hugh Kargon. Atomism in England from Hariot to Newton, 1966, Clarendon Press, Oxford.

Antonio Lolordo. Pierre Gassendi and the Birth of Modern Philosophy, 2007. Cambridge University Press.

Lucretius. De rerum natura. Tr. and commentary by Cyril Bailey. Clarendon Press, Oxford, 1947.

Stephen Menn. "The Intellectual Setting." In Daniel Garber and Michael Ayers, Eds. Cambridge History of Seventeenth Century Philosophy. Vol. 1, pp. 33–86, 2012, Cambridge University Press.

Henry Oldenburg. Correspondence of Henry Oldenburg.1965, University of Wisconsin Press, Madison.

Walter Pagel. Joan Baptista Van Helmont, 1982, Cambridge University Press.

Andrew Pyle. Atomism and Its Critics From Democritus to Newton, 1995, Thoemmes Press, England.

CHAPTER 9

C. Dobell. Antony von Leeuwenhoek and his "Little Animals." 1958, p. 117, 1958 (Leeuwenhoek. Letter 18) in Russell and Russell, New York. Dover reprint, 1960. An important source, in one book, of the letters Leeuwenhoek addressed to the Royal Society.

Brian J. Ford. The Story of the Simple Microscope, 1985, Harper and Row Publishers, New York. Leeuwenhoek's microscopes.

Howard Gest. "The Discovery of Microorganisms by Robert Hooke and Antoni van Leeuwenhoek, Fellows of the Royal Society." Notes and Records of the Royal Society of London 58(2):187–201, 2004.

Robert Hooke. Micrographia. First published by the Royal Society in 1665. 1960, Dover Publications, New York. First microscopic description of fungi.

Antony van Leeuwenhoek. "Observations, Communicated to the Publisher by Mr. Antony van Leewenheck, in a Dutch Letter of the 9th of October, 1676. Here English'd;

Concerning Little Animals by Him Observed in Rain-Well-Sea, and Snow Water; as Also in Water Wherein Pepper Had Lain Infused." Philosophical Transactions of the Royal Society, 12: 821–831, 1677–1678.

Antony van Leeuwenhoek. "An Abstract of a Letter from Mr. Antony Leewenhoek at Delft, dated Sep.17. 1683. Containing Some Microscopical Observations, about Animals in the Scurf of the Teeth, the substance call'd Worms in the Nose, the Cuticular Consisting of Scales." Philosophical Transactions of the Royal Society, 14: 568–574, 1684.

Antony van Leeuwenhoek. "Another Letter from the same Mr. Leewenhoek, concerning his observations on Rain Water." Philosophical Transactions of the Royal Society, 23: 1702–1703, pp-. 1137–1151, 1152–1155, 1304–1311.

CHAPTER 10

Thomas Bates. "A Brief Account of the Contagious Disease Which Raged Among the Milch Cows Near London, in the Year 1714." Philosophical Transactions of the Royal Society, 30: pp. 872–885, 1717–1719.

L. Bellini. A Mechanical Account of Fevers, 1720, London, The translator stated that the world was indebted to the "Italians for their Advancement of the most substantial Philosophy, leading into the only Means of arriving to the Knowledge of Nature, by Experiments and Mechanical Reasonings thereupon . . . and Malpighi, Redi Steno, Borelli . . . laid a sure Foundation in Anatomy . . . and Borelli in particular, by his application to Mechanicks, and the laws of Motion, taught him to account for the Powers of the Muscles . . . But it was this Scholar Bellini, the Author now before us, who first taught, upon the same Principles and Conduct, to reason demonstratively about the more minute and more unheeded Agency of the Animal Oeconomy" (p. iv, preface).

Richard Blackmore. A Discourse upon the Plague with a Preparatory Account of Malignant Fevers, 1722. London.

William Boghurst. Loimographia: An Account of the Great Plague in London in the Year 1665. Ed. J. F. Payne Shaw & Sons for the Epidemiological Society of London, 1894, London.

Robert Boyle. The Philosophical Works of the Honorable Robert Boyle; Causes of the Wholesomness and Unwholesomness of the Air: The Air consider'd With regard to Health and Sickness. Abridged, 2nd Ed., Vol. 3, Peter Shaw, 1738 , London.

Richard Bradley. The Plague at Marseilles Consider'd. 2nd Ed., 1721, London. "With Remarks upon the Plague in General, shewing its Cause and Nature of Infection, with necessary Precautions to prevent the spreading of the Direful Distemper. Publish'd for the Preservation of the People of Great Britain. Also some Observations taken from an Original Manuscript of a Graduate Physician, who resided in London during the whole Time of the late Plague. Anno 1665."

R. Bradley. "Some Microscopical Observations and Curious Remarks on the Vegetation, and exceeding quick Propagation of Moldiness, on the Substance of a Melon." Philosophical Transactions of the Royal Society of London, 29:490–492, 1714.

Richard Bradley. Preface to New Improvements in Planting and Gardening, 1717, London.

Richard Bradley. Chapter V in New Improvements of Planting and Gardening. 2nd Ed. pp. 80–100, 1718, W. Mears, London. It is clear what the term "putrefaction" meant for Bradley. It was a condition "which is always attended with Insects." It was the consequence of a disease process. It was part of a vocabulary ("decay, rotting, fermentation") used since antiquity for describing processes present in illnesses characterized by fevers. This topic is discussed by Don G. Bates in "Thomas Willis and the Fevers Literature of the Seventeenth Century" in "Theories of Fevers from Antiquity to the Enlightenment," Medical History Supplement No. 1, Ed. by W. F. Bynum and V. Nutton, Wellcome Institute for the History of Medicine, 1981, pp. 45–70.

Richard Bradley. Precautions Against Infection; Containing many observations necessary to be consider'd, at this time, on account of the dreadful plague in France. London. The date of this 38-page essay is not listed on the title page; it may be 1722. Bradley wrote: "we had the happiness to see a treatise concerning contagion . . . by a learned member of the College of Physicians [he was referring to Mead] wherein the author espouses the other opinion, viz, that infection is communicated from one person to another, by means of vitiated air; and that there are no insects in the case" (p. 2).

Richard Bradley. A General Treatise of Husbandry and Gardening. 1724, London.

Richard Bradley. Introduction to A Philosophical Account of the Works of Nature. 2nd Ed. 1739, London.

Theodore M. Browne. The Mechanical Philosophy and the "Animal Oeconomy": A Study of the development of English Physiology in the Seventeenth and Early Eighteenth Century. 1968.

University Microfilms, Inc., Ann Arbor, MI, 1968.

George Cheyne. A New Theory of Continual Fevers wherein Besides the Appearances of such Fevers, and the Method of their cure; occasionally, the structure of the Glands, and the manner of secretion, the operation of Purgative, Vomitive, and Mercurial Medicines, are Mechanically Explained. p. 2, 1701, London.

C. F. Cogrossi. Nuova idea del male contagioso de'buoi (New theory of contagious disease of oxen). Milan. Facsimile, with Eng. Trans. by Dorethy M. Schullian, 1953, Rome (6th Internat. Congr. Microbiol.).

William Coleman. "Mechanical Philosophy and Hypothetical Physiology." In Robert Palter, Ed. The Annus Mirablis of Sir Isaac Newton 1666–1966, pp. 322–332, 1970, MIT Press, Coleman, although very critical of Mead's mathematics and his unfounded speculations, acknowledged his importance in London medicine.

F. Egerton. "A History of the Ecological Sciences, Part 20: Richard Bradley, Entrepreneurial Naturalist." Bulletin of the Ecological Society of America, 87:117–127, 2006.

N. R. R. Fisher. "Robert Balle, Merchant of Leghorn and Fellow of the Royal Society (ca. 1640–1734)." Notes and Records of the Royal Society of London, 55(3):351–371, 2001. Balle was one of a wide circle of natural historians, botanists, and gardening enthusiasts that included Hans Sloane and James Petiver. Richard Bradley was a protégé of Balle, who was owner of the estate at Campden House.

A. Guerrini. "James Keill, George Cheyne, and Newtonian Physiology, 1690–1740." Journal of the History of Biology, 78: 247–266, 1985.

A. Guerrini. "The Tory Newtonians." Journal of British Studies, 25:293, 1986.

Anita Guerrini. "Archibald Pitcairne and Newtonian Medicine." Medical History, 31: 70–83, 1987.

A. Guerrini. "Isaac Newton, George Cheyne and the Principia Medicinae." In The Medical Revolution of the Seventeenth Century, Ed. Roger French and Andrew Wear. pp. 224, 228, 1989, Cambridge University Press.

A. Guerrini. "The Varieties of mechanical medicine: Borelli, Malpighi, Bellini, and Pitcairne." In Marcello Malpighi Anatomist and Physician, Ed. D. B. Meli, pp. 111–128, 2007. Leo S. Olschki Publisher, Florence. Borelli, a mentor of Bellini, produced a treatise, De motu animalium, contending that the movement of the animal body could be understood by a combination of mechanics and mathematics. He used this theoretical base to understand fevers and concluded they were due to chemical and physical effects that increased the motion of the blood, resulting in fever (p. 115). Marcello Malpighi, a contemporary of Bellini and a colleague of Borelli, was also a practicing physician and was confronted with fevers and developed a theory of cause based on understanding the processes of the body in mechanical terms. Fevers were the result of "corrosive particles," also described as "volatile and sulfurous" particles causing motion and breakdown in the blood (p. 118).

Stephen Hales. Vegetable Staticks. Macdonald, London, 1969. First published in 1727, the text was so named to declare that plant physiology was to be explained in quantitative physical and chemical terms. Hales contended, like Bradley, that there was a great analogy between plants and animals, and consequently the same mechanical methodology used to illuminate the "animal oeconomy" could be used to uncover the operation of plant processes (p. xxxii). Newton is the constant authority for many of Hale's discussions. There is a section on the nature of fire, which is discussed by Newton in Queries 7, 9, and 10 (p. 160). With regard to fermentation, Hales adopted the "rational account" by "That illustrious Philosopher Sir Isaac Newton" (p. 165). And in his experiments concerning "elastick air" he refers to Newton's Queries 29 and 30 (p. 177).

Blanche Henry. British Botanical and Horticultural Literature Before 1800. Vol. 2, p. 39, 1975. Oxford University Press. Bradley was of course widely known as a distinguished contributor to the gardening and horticulture community in Britain. He knew everyone in this group, which is amply documented by Blanche Henry.

Nathaniel Hodges. Loimologia: Or An Historical Account of the Plague in London in 1665. 1720, London. This book's subtitle is "With precautionary directions against the like contagion To which is added, An essay on the different causes of Pestilential diseases by John Quincy, M.D. With remarks On the infection now in France, and the most probable Means to prevent its spreading here." Hodge's book contains an example of the persistence of a divine cause of disease (p. 31). Although he presented a Galenic–mechanical explanation of the cause of the plague of 1665, he wished to counter "the suspicion of Atheism" and declared that "Pestilence is as much a Part of my faith." "The sacred Pages clearly and demonstratively prove that the Almighty, by his authority, and at his Pleasure, may draw the Sword, or shoot the Arrows of Death, and a Retrospection into Times past, shews many convincing Proofs of this terrible Truth; and in this Contagion before us, the Footsteps of an over-ruling Power are very legible, especially so far as concerns his divine Permission; But the great God's Purposes are Secret's too awful for mortals to pry into, although we know that he punishes as a Parent, and chides for our Good, which makes it our Duty to kis the Rod and submit." The term "Rod" refers to Exodus 7:17, The Torah Commentary: "I shall strike the water in the Nile with the rod that is in my hands and it will be turned into blood."

James Keill. Essays in Several Parts of the Animal Oeconomy, 2nd Ed., 1717, G. Strahan, London.

Nicolas Malebranche. The Search After Truth. Tr. Thomas M. Lennon and Paul J. Olocamp, 1980, Ohio State University Press.

Benjamin Marten. A New Theory of Consumptions, 1720, London.

Richard Mead. "An Abstract of part of a Letter from Dr. Bonomo to Signior Redi, containing some Observations concerning the Worms of Humane Bodies." Philosophical Transactions of the Royal Society, 23: pp. 1296–1299, 1702–1703.

Richard Mead. A Short Discourse Concerning Pestilential Contagion and the Methods to be Used to Prevent It. 3rd Ed., 1720, Buckley and Smith, London.

Richard Mead. "A Mechanical Account of Poisons." 1702, London. Mead had the support of the Royal Society for this view, expressed by the reviewer in the same issue that the Itch paper, see pp.152–3, appeared. In the summation, the reviewer assured the reader "that their learned author [Mead] was modest, when he said that it was not difficult to say something on these Heads [the circulation of the blood and pestilential fevers] more tolerable than authors had before said." The Royal Society was already a strong supporter of the mechanical philosophy, which had also been accepted as official doctrine by the London College of Physicians by 1700. Theodore M. Browne. "The College of Physicians and the Acceptance of Iatromechanism in England 1665–1695." Bulletin of the History of Medicine, 36:12–30, 1970.

Richard Mead. "A Treatise Concerning the Influence of the Sun and the Moon upon Human Bodies, and the Diseases thereby Produced." In The Medical Works of Richard Mead M.D. Ed. A. Donaldson and C. Elliot, 1775, Edinburgh: pp. 115–156, 1978, ASM Press, New York, 115–156. Originally published in 1704.

Sir Isaac Newton. Newton, Texts, Background, Commentaries, Eds. I. Bernard Cohen and Richard S. Westfall, 1995, W.W. Norton & Co., New York.

John Quincy. An Essay on the Different Causes of Pestilential Diseases and how they became contagious: With Remarks upon the infection now in France. 3rd Ed. 1721, London.

W. Roberts. "R. Bradley, Pioneer Garden Journalist." Journal of the Royal Horticultural Society of London, 64:164–174, 1939. Bradley was of course widely known as a distinguished contributor to the gardening and horticulture community in Britain.

Charles Singer. "The Dawn of Microscopical Discovery." Journal of the Royal Microscopical Society, pp. 317–340, 1915. Singer included Hooke, Grew, Malpighi, Leeuwenhoek, and Swammerdam as the great microscopists of the period from 1660 to the early decades of the eighteenth century.

S. M. Walters. The Shaping of Cambridge Botany. 1981, Cambridge University Press. Chapter 3 is devoted to Richard Bradley, documenting his "exceptional worth and reputation," including his foundational work on plant breeding.

R. Williamson. "The Germ Theory of Disease. Neglected Precursors of Louis Pasteur, Richard Bradley, Benjamin Marten, J.P. Goiffon." Annals of Science 2:44–57, 1955.

Catherine Wilson. The Invisible World, Early Modern Philosophy and the Invention of the Microscope, 1995. Princeton University Press,

Arnold Zuckerman. "Plague and Contagionism in Eighteenth-Century England: The Role of Richard Mead." Bulletin of the History of Medicine, 78:273–308, 2004.

CHAPTER 11

G. C. Ainsworth. Introduction to the History of Mycology, 1976, Cambridge University Press.

G. C. Ainsworth. Introduction to the History of Plant Pathology, 1981, Cambridge University Press.

Augostino Bassi. Del Mal del Segno (Disease of Silkworms). Tr. P. Y. Yarrow. Phytopathological Classics, No. 10, Ed. G. C. Ainsworth and P. Y. Yarrow, 1958. American Phytopathological Society, Baltimore.

Miles J. Berkeley. Observations, Botanical and Physiological, on the Potato Murrain, 1845. Phytopathological Classics No. 8, 1948, American Phytopathological Society, East Lansing, MI.

Bartolomeo Bizio. "Letter to the Most Eminent Priest, Angelo Bellani, Concerning the Phenomenon of the Red-Colored Polenta." Tr. C. P. Merlino. Journal of Bacteriology, 9:527–543, 1924.

R. Bradley. "Some Microscopical Observations and Curious Remarks on the Vegetation, and exceeding quick Propagation of Moldiness, on the Substance of a Melon." Philosophical Transactions of the Royal Society of London, 29:490–492, 1714.

C. Cagniard de la Tour. "Memoire on Alcoholic Fermentation." Annals of Chemistry and Physics, 68:206–222, 1838.

Ferdinand Cohn. "Untersuchurgen uber Bacterien IV. Beitrage zur Biologie der Bacillen." [Investigations into Bacteria IV. Contributions to the Biology of Bacilli.] Beitrage Biologie Pflanzen [Contributions to the Biology of Plants], 2:249–276, 1876.

Anton de Bary. Investigations of the Brand Fungi and the Diseases of Plants Caused by Them with Reference to Grain and Other Useful Plants. 1853. Tr. R. M. S. Heffner, D. C. Army, and J. D. Moore. Phytopathological Classics No. 11, 1969, American Phytopathological Society, Madison WI. Even after the demonstration by de Bary that plant disease were caused by microscopic living agents there was little interchange between individuals working on plant diseases and those working on animal and human diseases, as noted by Arthur Kelman and Paul D. Peterson in "Contributions of Plant Scientists to the Development of the Germ Theory of disease." Microbes and Infection, 4: 257–260, 2002.

Gerhart Drews. "The Roots of Microbiology and the Influence of Ferdinand Cohn on the Microbiology of the 19th Century." FEMS Microbiology Reviews, 24:225–249, 2000.

Mathieu du Tillet. Dissertation on the Cause of the Corruption and Smutting of the Kernals of Wheat in the Head Bordeaux, 1755. Tr. Harry Baker Humphrey. Phytopathological Classics No. 5, American Phytopathological Society, Ithaca, NY.

Felice Fontana. Observations on the Rust of Grains, 1767. Tr. P. P. Pirone. Phytopathological Classics No. 2, 1932, American Phytopathological Society, Washington, D. C.

F. Kutzing. "Microscopic Investigations on Yeast." *Journal of Practical Chemistry*, 11:385–409, 1837.

E. C. Large. The Advance of the Fungi. 1940, Henry Holt & Co., New York, 1940.

Joseph Lister. "On a New Method on Treating Compound Fracture, abcess, etc. with Observations on the Conditions of Suppuration." Lancet 1:326–329, 357–359, 387–389, 507–509; 2:95–96, 1867.

Joseph Lister. "On the Lactic Fermentation and Its Bearing on Pathology." Transactions of the Pathological Society of London, 29:425–467, 1878.

John T. Needham. "A Summary of Some Late Observations upon the Generation, Composition, and Decomposition of Animal and Vegetable Substances." Philosophical Transactions of the Royal Society of London, 45:615–666, 1748.

Benedict Prevost. Memoir on the Immediate Cause of Bunt or Smut of Wheat, and of Several other Diseases of Plants, and of the preventatives of Bunt. 1807. Tr. George Wannamaker Keitt. Phytopathological Classics No. 6, 1939, American Phytopathological Society, Menasha, WI.

Francesco Redi. Experiments of the Generation of Insects, Tr. from 1688 edition by Mab Bigelow, 1909, Open Court Publishing Co., Chicago.

J. Schroeter. Uber einige durch Bacterien gebildete Pigmengte Beitrage zur Biologie der Pflanzen 1875, Breslau J.U. Kern's Verlag, pp. 109–126.

L. Spallanzani. Nouvelles recherches sur les decourvetes microscopique et la generation des corps organizes, 1769. London and Paris, chez Lacombe.

J. Swammerdam. Histoire Generale des Insectes, C. G. Seyffert, 1682, Utrecht.

Giovanni Targioni Tozzetti. True Nature, Causes and Sad Effects of the Rust, The Bunt, The Smut, and Other Maladies of Wheat, and of Oats in the Field, 1766, Tr. Leo R. Tehon. Phytopathological Classics No. 9, 1952, Phytopathological Society, The Cayuga Press, Ithaca, NY.

CHAPTER 12

Thomas D. Brock. Robert Koch A Life in Medicine and Bacteriology, 1988, Science Tech Publishers, Madison, WI.

K. Codell Carter. "Koch's postulates in relation to the work of Jacob Henle and Edwin Klebs." Medical History, 29:353–374, 1985.

K. Codell Carter and George Tate. "The earliest-known account of Semmelweis's initiation of disinfection at Vienna's Allgemeines Krankenhaus." Bulletin of the History of Medicine, 65:252–257, 1991.

Charles Cowdell. A Disquisition on Pestilential Cholera being An Attempt to Explain its Phenomena, Nature, Cause, Prevention, and Treatment, by Reference to an Extrinsic Fungous Origin, 1848, Samuel Highly, London.

Rene Dubos and J. Dubos. The White Plague: Tuberculosis, Man and Society, 1952. Little Brown, Boston.

John Farley. The Spontaneous Generation Controversy from Descartes to Oparin, 1977. Johns Hopkins University Press.

Gerald L. Geison. The Private Science of Louis Pasteur, 1995. Princeton University Press.

A. Gordon. A Treatise on the Epidemic Puerperal Fever of Aberdeen, 1795, Robinson, London.

Heinrich Heine. "The Cholera in Paris." In Medicine: A Treasury of Art and Literature, Eds. Ann G. Carmichael and Richard M. Ratzan. Hugh Lauter Levin Assoc. Inc., 1991, pp. 148–152. The presence of Cholera of 1832 in Paris.

Sandra Hempel. The Strange Case of the Broad Street Pump, John Snow and the Mystery of Cholera, 2007. University of California Press, Berkeley.

Jacob Henle. On Miasmata and Contagia [1840]. Tr. George Rosen, 1938. Johns Hopkins Press. Henle argued that contagious diseases were caused by living agents; however, he believed it was impossible at the time to prove this hypothesis because the methodology was unavailable.

Oscar Hertwig. The Cell: Outlines of General Anatomy and Physiology [1892]. Tr. Henry J. Campbell, 1909, Macmillan, New York. An introductory chapter contains a brief history of the cell theory, the history of the protoplasmic theory, and representative references to important work from the late eighteenth through the nineteenth centuries.

Oliver W. Holmes. "The Contagiousness of Puerperal Fever." New England Quarterly Journal of Medicine and Surgery, 1:518, 1843.

Thomas H. Huxley. "The Physical Basis of Life." In Collected Essays, Vol. 1, pp. 130–165, 1894, D. Appleton and Co., New York. According to this 1868 essay, there is a unity in nature. All plants and animals and the cells of microscopic entities are composed of protoplasm, a nitrogen-containing material.

Robert Koch. "The Etiology of Tuberculosis." Berliner Klinischen Wochenschrift, April 10, 1882, pp. 221–230. The first demonstration that a specific bacterium is the cause of tuberculosis.

Margaret Pelling. Cholera, Fever, and English Medicine 1825–1865, 1978. Oxford University Press. This is an authoritative review of the various theories for the cause of cholera, including the microscopic studies that implicated a fungal origin of the disease of 1848.

Terrie M. Romano. Making Medicine Scientific: John Burden Sanderson and the Culture of Victorian Science, 2002. Johns Hopkins University Press. The development and use of laboratory medicine in Britain to study the cause of contagious disease, featuring the life and work of Sanderson.

Theodore Schwann. Microscopical Researches Into the Accordance of the Structure and Growth of Animals and Plants. Tr. Henry Smith, 1847, Sydenham Society, London. With a contribution to Plant Development by M. J. Schleiden. The major document correlating facts from which a cell theory was produced which contended that all plants and animals are composed of cells.

Ignaz Semmelweis. The Etiology, Concept, and Prophylaxis of Childbed Fever. Tr. and Ed. K. Codell Carter, 1983. University of Wisconsin Press. A translation of the writings of Semmelweis, which reveal the origins of his theory of the cause of childbed fever.

John Simon. "An Essay on Contagion: Its Nature and Mode of Action." The British Medical Journal, Dec. 20, 1879, pp. 973–975. Simon's mature speculations on the cause of contagious diseases.

James E. Strick. Sparks of Life: Darwinism and the Victorian Debates over Spontaneous Generation, 2000. Harvard University Press.

J. A. Villemin. Studies on Tuberculosis: Experimental Proof of its Specificity and its Innoculability, 1868. J. B. Bailliere, Paris.

R. Virchow. "Cellular Pathology." Virchows Archives of Pathology, Anatomy, Physiology, 8:1–15, 1855.

Owen H. Wangensteen. "Nineteenth Century Wound Management of The Prurient Uterus and Compound Fracture: The Semmelweis–Lister Priority Controversy." Bulletin of the New York Academy of Medicine, 46(8):565–596, 1970 (see p. 566). Contains many references in support of Semmelweis' views from medical scientists in Vienna but not from Rudolph Virchow.

Lise Wilkinson. Animals and Disease: An Introduction to the History of Comparative Medicine, 1992. Cambridge University Press. A quick survey of animal and human diseases and theories of disease causation from antiquity to the twentieth century.

Human and animal medicine became equated when it was recognized that epidemic diseases of domestic animals and humans had similar causes.

Michael Worboys. Spreading Germs: Disease Theories and Medical Practice in Britain, 1865–1900. 2000. Cambridge University Press.

CHAPTER 13

Angela N. H. Creager. The Life of a Virus: Tobacco Mosaic Virus as an Experimental Model, 1930–1965. 2002. University of Chicago Press.

Oswald T. Avery, C. M. McLeod, and M. McCarty. "Studies on the chemical nature of the substance inducing transformation of pneumococcal types; induction of transformation by a desoxyribonucleic acid fraction isolated from pneumococcus type III." Journal of Experimental Medicine 79:137–158, 1944.

M. Delbruck. "A Physicist's Renewed Look at Biology: Twenty Years Later." Science, 168: 1312–1315, 1970.

Felix d'Herelle. "Sur un microbe invisible antagoniste des bacilles dysenteriques." [About an invisible microbe antagonistic to the dysentery bacillus.] Comptes Rendus, 165:373–375, 1917.

Felix d'Herelle. "Discussion of the Bacterophage (Bacteriolysin)." The British Medical Journal, Aug. 19, 1922, pp. 289–293.

Emory L. Ellis and Max Delbruck. "The Growth of Bacteriophage." Journal of General Physiology, 22:365–384, 1939.

H. Fraenkel-Conrat. "The Role of the Nucleic Acid in the Reconstitution of Active Tobacco Mosaic Virus." Journal of the American Chemical Society, 78:882–883, 1956.

Horace Freeland Judson. The Eighth Day of Creation: Makers of the Revolution in Biology, 1979. Simon & Schuster, New York. A brilliant summary of interviews with scientists about the theories and practices that led to the field of molecular biology and the elucidation of the structure of DNA.

Alfred D. Hershey. "The Injection of DNA into Cells by Phage." In Phage and the Origins of Molecular Biology, Ed. J. Cairns, G. S. Stent, and J. D. Watson, pp. 100–108, 1966. Cold Spring Harbor Laboratory of Quantitative Biology.

Lily E. Kay. "W. M. Stanley's Crystallization of the Tobacco Mosaic Virus 1930–1940." Isis, 77(3): 450–472, 1986.

S. E. Luria. General Virology, 1953. John Wiley & Sons, New York.

Andre Lwoff. "Lysogeny." Bacteriological Reviews, 17:269–337, 1953.

Martinus Beijerinck, Concerning a Contagium Vivum Fluidum as a Cause of the Spot-Disease of Tobacco Leaves, 1898. Tr. James Johnson, Phytopathological Classics No. 7, 1942, The American Phytopathological Society, The Cayuga press, Ithaca, NY.

Dimitri Ivanovski, Concerning the Mosaic Disease of the Tobacco Plant, 1892. Tr. James Johnson, Phytopathological Classics No. 7, 1942, The American Phytopathological Society, The Cayuga press, Ithaca, NY.

John Northrup, Moses Kunitz, and Roger Herriott. Crystalline Enzymes, 2nd Ed., 1948. Columbia University Press, New York, .

Richard E. Shope. "Swine Influenza III Filtration Experiments and Etiology." Journal of Experimental Medicine, 54(3):373–385, 1931.

Richard E. Shope. "Swine Influenza." Harvey Lectures 31:183–213, 1936.

Wilson Smith, C. H. Andrewes, and P. P. Laidlaw. "A Virus Obtained From Influenza Patients." Lancet, 222(5732): 66–68, 1933.

W. M. Stanley. "Isolation of a Crystalline Protein Possessing the Properties of Tobacco-Mosaic Virus." Science, 81:644–645, 1935.

Mary Wortley Montagu. Letters from Mary Wortley Montagu 1709–1762. Ed. Ernest Rhys. 1906, Everyman's Library #69, J. M. Dent & Sons, London. Montagu's enthusiastic report of the immunization of Turkish children using material from the pustules of small pox. Later in the century Edward Jenner, "on the 14th of May, 1796,"scratched the surface of the arm of an eight-year-old boy with matter "taken from a sore on the hand of a dairymaid who was infected by her master's cows." He exhibited mild symptoms about a week later but on the eighth day was "perfectly well." Two months later, in July, in order to ascertain whether the boy was secure from the contagion of Small-pox, he was inoculated with variolous matter, immediately taken from a (small-pox) pustule. The boy remained unaffected. This procedure is labeled vaccination ("vacca" is the Latin name for cow).

F. C. Bawden and N. W. Pirie. The isolation and some properties of liquid crystalline substances from solanaceous plants infected with three strains of tobacco mosaic virus." Proceedings of the Royal Society, B, 128:274–320, 1937.

PERMISSIONS

THE AMERICAN ASSOCIATION FOR THE Advancement of Science gave permission to quote from the following:

1. M. Delbruck, "A Physicist's Renewed Look at Biology: Twenty Years Later." *Science*, June 12, 1970, p. 1315.
2. Wendell M. Stanley, "Isolation of a crystalline protein possessing the properties of tobacco-mosaic virus." *Science*, 81:644–645, 1935 (citation from p. 645).

The author wishes to thank The American Phytopathological Society for permission to use quotes from the following publications:

1. Mathieu du Tillet, *Dissertation on the Cause of the Corruption and Smutting of the Kernals of Wheat in the Head, Bordeaux*, 1755. Tr. Harry Baker Humphrey. Phytopathological Classics No. 5, American Phytopathological Society, Ithaca, NY, 1937.
2. Benedict Prevost, *Memoir on the Immediate Cause of Bunt or Smut of Wheat, and of several Other Diseases of Plants, and on Prevention of Bunt.* 1807, Tr. George Wannamaker Keitt. American Phytopathological Society, Menasha, WI, 1939 (1970).
3. Anton de Bary, *Investigations of the Brand Fungi and the Diseases of Plants Caused by Them with Reference to Grain and Other Useful Plants,* 1853. Tr. R. M. S. Heffner, D. C. Army, and J. D. Moore. Phytopathological Classics No. 11, American Phytopathological Society, Madison, WI, 1969.
4. Alfred Mayer, *Concerning the Mosaic Disease of Tobacco*, 1886. Tr. J. Johnson, Phytopathological Classics No. 7, American Phytopathological Society, St. Paul, MN, 1942.
5. Dmitrii Ivanowski, *Concerning the Mosaic Disease of the Tobacco Plant*, 1892. Tr. J. Johnson. Phytopathological Classics No. 7, American Phytopathological Society, St. Paul, MN, 1942.

The American Society for Microbiology Publications (copyright assigned to ASM) gave permission to quote from the following:

1. Milestones in Microbiology: 1546 to 1940, Tr. and Ed. Thomas D. Brock. Report of the Commission for Research on the Foot-and-Mouth Disease, Friedrich Loeffler and P. Frosch, pp. 149–153, 1898. ASM Press, Washington, DC. 1961.
2. Milestones in Microbiology: 1546 to 1940, Tr. and Ed. Thomas D. Brock. Report of the lactic fermentation, p. 28; Influence of oxygen on the development of yeast and on alcoholic fermentation, Louis Pasteur, p. 44, 1861. ASM Press, Washington, DC, 1961.
3. Milestones in Microbiology: 1546 to 1940, Tr. and Ed. Thomas D. Brock. The etiology of tuberculosis, Robert Koch, p. 111, 1882. ASM Press, Washington, DC. 1961.

Robert G. Frank gave permission to quote from Robert G. Frank, *Harvey and the Oxford Physiologists*, University of California Press, Berkeley, 1980.

Brill gave permission to quote from Saul Fisher, *Pierre Gassendi's Philosophy and Science: Atomism for Empiricists*, Brill, Leiden, 2005.

Cambridge University Press gave permission to quote from the following:

1. K. Codell Carter, "Koch's Postulates in Relation to the Work of Jacob Henle and Edwin Klebs," *Medical History* 29: 353–374 (citation p. 365), 1985.
2. P. Micheli, p. 86 in G. C. Ainsworth, *Introduction to the History of Mycology.* Cambridge University Press, Cambridge, 1976.
3. Andrew Wear, Chapter 6, pp. 275–276, in *Knowledge and Practice in English Medicine, 1550–1680*. Cambridge University Press, Cambridge, 2000.
4. Antonia Lolordo, pp. 142–143 in *Pierre Gassendi and the Birth of Modern Philosophy*. Cambridge University Press, Cambridge, 2007.
5. Steven Nadler, Doctrines of Explanation in Late Scholasticism and in the Mechanical Philosophy, Chapter 17, p. 519, in *The Cambridge History of Seventeenth Century Philosophy,* Vol. 1, ed. Daniel Garber and Michael Ayers. Cambridge University Press, Cambridge, 1998.
6. Walter Pagel, *Joan Baptiste Van Helmont* Cambridge University Press, Cambridge, 1982, p. 7.
7. D. Garber, Part 1, Chapter 2, p. 26, Cambridge History of Science, Vol. 3, *Early Modern Science*, Eds. Katharine Park and Lorrine Daston. Cambridge University Press, Cambridge, 2006.
8. William R. Newman, Part 3 Chapter 21, p. 497
   Cambridge History of Science, Vol. 3, *Early Modern Science*, Eds. Katharine Park and Lorrine Daston. Cambridge University Press, Cambridge, 2006. The Cambridge History of Science Vol. 3 Early Modern Science. Ed. Katharine Park and Lorrine Daston. Cambridge University Press, Cambridge. Ch. 21
9. Jean Theodorides, "Casimir Davaine (1812–1882): A Precursor of Pasteur." *Medical History* 10(2):155–165, 1966 (Cambridge).
10. M. R. Johnson and Catherine Wilson, Chapter 8, Lucretius and the History of Science, pp. 131–148 in *Cambridge Companion to Lucretius*, Eds. Stuart Gillespie and Philip Hardie. Cambridge University Press, Cambridge, 2011.

11. R. Hooykaas, "The Rise of Modern Science: When and Why?" *British Journal for the History of Science*, 20:453–473, 1987.

Dover Publications, Inc. gave permission to quote from Clifford Dobell, *Antony von Leeuwenhoek and His "Little Animals,"* Dover, 1960, p. 42.

Elsevier gave permission to quote from the following:

1. *The Works of William Harvey*, Tr. Robert Wills. Johnson Reprint Corp. The Sources of Science #13, New York, 1965.
2. Fracastoro, *Syphilis or the French Disease*, Tr. Heneage Wynne-Finch. London, 1935, pp. 46, 54, 61, 69, 75, 77.

Harper-Collins gave permission to quote from Brian Ford, *Single Lens: The Story of the Simple Microscope*, Harper and Row Publishers, New York, 1985, p. 1.

Harper Collins Publishers Ltd. gave permission to quote from Dava Sobel, 1999, *Galileo's Daughter* and Walker Books, an imprint of Bloomsbury Publishing Plc. *and* copyright © 1999 Dava Sobel. Reprinted by permission of Harper Collins Publishers Ltd.

Reprinted from the Tanakh: The Holy Scriptures by permission of the University of Nebraska Press. © 1985, 1999 by the Jewish Publication Society, Philadelphia.

Harvard University Press gave permission to quote from A. C. Crombie, *From Augustine to Galileo,* p. 45. Harvard University Press. © 1952, 1959 by A. C. Crombie.

Johns Hopkins University Press gave permission to quote from the following:

1. Jacob Henle, *On Miasmata and Contagia*, 1840. Tr. George Rosen, pp. 6, 36, 42. © 1938, The Johns Hopkins Press, Baltimore, MD.
2. Terrie M. Romano, *Making Medicine Scientific: John Burden Sanderson and the Culture of Victorian Science*, p. 68. © 2002, Johns Hopkins University Press.
3. Bertolini Meli, *Mechanism, Experiment, Disease,* p. 44. © 2011, Johns Hopkins University Press.
4. Carlo Ginzburg. Translated by John and Ann C. Tedeschi, *The Cheese and the Worms: The Cosmos of a Sixteenth-Century Miller*, pp. 56–57. © 1980, Johns Hopkins University Press and Routledge Kegan Paul Ltd.
5. Aristotle, *Aristotle's Politics.* Tr. Benjamin Jowett. New York Modern Library, Johns Hopkins University Press, 1943.

Johns Hopkins University Press gave permission to reprint some material in Chapter 10 that appeared in Melvin Santer, "A unified, living agent theory of the cause of infectious diseases of plants, animals, and humans in the first decades of the 18th century." *Perspectives in Biology and Medicine* 52:566–578, 2009.

Olivia Judson, Cold Spring Harbor Laboratory Press, gave permission to quote from Horace Freeland Judson, *The Eighth Day of Creation.* Simon & Schuster, New York, 1979, p. 51.

S. Karger AG, Basel, gave permission to quote from Walter Pagel, *Paracelsus: An Introduction to Philosophical Medicine in the Era of the Renaissance* Karger, Basel, 1958, p. 50.

Kazi Publications gave publication to quote from *852 Avicenna, The Canon of Medicine,* Adapted by Laleh Bakhtiar, 1999. Tr. O. Cameron Gruner and Mazar H. Shah. Great

Books of the Islamic World, Inc. Distributed by Kazi Publications 3023 Belmont Ave. Chicago, IL 60618, email:kazibooks@kazi.org

Professor G. E. R. Lloyd gave permission to quote from *Hippocratic Writings*, Tr. J. Chadwick and W. N. Mann, Ed. G. E. R. Lloyd. Penguin Books, London, 1978.

Zelle Luria gave permission to quote from S. E. Luria, *General Virology*. John Wiley & Sons, New York, 1953, pp. 157–158 (citation is from p. 100).

Manchester University Press gave permission to quote from *The Black Death*, Tr. and Ed. Rosemary Horrox. Medieval Sources Series, Manchester University Press, Manchester, 1994. Copyright Rosemary Horrox.

Spencer Pearce. *The Cultural Heritage of the Italian Renaissance: Essays in Honour of T. G. Griffith*, ed. C. Griffiths and R. Hastings. Edwin Mellen Press, Lewiston/Queenston/Lampeter, 1993, p. 236. Edwin Mellen Press gave permission to quote.

Oxford University Press gave permission to quote from the following.

1. Hesiod, *Theogony; and Works and Days*. Tr. West. Oxford University Press, 1988, p. 40.
2. *The First Philosophers: the Presocratics and Sophists*. Tr. R. Waterfield. Oxford University Press, 2000, pp. xiv, 26–27, 29, 41, 59, 124, 171–172.
3. Lucretius, *De rerum natura*. Tr. Cyril Bailey. Oxford University Press, 1922, from Books 1, 2, 4, 5, 6.
4. Robert H. Kargon, *Atomism in England from Hariot to Newton*. Oxford University Press, 1966.

Science History Publications/USA, a division of Neale Watson Publishing International LLC, gave permission to quote from Allen G. Debus, *The Chemical Philosophy*, Vol. 1. Neale Watson Academic Publications, Inc., New York, 1977.

Carl Hempel, *Philosophy of Natural Science*. Prentice Hall, 1966, p. 47. Permission granter by Pearson Education (PE Ref #179,331)

Princeton University Press gave permission to quote from the following:

1. Plato, *The Collected Dialogues*. Eds. Edith Hamilton and Huntington Cairns. Princeton University Press, 1971, pp. 78, 515–516. (Bollingen Series LXXI, 1971)
2. Plato, *Timaeus* (49 bc). In *The Collected Dialogues*. Eds. Edith Hamilton and Huntington Cairns. Princeton University Press, 1971. (Bollingen Series LXXI, 1971)
3. Plato, *Cratylus* (402a8). In *The Collected Dialogues*. Eds. Edith Hamilton and Huntington Cairns. Princeton University Press, 1971, p. 339 (Bollingen Series LXXI, 1971)
5. Gerald L. Geison, *The Private Science of Louis Pasteur*. Princeton University Press, 1995, pp. 95–96.
6. C. E. A Winslow, *The Conquest of Epidemic Disease*, Princeton University Press. 1943, p. 143.

Andrew Pyle gave permission to quote from Andrew Pyle, 1997, *Atomism and its Critics*, Thoemmes Continuum, an imprint of Bloomsbury Publishing Plc. Copyright © Andrew Pyle, 1997.

Rockefeller University Press gave permission to quote from Oswald T. Avery, Colin M. MacLeod, and Maclyn McCarty, "Studies on the Chemical Nature of the Substance Inducing Transformation of Pneumoccocal Types: Induction of Transformation by a Desoxyribonucleic Acid Fraction Isolated from Pneumoccocus Type 3." *Journal of Experimental Medicine* 79:137–159, 1944 (citations from pp. 152, 155). © 1944.

The Royal Society gave permission to reprint some material in Chapters 10 and 12 that appeared in two issues of *Notes and Records of the Royal Society*:

1. Melvin Santer, "How it happened that a portion of a treatise entitled *New Improvements of Planting and Gardening Both Philosophical and Practical* by Richard Bradley FRS, which dealt with blights of trees and plants, provided the first report of an environment that contained green sulphur photosynthetic bacteria." *Notes and Records of the Royal Society* 61:327–322, 2007.
2. Melvin Santer, "Joseph Lister: First use of a bacterium as a "Model Organism" to illustrate the cause of infectious disease of humans." *Notes and Records of the Royal Society* 64:59–65, 2010.

Montaigne *The Complete Essays of Montaigne*. Tr. Donald M. Frame. Stanford University Press, pp. 429–430, 1958. Stanford University Press gave permission to quote

Gerrad Naddaf, *The Greek Concept of Nature*, SUNY Press, 2005, p. 33. SUNY Press gave permission to quote.

The Cell in Development and Heredity, 3E by Edmund B. Wilson. Copyright © 1925 by Edmund B. Wilson; copyright renewed 1952 by Anne M.K. Wilson. Permission given by Scribner Publishing Group, a division of Simon & Schuster.

James Longrigg, *Greek Rational Medicine: From Alcmaeon to the Alexandrians*, Routledge, London/New York, 1993. Taylor and Francis Books (UK) gave permission to quote

R. Williamson, "The Germ Theory of Disease. Neglected Precursors of Louis Pasteur. Richard Bradley, Benjamin Marten, J.P. Goiffon." *Annals of Science* 2:44–57, 1955. Taylor and Francis Ltd. (via Copyright Clearance Center) gave permission to quote.

Taylor and Francis Ltd. gave permission to quote from the following:

1. Phillip R. Sloan, The Origins of the Science of Natural History, p. 304 in *Companion to the History of Modern Science*, Eds. F. R. C. Olby, G. N. Cantor, J. R. R. Christie, and M. J. S. Hodge. Routledge, London and NY, 1990.
2. M. J. S. Hodge, Chapter 24, An Alternative to the Dominant Historiographic Tradition, pp. 374–395 (quote from pp. 378–379) in *Companion to the History of Modern Science*, Eds. F. R. C. Olby, G. N. Cantor, J. R. R. Christie, and M. J. S. Hodge. Routledge, London and NY, 1990.

Ughetta Fitzgerald Lubin gave permission to quote from the following:

1. Homer, *The Iliad*. Tr. Robert Fitzgerald. Anchor Press, Doubleday, New York, 1975. Book 19.
2. Homer, *The Odyssey*. Tr. R. Fitzgerald. Anchor Doubleday, 1963. Book 11:167 (p. 190).

INDEX

Fracastoro, Girolamo, 66, 72, 74–75, 97, 315
  on agents of contagion, 88–91
  disease theory of, 67, 79–83, 316
  and seeds, 314
Fraenkel-Conrat, Heinz, 311
Frazer, R.M., 5
Frosch, Paul, 283, 285, 286
fumes, 63, 124, 163–165, 165
fungi, 13, 139–140, 161, 176, 205
  biology of, 206–207
  bunt disease, 201
  as cause of disease, 206, 210
  as cause of plant disease, 216–217, 228
  and cholera, 227
  culturing of, 204
  on decayed plants, 180, 205
  experiments with, 220–221, 271
  and fermentation, 207–209, 246
  growth patterns of, 218
  lycoperdon, 198
  as microorganisms, 176
  microscopic, 257
  as microscopic agents, 161, 214
  microscopic studies of, 179–180, 183
  Neurospora crassa, 306
  origin of, 178–179, 181
  and plant diseases, 175–176, 193–194, 196–201, 204, 214–217
  reproduction of, 181, 219, 277
  Robert Hooke on, 179–180
  and spontaneous generation, 176, 185, 198, 202, 219–220
  spores of, 178, 182
  spread of, 210
  taxonomy of, 208
  uredo, 203
  and yeast, 207–209, 218, 235

Galen, xvi, xvii, 44–45, 55, 114
  disease theory of, 46–48, 50, 54, 62, 64
  and four-element system, 77, 151
  medicine of, 51, 67, 73–76, 90–93, 127, 171
  model of circulation, 102
  on natural causes, 57

  on seeds, 49, 87
  theory of faculties, 105
Galileo, 96, 145
Garrison, Fielding H., 313
Gassendi, Pierre, 105–111, 113, 115–116, 118, 127
  on atomism, 103–105
Geison, Gerald, 253–254
germs, 192, 201, 244, 265–266, 293.
  see also seeds; seminaria
  conveyed by air, 236, 240
  of disease, 236
  and fungi, 219, 227
  heat sensitivity of, 251
  as living organism, 315
  in model of epigenesis, 191
  self-reproduction of, 211
  as term, 80, 87, 313, 314, 315
  theory of disease, 177, 256
Glanville, Joseph, 170
God. *see also* gods
  anger of, xvi
  as cause of disease, 55
  as cause of plague, 57
  disease transmission by, 12
  divine wrath, 64
gods, 6
  as actors in causing world phenomena, 17
  as cause of disease, 6, 57–58
  as cause of natural phenomena, 1
  contractual relations with humans, 9–10
  in control of physical and biological world, 4
  as healers, 11
Goiffon, Jean-Baptiste, 143, 171, 172–173, 174, 241
Gordon, A., 237–238
gravity, 149
Great Hunger, 214–217
Great Outdoor Field Experiment, 198–199
Greeks, 1
  medicine of, 31
  plays of, 3, 10
Gregory, David, 146
Grew, Nehemiah, 155–156, 169
grotto, 163–166

Hales, Stephen, 155, 184–185
Hartlib, Samuel, 117–118
Harvey, William, xv–xvi, 101–103, 118, 123
    on infectious disease, 224
    on ovism, 186–187
healers, 11–12
Hebrews, 1, 7, 12, 68, 177. *see also* Israelite writings
Hempel, Carl, xviii, 198
Hempel, Sandra, 229
Henle, G. Jacob, 209–212, 222, 223
    disease theory of, 222–223, 263, 265, 267
    on experimental method, 270
Henry, Blanche, 167
Heraclitus, 18, 19, 21, 22
Heriot, Roger, 98, 99, 101
Hermeticism, 66
Herodotus, 1, 13, 18, 36
Hershey, Alfred D., 309–310
Hesiod, 1, 3–4, 9
Hippocrates, xv, xvi
    Hippocratic Writings 16, 17, 31–43, 54–55, 61, 171
    medicine of, 54–55, 61, 171, 314
Histories, The, 1, 13. *see also* Herodotus
Hobbes, Thomas, 101
Hohenheim, Theophrastus von. *see* Paracelsus
Holmes, Francis, 296, 298
Holmes, Oliver Wendell, 237–238
Homer, 1–2, 4, 12, 14
    epics of, 7–10, 16, 19–20, 59
    and gods, 18
    period of, 27, 177
Homeric–Hesiod epics, 1, 4, 5, 16, 18, 19
Hooke, Robert, 132, 135–136, 156, 159–162, 169
    cell theory of, 239
    and equivocal generation, 139–140, 192
    mechanical philosophy of, 113, 118, 121–122, 185
    and microscopic studies, 143, 179–180, 184
Hooykaas, R., 156

human cause of disease, 57, 64–65
"Humors", 39. *see also* Hippocrates
Huxley, Thomas Henry, 255–258

Ibn el-Khatib, 62–64
Ibn Khatima, 62–64
Iliad, 2–3, 6–9, 14, 43. *see also* Homer; Homeric–Hesiod epics
    and disease theory, 59
    human-god contractual agreement in, 10–11
inanimate agent, 142
influenza, 286–291
Ionia, 17–18
Isidore of Seville, 52
Israelite writings, 1, 5–6
Italian School mechanical philosophy, 145, 153
Italy, 17–18
Itch (disease), 151–152
Iwanowski, Dimitri, 281–282

Jacob Henle for justification, 212
Jenner, Edward, 279–280
Jews, 1, 57. *see also* Israelite writings
John Snow, 236
Johnson, M.R., 315
Judson, Horace Freeland, 303, 306

Kargon, Robert Hugh, 98–99
Keill, James, 149
Kepler, Johannes, 96, 98, 145
Kircher, Athanasius, 151
Klebs, Edwin, 266–267, 269–270, 274
Koch, Robert, 262, 269, 270–275
    rules for establishing causality on tuberculosis, 275–276
    on tuberculosis, 276, 280
Kützing, Friedrich, 207–208, 246

laboratory methodology, 223
Laidlaw, P.P., 289, 291
Large, E.C., 216
Laws, 16. *see also* Plato
Lechevalier, H.A., 314
Lederberg, Esther, 308

vs. Robert Boyle, 153
treatises of, 145
mechanical philosophy, 114–120
corpuscularian version of, 163
as explanation for contagious diseases,
141
Isaac Newton's acceptance of, 127,
128, 145–146, 148, 195
medicine, 33, 38, 39, 95, 146
Hippocratic–Galenic, xvi, 54, 55
influence of Isaac Newton on, 127
Western, 33
Melo, Pietro, 242
metaphysics, xvii, 17, 22, 105fig, 170, 199
in ancient theories, 186
and Cartesian mechanics, 104
and Descartes' corpuscularian theory,
107
vs. medicine, 75
Parmenidean, 17, 23, 27
miasms, 90, 126, 211–212, 222, 238,
314–315. *see also* fumes
Micheli, Pier Antonio, 180–182, 184, 202,
271
microbiology, 131, 314
micro-composition, 97
Micrographia, 132, 159
described, 139–140
influence on Antony van Leeuwenhoek,
136–137
microscopic observations in, 121, 179
microscopic studies discussed in,
169–170
micro-mechanics, 96, 153
micro-mechanisms, 97–101
microscopes, 108, 139, 271
creation of, 130
electron, 309, 311
fabricated by Antony van Leeuwenhoek,
131, 135–138
to study atoms, 107, 121
to study bacteria, 272, 303, 308
to study cells, 239–240
to study contagious agents, 206
to study corpuscular basis of living
matter, 170

to study fungus, 178, 179–180, 183,
210
to study insects, 187
to study plant disease, 158, 197, 243
to study yeast, 208
used by Antony van Leeuwenhoek,
132–134
used by Joseph Lister, 265
used by Richard Bradley, 160–161
used in medical laboratory work, 230
microscopic agents. *see* microscopic
entities
microscopic entities, 158–159, 162, 204,
257. *see also* ultramicroscopic
entities
in air, 181, 252
as animalcules, 129, 134, 144,
151–152, 239
and anthrax, 261–262
bacteria, 223, 255, 270
bacterides, 262
in blood, 261, 272
as cause of disease, xii, xvi, 141, 166,
206
in cholera, 230
as contagious matter, 269
culturing of, 193
development of, 250
as disease agent, 270
in disease of oxen, 168
existence of, 240, 249
and fermentation, 177, 254, 266, 268
fungus, 140, 178, 210, 214, 221
generation of, 193–194
growth of, 286
as insects, 157, 161
in living agent theory of disease, 139,
143, 155–156, 209, 276
as living cells, 277
and lysis, 294
measurement of, 108, 136, 170
nature of, 277
parasitic plantlet, 202
in plague, 171
in plant disease, 174
protoplasm, 256

microscopic entities (*continued*)
  reproduction of, 184, 251, 280
  size of, 172
  in spontaneous generation, 194, 212, 260, 277
  spores, 203
  in studies by Louis Pasteur, 249–251, 253, 254–255
  yeast, 248
microscopic life. *see* microscopic entities
microscopic observations, 139, 140, 169–171, 172, 286
  of animalcules, 169, 192, 245
  of Antony van Leeuwenhoek, 131, 137
  of bacteria, 275
  of bacteriophages, 294
  of Bartolomeo Bizio, 242
  of Benjamin Marten, 172
  of cattle plague, 264
  of cells, 206
  of cholera, 230
  of development of insects from eggs, 191
  of disease, 131
  of fermentation, 209
  of foot and mouth disease, 283–285
  of fungus, 158, 179, 220, 228, 231
  and God, 187
  of Isaac Benedict Prevost, 204
  of Jean Victor Audouin, 210
  of Louis Pasteur, 248, 255
  in medical laboratories, 230
  and microscopic agent theory of disease, 272
  of plant material, 207
  published by Robert Hooke, 121, 139, 179
  of Richard Bradley, 156
  of Robert Koch, 271
  of Rudolph Virchow, 239
  of snake venom, 149
  of yeast, 208, 209
microscopic organisms. *see* microscopic entities
microscopic studies. *see* microscopic observations

Milesia, 17–18
molds, 131, 157–158, 177, 181–184, 192, 202
Montagu, Lady Mary Wortley, 278
Monte, Giambattista da, 75, 90, 313
moon, 1, 16, 71, 130, 149
mosquitos, 285–287
Muller, Herman J., 300, 302
murrain, 124, 168, 175
mycology, 175

Nadler, Steven, 100
natural causes, 16–17, 17, 31, 34–37, 40, 58–62
"Natural History", 52. *see also* Pliny
naturalism, xvii, 25, 55, 64, 75, 155
"Nature of Man, The", 32, 34, 39–40, 44. *see also* Hippocrates
  four-humor theory in, 37–38
Needham, John Tuberville, 191–195, 239, 240, 250
neo-Platonism, 66
Newton, Isaac, 142, 145, 147, 195
  contributions to science of, 127–129
  disease theory of, 154
  mechanical philosophy of, 145, 166, 184
  Newtonian philosophy, 149
  Newtonian theory of matter, 146
  on particulate matter, 150
  physico-theological argument, 190
  physics of, 143, 144, 149, 192
  in the Scientific Revolution, 96
Northrup, John, 299–300
nucleic acid, DNA, RNA, 298, 301, 303–311
  of bacterial viruses, 310
  and tobacco mosaic viral heredity, 311
Nutton, Vivian, 314–315

occult, 66, 69, 84–85, 90–91, 121, 314
Odyssey, 9–11. *see also* Homer; Homeric–Hesiod epics
Oedipus Rex, 6. *see also* Sophocles
Oldenburg, Henry, 121, 131–132, 134, 136–138, 155

Redi, Francesco, 151, 154, 186, 188–190
Reed, Walter, 285–286
"Regimen", 32. *see also* Hippocrates
religion, xii, xix, 41, 53, 66
  Christianity, 51
  and contagious disease theory, 66
  Eastern vs. Western, 52
  influence on Louis Pasteur, 253–254
  of Paracelsus, 75–76
  religious speculation, hermeticism,
    neo-Platonsism, 69
  in the Roman Empire, Christianity, 51
  St. Augustine on, 53
  and study of chemistry, 91
  of Walter Charleton, 115
  in writings of Hermes Trismegistus, 68
Robert Boyle, 125–126
Roman Empire, 51
Royal Society, 112, 132
  founding of, 117, 120–122
  support of mechanical philosophy, 121,
    139–140

Sachs, Julius, 156
"Sacred Disease, The", 34, 36–37. *see*
  also Hippocrates
Sanderson, J. Burden, 264, 269, 270
scabies, 45–46, 48, 211
scarlet fever, 263, 285
Schleiden, Matthew, 206, 210
Scholasticism, 66–67
Schroeter, J., 244, 273–274
Schwann, Theodore, 176–177, 206–210,
  250–251
  on fermentation, 213, 235, 246, 248,
    263
Scientific Revolution, 96
seasons, influence on disease, 41, 54
seeds, 81, 88, 144, 150, 178, 264. *see also*
  germs; semina
  in chemical philosophy, 92
  compared to spirit, 69
  conveyed by air, 80
  of disease, 48, 50, 78, 87, 167
  Epicurus on, 27
  equivalent to atoms, 84

  in experiments, 193, 197–200, 220
  in experiments of Lazzaro Spallanzani,
    240, 242
  and fungus, 140, 158–159, 179–181,
    203
  generation of, 190–195, 211
  in Girolamo Fracastoro, 89–90, 313
  in Jan Swammerdam, 187–188
  as living agents, 315
  in Lucretius, 100
  metallic, 79
  as metaphor, 315
  millet, 132–133,136
  and mold, 184
  molecular changes in, 235
  in "On Contagion", 88–89
  origin of, 110
  in Pagel, 314
  of poison, 124
  in Richard Bradley, 182–183
  role in reproduction, 49
  transported by air, 202, 204
  in William Harvey, 186
  and worms, 189
semina, 72, 79, 90–91, 94. *see also* seeds
  as agent of contagion, 89
  and archei, 94
  in Lucretius' model for contagion,
    84–88
  as term, 313
seminaria. *see* germs
Semmelweis, Ignaz, 236–239
Shope, Richard E., 289, 291
Sicily, 17–18
Simon, John, 229, 236, 263–264, 269–270
simulacra, 27
Singer, Charles, 313
Sloane, Hans, 162, 166
smallpox, 223, 260, 263, 278–280
  caused by bacteria, 285
  caused by poison, 235
  fever in, 123
Smith, Wilson, 289
snake venom, 149
Snow, John, 232–236
Snow, Stephanie, 231